Also by Michael Novacek

Time Traveler
The Biodiversity Crisis (editor)
Dinosaurs of the Flaming Cliffs
Extinction and Phylogeny (editor)

TERRA

TERRA

Our 100-Million-Year-Old Ecosystem—

and the Threats That Now Put It at Risk

MICHAEL NOVACEK

Farrar, Straus and Giroux • New York

Farrar, Straus and Giroux
18 West 18th Street, New York 10011

Copyright © 2007 by Michael Novacek
Distributed in Canada by Douglas & McIntyre Ltd.
Printed in the United States of America
Published in 2007 by Farrar, Straus and Giroux
First paperback edition, 2008

The Library of Congress has cataloged the hardcover edition as follows:
Novacek, Michael J.
    Terra : our 100-million-year-old ecosystem—and the threats that now put it at risk /
Michael Novacek.
        p.   cm.
    Includes bibliographical references (p.   ) and index.
    ISBN-13: 978-0-374-27325-5 (hardcover : alk. paper)
    ISBN-10: 0-374-27325-1 (hardcover : alk. paper)
    1. Extinction (Biology)   2. Evolution (Biology)   3. Nature—Effect of human beings on.
4. Paleontology.   5. Environmental degradation.   I. Title.

QE721.2.E97 N68 2007
576.8'4—dc22

                                                                    2007009126
                    Paperback ISBN-13: 978-0-374-53141-6
                    Paperback ISBN-10: 0-374-53141-2

                    Designed by Jonathan D. Lippincott

                        www.fsgbooks.com

            1   3   5   7   9   10   8   6   4   2

In loving memory of my father,
who found beauty in his music,
all people, and all nature

CONTENTS

PROLOGUE: THE HYENA

On a summer morning, just like summer mornings for hundreds of millennia, the Sahara turned its scorched earth toward the sun. We tentatively stepped out of the shadow of our Land Cruiser, from a cool dark place to the white hot of bleached cliffs. Earth is a planet of bountiful life, but the Sahara seemed to us the antithesis of anything living. Rock and sand stretched to the horizon, interrupted by only the bristle of a dead thornbush, enlivened only by the distant shriek of a shadowy bird. We were in this dead place not because of what it is but because of what it once was. As we crept along in the awful stillness of the heat, we bent over a heterogeneous surface of pebbles, rocks, and sand, picking out alien items: spool-shaped vertebrae of giant crocodiles, strange, polished combs of ray teeth, and pieces of turtle shells that were weathered to the texture of ancient bronze coins. Even with a dusty fossil bone fragment clenched in my hand I could not conjure at that moment the 60-million-year-old world it came from. I could not appreciate that I was in fact walking the ancient shoreline of a warm sea whose surface had once been wrinkled with rising fish and meandering crocodiles. I could not pick up the scent of primeval magnolias in a cool breeze. That world was gone, burned out of the intervening epochs. The sea had long ago shriveled away, and although in a subsequent stretch of time this place still harbored forest and stream, the desert had at last come and buried it all.

And this was not even the emptiest core of the Sahara. Our expedition was crisscrossing the edge land of northern Mali, the zone between sand seas and the brush-covered hills of the Sahel (pronounced Sa-HELL). We were a small group led by the paleontologist Dr. Maureen O'Leary, whose fluent French and organizational acumen had succeeded in providing us with six Kalashnikov-armed, blue-turbaned Tuaregs. She had made the deal with local militia at the town of Gao, our jumping-off point, some 129 kilometers (80 miles) to the southwest. Over the years I had worked in many hot places—southern New Mexico's badlands, the cactus gardens of Baja California, the stifling humid clime of the Tehama along the Red Sea, and the solar oven formed by Mongolia's Gobi Desert—but I had never felt heat so intense, so immobilizing. I climbed a pediment slope and stopped suddenly, panting in the thick air. I looked down the hill to the flats about a hundred feet below, where I could see our escorts lying below the bellies of their cars, only their sandaled feet exposed to the sun. It seemed that the only water left on the planet was being relentlessly sucked out of my tissues and would hardly be replenished by the two liters in my backpack. It was only 10:00 a.m., two hours until I too would be allowed to slither below the oily chassis of a Land Cruiser. I thought with some shame of the herdsmen we had seen two days before, driving their spectral goats through the ashen thorn forest to a tiny well, where their camels slurped. They greeted us, laughing generously. Not one of them mentioned the heat. Adaptation is an amazing thing.

Of course one does not wage the struggle for existence on spirit alone. The conditions of the southern Sahara for human habitation are marginal at best. At worst, in places devoid of resource, community, and humane treatment, they are simply a death sentence. Seventeen hundred miles to the east in the Darfur region of Sudan, hundreds of thousands of people were being driven from their homes and arable lands. Many of these victims were killed, wounded, or raped. Others were left starving, ravaged by disease, collapsing in the heat. They were casualties of a genocide being carried out by the Janjaweed militia, which conveniently relied on the forces of nature to inflict pain and suffering. The root of this tragedy involved the usual motivations for "ethnic cleansing," but long-standing strife between farmers and nomadic herdspeople over dwindling water and productive land doubtless fueled the conflict. The Sahara is part of a global environmental disaster, desertification, which is spreading rapidly southward with the aid of a fickle climate and the mere trickle of a shrinking water table, incapable of supporting the 5 million people who live here. Indeed, of all natural resources, the most tenuous, most threatened, and most precious worldwide, freshwater ranks number one. In the Sahara the age of drought and devastation has already come.

Humans had been through hard times before. There is vivid fossil evidence of the harsh conditions faced by 2-million-year-old human populations in Africa. Not only did these small clusters of people have to track sporadic food and water sources through marked changes in seasons, but leopards, lions, and hyenas frequently preyed upon them. At Sterkfontein and Swartkrans, famous fossil cave sites in southern Africa, the gnawed bones of ancient australopithecine humans have been found strewed about the entrance to caves that were their shelter. But humans eventually found themselves sufficiently numerous and well armed to turn the tables, becoming hunters instead of the hunted.

Even with this accession of power, human predation on Africa's wildlife seemed oddly well managed. Hunting did not lead to rampant extermination of large animals, as it did when humans arrived much later in Australia, Eurasia, the Americas, and numerous islands. Perhaps this difference was due to the long history of cohabitation of humans and large animals in Africa. Coevolution effected a balance in the alternating roles of prey and predator. African animals over millions of years evolved into forms more wary of humans; they kept their distance and survived.

Despite the vast evolutionary time invested in attaining this equilibrium, things do go wrong. During the recent centuries of its colonial imprisonment Africa was invaded by Europeans intent on farming, mining, enslaving indigenous human populations, and rampantly hunting out vast tracts of wildlands. During a single year, 1911, one safari company killed 700 to 800 lions. The carnage promoted by both foreigners and residents alike goes on. In defiance of conservation efforts of more recent decades, poachers have killed thousands of African animals. By 2003, the same year we were scrambling over the bleached cliffs of Mali, the numbers of elephants in Virunga National Park in the Democratic Republic of Congo (DRC), the most species-enriched reserve in Africa, were at an all-time low of fewer than 250 individuals. Poachers had relentlessly culled the elephant populations in the region from 1960 levels numbering more than 4,000. Less malevolent but equally destructive forces that come with human population expansion and development have either destroyed or marginalized many natural sanctuaries.

At the same time, our species has not totally domesticated any continent or indeed the planet. There are still wild places where humans are interlopers, where they must tread lightly. This has always been the case since humans first appeared. Even today wild animals kill hundreds of humans. In many cases, these deaths simply result from unhappy accidents, where the victims find themselves in the path of an aggressive and protective hippo, elephant, or African buffalo. In other cases, humans have once again found themselves repeating the roles of their ancestors: they are simply prey items. Since 1990 in Tanzania, lions have killed nearly six hundred people and injured another three hundred.

Sometimes wild predators terrorize in tandem. A 2005 World Wildlife Fund dispatch tells us that scientists visited an area in a Mozambique village near the Tanzanian border, where they discovered a nightmarish scene. Wild elephants had raided the farmers' fields, absconding with an estimated two-thirds of the crops. Farmers were not allowed to shoot the recently protected elephants, so instead, they slept in their fields, only awaking to bang on pots in a feeble attempt to chase away the noisy invaders. Unfortunately, stealthy lions and hyenas soon learned that sleeping farmers were easy prey. Along a twenty-kilometer (twelve-mile) stretch of road thirty-five deaths were recorded, largely because of the gluttony of a single rogue lion. During the same year hyenas attacked and seriously injured fifty-two people, resulting in twenty-eight deaths. Many inhabitants were so terrorized by these predators that they fled the area; certain villages were rapidly transformed into deserted ghost towns. In the Sahara, where densities of lion populations vary from low to nonexistent, the hyena is often the top predator, feared by wild species, domestic animals, and humans alike.

As I walked along the scarred Saharan cliff, I picked up an evil scent, an incisive rot of decaying flesh. In a ravine descending from a dark overhang in a rock wall, I saw scattered bone. As I approached the ravine, I could inspect this graveyard with a paleontologist's eye. There were small, slender bones of a hoofed animal, perhaps an antelope, and other small bones that clearly belonged to a goat or two. There were even some robust ribs, limb bones, and vertebrae of a large animal, most likely a camel. For someone like me who spends much time looking at the ground when walking the earth, piles of bones are not an unusual sight. Nonetheless, this isolated monument to carnage should have sparked some sense of trepidation in me. Why it did not I have pondered many times since. Perhaps the heat had dulled my brain. Perhaps for a moment I forgot I was in Africa, where the same formidable prowlers of the Serengeti can be encountered even in a lifeless desert.

Then I heard a heavy thud of a footfall, not typical of a two-legged human but of something four-footed and largish. In the dead air I picked up the sound of a low exhale, almost like the chuff of a locomotive. I stopped. A beast emerged from behind a boulder only eight feet from where I stood. The first thing that struck me was its mass; it seemed more the size of a pony than a typical African carnivore. Its body was as dull gray as the boulder between us. I could see shaggy, limp hairs, wet with sweat, hanging like entrails from its flanks. I could count ribs. The animal's muscular neck ended in a head that was huge, ugly, and misshapen, with a tiny eye perched above an oversize jaw. The canines jutted from a rippling line of black gums. For a moment I was completely transfixed by the weirdness, the grotesque nature of the thing. My experience as a biologist counted for nothing. The beast did not look of this world. Then something more conscious and analytical came to the surface. This was in fact a very large hyena.

The creature gave me an evil and ambiguous beady eye and a snarl. I was too frozen in shock to tell whether it was standing still or slowly drifting to my right. Then I could perceive the hyena at last receding, with agonizing slowness. Suddenly it chuffed again and lunged down the ravine. Several yards away the hyena made a complete turnabout, and for one terrible instant I thought it had resolved to charge me directly. But it suddenly shifted again to the far end of the ravine and started loping up to the top of the cliffs. Even from that distant summit its silhouette had a striking mass to it.

My hand radio crackled. "Did you see that? That's the biggest hyena I've ever seen!" Eric, one of the team geologists, was calling.

"I saw it," I replied.

I could hear excited shouts in French and Tamashek, the high-pitched dialect of the Tuareg soldiers, rising from somewhere below. Please don't unload your automatics, gentlemen, I thought. I'm up here.

In the end this unexpected encounter in the Sahara failed to distract us from our morning's prospecting for fossils. By noon I was lying under the Land Cruiser, only consumed with avarice as I watched Leif, one of the young geologists, pour his private stash of Gatorade into a canteen full of hot, alkaline water. My meeting with the hyena seemed worthy of only brief conversation; it was too hot to talk much. Compared with the grim statistics on animal attacks and the horrific tragedies that were part of life in this harsh place, the whole affair was hardly dramatic. Later, though, my memory of that encounter came back, and I relived the few seconds of that Saharan rendezvous. I have many times seen that gray, ugly head and that beady eye and picked up the foreboding stench from a hyena's cave. It is a memory that shocks me with its oddity and its potential primal violence. It seems an event reincarnated from someplace deeper in my own history, from a time when my ancestors walked the plains of Africa in trepidation. Ecologists often point out that the image of Earth still harboring unspoiled, pristine wild places is a myth. We live in a human-dominated world, they say, and virtually no habitat is untouched by our presence. Yet we are hardly the infallible masters of that universe. Instead, we are rather uneasy regents, a fragile and dysfunctional royal family holding back a revolution. The relationship between what we call the civilized and the untamed can be mysterious, changeable, and unpredictably violent. Sometimes nature can show its awesome, intolerant power, a power that emanates from a history much longer than our own. Where does that history lead us now?

About 7 million years ago—58 million years after the great dinosaur extinction event, nearly 100 million years after the modern land ecosystem of flowering plants and pollinating insects emerged, and 3.493 billion years after life first appeared on Earth—humans became the primate group to watch. At first there was not much to see. Small bands of primitive, human-like species lived out desperate lives, dividing their time between being either slow-moving, poorly armed predators or, as I have noted, prey items themselves. Six million years later or more, tens of thousands of years ago, humans made deeper incisions on the land. They decimated large animals on many islands and most continents and, later, created their own biofactories through plant cultivation and animal domestication. The advent of agriculture fueled an unprecedented acceleration of human populations, but one that seemed sustainable in a wild and bounteous world. It was not until 1800 that global human population reached 1 billion. And still, even in the face of this horde of humanity, Earth seemed an expansive, mysterious, even intimidating place. Major tracts of rainforest in Africa and South America were inhabited only by small clusters of people who carried on their secretive hunter-gatherer traditions amid a staggering cornucopia of life. Huge areas were poorly known or not known at all. Antarctica, an entire continent larger than Europe and only slightly smaller than South America, was not seen by humans until 1820. A well-known European map from that same year shows western North America, interior Australia, and central Africa in blank white. These were spaces of terra incognito, whose landforms, animals, plants, and indigenous peoples were all but unknown to Westerners. Not until well along in the nineteenth century did Western scientists encounter, describe, and name the Sumatran rhino, the blue whale, the two living species of tree sloths, the wombat, the pigmy hippo, the spider monkey, and huge numbers of insects and plants. Meanwhile, the passenger pigeon, one of the emblematic casualties of the decades to come, numbered an estimated 3 to 5 billion birds, its flocks engulfing the sky of North America in deep shadow.

How things can change in a few short years! Joel E. Cohen has de-scribed the change in terms of the human population explosion:

It took from the beginning of time until about 1927 to put the first 2 billion people on the planet; less than 50 years to add the next 2 bil-lion people (by 1974); and just 25 years to add the next 2 billion (1999). The population doubled in the most recent 40 years. Never before the second half of the 20th century had anyone lived through a doubling of global population. Now some have lived through a tripling. The human species lacks any prior experience with such rapid growth and large numbers of its own species.

The matter of sustaining such an unimaginably huge human popula-tion today becomes more challenging. Our voracious use of land for food production, energy, and habitation is necessary for our survival. As Norman Myers has pointed out, in less than thirty years we shall need to feed an es-timated 8.2 billion people, 32 percent more than exist today. This requires a boost of 50 to 60 percent in food production from current levels, or about a 2 percent increase per year. The immensity of the challenge and what it will cost can be appreciated by considering the so-called miracle of bioagri-culture in our time. The agricultural breakthroughs of recent decades have produced only a 1.8 percent *cumulative* increase in food in the decade be-tween 1985 and 1995, for example. To make matters worse, the agriculture that has thus far sustained human populations is hardly itself sustainable enough for future needs. Land use over the past two decades presents a

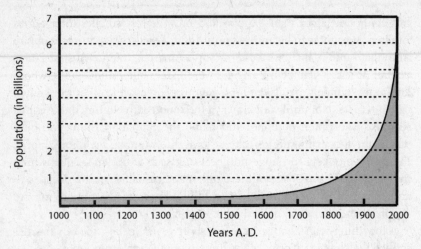

The human population curve

disturbing picture of degradation. In that short span of time 5 billion tons of topsoil have been removed from formerly arable lands. During the past forty years at least 4.3 million square kilometers of cropland (more than twice the size of Alaska) have been abandoned because of soil loss.

Of course, to have degraded soils, you must first have converted land that was once untamed and used it for cultivation. From the advent of agriculture eleven thousand years ago humans have converted more than half of Earth's once pristine forestland. Today's forests occupy only about 30 percent of the total land surface (about 4 billion hectares, or 8 billion football fields' worth). Deforestation is carried out at a global annual rate of 7.3 million hectares (17.5 million acres); in the last five years alone the world has lost 37 million hectares (91 million acres) of forest, about the size of Germany. This number is also somewhat deceptive; the real loss of original forest has been about 64.4 million hectares, about 13 million hectares a year, but 27. 8 million of those hectares are either naturally regenerated or have become converted, industrialized tree farms, such as the timber plantations in North America and oil palm plantations in Southeast Asia. There is no guarantee that regenerated forests or especially those tree farms have recovered many of the species that once lived there.

The situation for tropical rainforest, the habitat that is estimated to account for 40 to 50 percent of all species on Earth, is particularly distressing.

Despite impassioned pleas and elaborate strategies for conserving rainforests in places like the Amazon or the Congo River basins, the rate of loss
has hardly abated. Brazil lost 16 million hectares between 2000 and 2005,
at a rate that, if unchecked, will reduce its forestland to 60 percent of its current size by 2050. In fact, Brazil's annual rate of loss in the first decade of
this century actually increased over that for the 1990s. During the same
2000–2005 interval South America as a whole lost 21 million hectares.
Africa, with a significantly smaller amount of forest cover, lost a comparable
21 million hectares. If the pace of deforestation continues, many of the
world's tropical rainforests will be gone in a few decades. Even the great
tracts of forest in the Amazon and the Congo seem destined not to outlast
this century.

Relentless conversion of land for habitation and agricultural product is
not the only dangerous driver of environmental destruction. Humans have
promoted some other contributors, dark forces, along with land conversion,
we might analogize with the Four Horsemen of the Apocalypse: oppression,
war, famine, and pestilence. Like the Four Horsemen (Jared Diamond has
called a very similar list the Evil Quartet) bearing down on us, these forces
are assaulting the planet. Deforestation or other forms of habitat conversion
and destruction are only the first; overharvesting of certain animals whether
through hunting and fishing, the introduction of invasive species into new
habitats, and pollution are others. Fisheries have erased more than 25 percent of the productivity in upwelling ocean regions and 35 percent of the
productivity in shallow temperate waters along coastlines and over continental shelves. Invasive species have wiped out resident birds on Pacific islands, destroyed the original fish communities in African lakes, and choked
the waterways of eastern North America. Most of the major coral reef systems in the world have been more than 50 percent degraded through a
combination of overfishing, pollution, disease, and bleaching caused by
warmer waters. The sixteenfold increase in energy use during the twentieth
century created 160 million tons of sulfur dioxide emissions per year, twice
the levels of emissions from natural sources. The last decade marks the time
when air pollution became global in extent, when we humans freely ex-
changed our polluted air across oceans and continents.

Some of these actions have cascading effects. Fossil fuel burning and
agriculture have together created an ultragreenhouse atmosphere, with an
increase in carbon dioxide of 30 percent and in methane by more than 100

percent. More CO_2 in the air means a warmer planet, and here the historical facts are riveting. CO_2 levels in the atmosphere have not been so high in 10 million years, when climates were warmer, sea levels were higher, and large portions of land now occupied by millions of people were under water. A significant chunk of what we now call land we may soon call ocean. Models that account for the complexities of climate change over time predict an increase in global mean surface temperatures by 2030–2050 of anywhere between 1.5° and 4.5°C, depending on the region. This range of increase in temperature will not only dramatically warm the climate but disrupt current weather patterns and threaten the survival of many lifeforms. Adapting to this profound climatic change will be enormously difficult for many species, since the loss and fragmentation of natural habitats will make it impossible for them to escape the heat and expand into more hospitable areas.

Some people have the temerity to act as if this destruction were the inevitable result of progress, the necessary by-product of the human path to world domination. But if that is so, then our ecological conquest is a pyrrhic victory. Not only do these destructive forces erode the potential of the biota to sustain humans more efficiently, but they have massively negative impacts on species living in those destroyed habitats that are critical to maintaining the ecosystems and of enormous benefit to humans. To biologists, a record of species loss is a particularly important measure of environmental destruction. Species, the fundamental units in biology, are groups of individuals, reproductively isolated from other similar groups, that share a unique evolutionary history. They are the points on the map of the biological world. With the upsurge of the human invasion, some species have become marginalized in fringe habitats, hanging on in small highly vulnerable numbers. In other cases we've taken no prisoners; species have been entirely eliminated.

The casualty list of extinguished or endangered large animals likely resulted from hunting by humans started about fifty thousand years ago. Those casualties are easily recognized, offering vivid images of the mighty and the fallen—cave bears, woolly mammoths, giant kangaroos, and "Irish elks." Most of the species currently under threat or facing extinction would hardly be recognizable to all but a few specialists; some are not even known to science. We can estimate that the destruction of natural habitats along with the dark forces of overexploitation, invasive species, pollution, and human-induced climate change means that every year thousands, perhaps tens of thousands of species are going extinct. This means that 30 percent and perhaps as much as 50 percent of all species might be extinct by the middle of this century. Not all these endangered species are the familiar and the cherished—the tiger, the polar bear, the right whale, and numerous birds. They are also certain moths, midges, wasps, flies, worms, flowers, fungi, protists, algae, and bacteria that keep the energy flowing through the global ecosystem.

The statistics on current and projected extinction of species can be rolled out, and they have been in many papers and books. But what do all the horrific numbers mean? After this massive destruction will life starkly change for the human and other remaining denizens of the planet? Various scenarios are proposed for the future of the biota and the future of our world resulting from the current catastrophe. Many of these are far from precise; there are many things we do not know and many events we cannot forecast. Nonetheless, there is also a sobering baseline certainty about the nature of the Earth shock we are experiencing.

How did we get to this place, this pivotal and troublesome moment? Perhaps we have relied too much on a conviction, whether explicit or subconscious, that nature will sustain itself in spite of our meddling. The fossil record tells us otherwise, however. We learn from it that life, notwithstanding its vast diversity and apparent robustness, has an unpredictable and tenuous quality. Life's history is tumultuous. The present-day diversity of life—some 1.75 million named species and doubtless millions more yet to be discovered—is but the surface coating of the multitudinous life-forms on this planet that have appeared and disappeared over eons. There are no more arthrodiran fishes with their robotic-looking armored heads, nor are there trilobites sporting spines and hundreds of beady eyes. Gone are the dinosaurs as tall as a three-story town house. Gone are the beautiful coiled-shelled ammonites, the brachiopods or lampshells, dragonflies with the wingspans of seagulls, ferns as tall as date palms, trees with barks like crocodile skin, twelve-foot-long dragons with sail fins along their backs, Australian mammals as big as rhinoceroses, lemurs as big as gorillas, giant hairy elephantine mammoths, countless small fungi, bacteria, pond organisms, insects, and all but one of the nearly twenty species of human and human-like apes.

So, during this long history of appearance, flourish, and decimation, what spawned the living world so familiar to us? What are the roots of our lively planet, the wild, the cultivated, the harvested, the stripped, and the scorched habitats of us landlubber humans? To answer these questions, we must focus on entire ecosystems, the name we give for the complex, inter-connected networks of energy exchange among huge numbers of different species in a single habitat. It would seem that the modern land ecosystems, whether pristine or altered, contain only an echo of an ancient world where long-necked seventy-ton swamp monsters cropped the foliage off huge, grotesquely twisted trees. Indeed, even scientists have long empha-sized the radical transformation of ecosystems in the aftermath of species extinction, climate change, and the jostling of landmasses over the past mil-lions of years.

But another history is emerging, one more accurate and illuminating for contemplating our evolutionary past as well as our environmental future. This book is based on the late-breaking evidence for that revised history. It covers not only new discoveries and new ideas but major questions that still challenge us. The quest of science is never completely fulfilled; science would indeed degrade to ennui if there were no remaining puzzles, dilem-mas, or debates. The book is meant to provide a statement of the ancient-ness, resiliency, and, at the same time, the vulnerability of the biological world we call modern. I argue that this modern world took shape certainly much earlier than the last few thousand years that have marked human do-minion over the land. It indeed long preceded the 7-million-year-old ap-pearance of humanlike species in gallery forests of the African continent. The essential architecture of our modern land habitats—in scientific par-lance, the present-day terrestrial ecosystem—was built in the very distant past, before *Tyrannosaurus rex* stalked the Earth and before our furry mam-malian forerunners exceeded the size of a house cat.

This book is a chronicle of 100 million years of modern living. Roughly that long ago, while dinosaurs were the big animals on Earth, our planet's terrestrial ecosystem began its metamorphosis into one like our own. The chapters in Part One, "The Way of the World," provide the diagnostics: the fleeting and threatened wildlife in the crypts of nature; our knowledge and our ignorance about the diversity of life; the dominance of diverse insects, plants, and microbial life-forms; the stark statistics for the current mass ex-tinction event, which is strikingly reminiscent of some mass extinction events we have learned about in the past; the chugging energetics of modern ecosystems and their instability in the face of human expansion. The final chapter of this section describes how science has developed the theory of evolution, the great explanation for the diversification of all present and past life, and how denying the power of this theory through ignorance or antipathy not only impedes enlightenment but threatens any practical strategy for our own survival.

In Part Two, "The World Becomes Modern," I trace the roots of the modern ecosystem, from 475 million years ago, when creatures from the sea invaded the land, to some 300 million years later, when the dark pine forests and fern gardens of the early dinosaurs dominated the landscape. To understand the scale of this history, we must dissect another dimension, time, and learn how the calendar of Earth's history marks both the birth and death of species and their replacement by new species as well as calibrates the milestones in the history of the modern ecosystem. This is an epic about a topsy-turvy world, the saga of the rise and fall of biological empires that culminates in a 100-million-year-old realm teeming with florid color and texture. I make the case for the intimate likeness between the present and past on the basis of many factors, not the least the appearance and enrichment of flowering plants and the coincident diversification of pollinating insects such as bees, wasps, and butterflies.

Part Three, "Death and Resurrection," deals with the destruction—possibly the incineration—of life coincident with the asteroid impact 65 million years ago, a catastrophe that wiped out all those dinosaurs except their living descendants, the birds, and erased more than 70 percent of all the species then living on Earth. Nonetheless, the basic architecture of what I call the modern ecosystem was, though battered, not completely shattered. Within a few million years following that dinosaur apocalypse, the garden was restored, albeit with a few changes. It bore more fruit, grew grassy around its edges, and was subject to vigorous cropping by big mammals that filled the ecological roles of the extinguished dinosaurs. Still, the prolonged recovery of the ecosystem was not at all complete.

What are the current state and prognosis of this 100-million-year-old ecosystem? Part Four, "Terra Humana," describes how humans evolved and took possession of the Earth and how the biota from the dinosaur age has been plowed, harvested, exploited, destroyed, and, in some rare salutary instances, even sustained. Chapters in this section deal with the evolution of the human family and our ancient roles as exterminators and then cultivators. Subsequent chapters describe the dark forces of habitat destruction, overhunting and overfishing, invasive species, and pollution, all drivers of extinction promoted by our own species. The penultimate chapter looks at the arresting new evidence that industrial emissions of greenhouse gases are changing the climate in fundamental ways, often to the detriment of species and ecosystems. The final chapter builds on the past and the present to forecast, with understandable qualification, the evolutionary future of our own species and other species. Here I also cite some examples of our remarkable capacity for mitigating some of the destruction engulfing the planet, efforts that must catch on in a much more comprehensive, global-scale way if we are to improve the odds for a sustainable future.

The mark of history that shaped the present-day planetary biota and its diverse species—the source of food, medicine, materials, and aesthetic pleasures so important to human lives—is indeed a profound one. But the mere last slice of this chronicle, the last few decades of our history, is perhaps the most dramatic and, from the perspective of our own self-interest, the most alarming. A forecast of how those systems may change brings us back to the central argument of this book. What has happened since the Industrial Revolution, especially during the last few decades, and what will happen in the next few decades are not simply degradations of various habitats that are accustomed to disruption and ready-built for response. We are seeing an unprecedented transformation of an ecosystem that evolution has refined to wondrous complexity and powerful function over 100 million years. As a function of its wealth and its countless benefits to ever more needy and voracious humankind, the modern ecosystem may be today facing its greatest threat in its 100-million-year life span. The wild garden that embraces us, and compels us to strike a balance between our perceived needs and our existence in a sustainable biological world, indeed has a long history of cultivation that is again under grave threat.

Many of the classic and powerful disclosures on environmental deterioration by Rachel Carson and others eventually in the 1990s inspired a much broader perception that something was actually amiss. Perhaps pivotal dates came with two events, the UN Convention on Biological Diversity in 1992 and the Kyoto Conference in 1997, events that galvanized many of the world's nations (but sadly not all, including the United States) and prompted the first massive efforts in developing international collaborative strategies that involved conservation and self-restraint. Thus in the last decade of the twentieth century a sense of urgency about the destruction of the natural world inspired a scientific effort to identify, once and for all, what we have and what we are losing. The mission was outlined by some key scientific leaders, including Edward O. Wilson of Harvard University, the indefatigable ant specialist and evolutionary biologist. Wilson was the first to publish the word *biodiversity* in the 1988 proceedings from a conference organized by W. J. Rosen, who originally coined the term. *Biodiversity* refers to the totality of diverse species and the myriad interconnections of those species. When it first appeared in print, the word was recondite, even esoteric. Now it is common parlance for scientists, environmentalist, economists, and policy makers.

The current decimation of species is commonly called the biodiversity crisis. Wilson himself wrote a book about this crisis entitled *The Diversity of Life*, published in 1992, a clarion call for urgent, high-priority work to be done by biologists and for swift and effective conservation action based on their results. Subsequently there have been many books, including one that I edited (*The Biodiversity Crisis: Losing What Counts*), that have addressed this problem. Yet it seemed to me, from my paleontological point of view, there was another dimension to this problem and to the saga of our biota, a history with a rich narrative, that needed further probing. Much of that history has been revised in the last five years with fresh discoveries about the fossil record. We are now, for example, in a new phase of exploration, and we better understand how the evolution of species and ecosystems has dovetailed with the physical transformation of the planet—the changes in atmosphere, oceans, and landforms. Such new insights have great implications for our prognostications about such matters as land fragmentation, invasive species, global climate change, and the current mass extinction event.

This book draws heavily on the past but in many ways it is more about our present and especially our future. I am a paleontologist by profession, but I must confess that beneath my professional interest lies a strong emotional attachment to the wonder of that past world. It is important to me and to us all. When I was younger and newer to science, it seemed that the fossils we found and the sites we uncovered generated their own isolated, autonomous awe, one remote, even detached from the present. But I have found it increasingly difficult to separate that past from the present. In relating this history of our modern ecosystem, I hope to deepen a sense of the extraordinary biological wealth that was cultivated through the epochs and that we have inherited. When an item is up for auction, its value is often elevated by knowledge about the source and the intricate history behind the piece. My unabashed bias is that a great history lends value to something and that this increase in value provides a truer sense of what we have and what we risk not having. Perhaps, I hope not too hubristically, this might help cultivate yet more motivation for stewardship at a time of great crisis.

Part One

THE WAY
OF THE WORLD

There are more things in heaven and earth, Horatio,
Than are dreamt of in your philosophy.

—William Shakespeare, *Hamlet*

A CREATURE IN THE FOREST

The most precious things in the world may be the things we never see. Perhaps at this very moment in a rain-doused stand of pristine forest in the Truong Son Mountains in central Vietnam, one of the last and least explored wildernesses on Earth, a creature with long, slightly recurved horns is emerging from an undergrowth of bamboo, palms, and saplings in the shadow under the forest canopy. A fleeting sunburst between the trees highlights an elegant black stripe on a chestnut back that looks like a signature of ancient calligraphy. As the 220-pound animal bends its thick neck down to a stream, it plants the cloven hoof of each foot on the bank. The sun catches the sharp etch of a white band above the feet, an anatomical accoutrement that looks like a bad practical joke, as if the animal were sporting black-and-white spats. In the shadows a tricolor tail of brown, cream, and black whisks against the first battalion of morning flies.

This scenario is entirely plausible but has probably never been witnessed. Yet the beast by the stream is not a fiction. This is the mysterious saola, scientifically known as *Pseudoryx nghetinhensis*, of Vietnam and Laos. The name saola is a local one, referring to the beast's horns (*sao*, spindle; *la*, post), which resemble the parallel posts on the spinning wheels used by people in the region.

The saola is one of the rarest creatures on Earth; few specimens have been collected, and scientists have never seen this animal in the wild. Most biologists, like George Schaller of the Wildlife Conservation Society, have tracked it by surveying the horns and other remnants of hunting forays in villages along the border between Laos and Vietnam. Most remarkably, the

The saola, *Pseudoryx nghetinhensis*

saola was first discovered by the Vietnamese ecologist Do Tuoc during a field survey of central Vietnam's Vu Quang Nature Reserve in 1992. In 1812 the great naturalist Baron Georges Cuvier had declared that no more large, hoofed, herbivorous mammals would be discovered in any part of the world. But here is this creature whose discovery did not come until the last decade of the twentieth century.

Aside from a few color markings and its long recurved horns, the saola looks like a typical hoofed mammal—say, a deer or an antelope. Indeed, its appearance is a helpful clue to its reasonable scientific identification. The saola is related to deer and antelope, but it is also very different from these animals. The scientific name *Pseudoryx nghetinhensis* is from the Greek, meaning "false oryx," suggesting the saola is deceptively similar in appearance, especially in its long curved horns, to the better-known oryxes; the second part of the name simply means "of nghetinh," a regional name that combines the names of two provinces in Vietnam, Ha Tinh and Nghe An, where the animal has been seen. *Pseudoryx* denotes the group, in traditional classification the genus, to which it belongs. The second name is meant to distinguish this species from all other species of *Pseudoryx*. Thus the horse, *Equus caballus*, and the donkey, *Equus asinus*, are separate but similar species that both belong to the genus *Equus*. The saola is so distinct, however, that it gets its own genus; there is as yet no other species of *Pseudoryx*.

As for the broader affinities of *Pseudoryx*, the cloven hooves, the shapes of its grinding teeth, and the structure of its anklebones are giveaways, demonstrating that the saola belongs to a diverse order of mammals known as artiodactyls, the cloven-hoofed mammals that include antelopes, bovids, giraffes, deer, camels, pigs, and hippos. Surprisingly, there is new evidence from DNA and anatomical features that whales might have diverged from some very early and primitive artiodactyl group, perhaps from a lineage that also led to pigs and hippos. Artiodactyls, the dominant large plant-eating mammals of today, were even more diverse in the past. The saola fits clearly within the artiodactyl family Bovidae, the group that contains cows, bison, many African antelopes, goats, and sheep, but just where it fits within this family is a trickier problem. Some students of saola anatomy have assigned it to the tribe Caprini, which includes goats, chamois, musk oxen, and relatives, but recent studies based on DNA put it within the tribe Bovini, which includes cattle and buffalo. For now this alliance seems to stick, and more comprehensive studies of both genes and anatomy are anticipated.

Since the saola was first discovered, researchers have accumulated only a small collection of twenty partial specimens, including three complete skins and two skulls. The saola has even taken pictures of itself, images snapped when the animal unknowingly tripped a camera as it trudged through the steep thick forest of the Truong Son. Yet these animals have not been observed up close by anyone except the people of the forest who have hunted them. Today locating the whereabouts of the saola is both a scientific mission and a sacred one, for this precious animal is making its last stand. It is estimated that only a few hundred saolas may be left in the forests of Vietnam and Laos, in a total area of about two thousand square miles (five thousand square kilometers). Sadly ironic is the increased local effort to find and kill these animals because of their great interest to outsiders and their presumed monetary value. In the Vu Quang Reserve twenty-one saolas were killed and three were taken alive and brought to Hanoi between 1992 and 1994. In Laos seven saolas were caught, but only one survived, and then just for three weeks. This individual, a female, was shy and docile, allowing herself to be petted and hand-fed. The Hmong people of Laos call the saola *saht supahp*, "the polite animal." Unfortunately, it may be too polite for its own good; its numbers are dwindling as it falls prey to hunters and its habitat is winnowed away by loggers and farmers. In recent years measures have been taken to establish more nature reserves and to strengthen their protection; a new initiative extends some of these protected areas across the international border between Vietnam and Laos.

In the spring of 2002 I had a chance to observe at close range the efforts to preserve Vietnam's wildlife, though that was not my primary assignment. At the behest of my employer, the American Museum of Natural History, I traveled to Hanoi with the museum's president, Ellen Futter, and several other officials to announce the opening of the first major U.S. exhibition on Vietnamese culture since the Vietnam War. The exhibit, to be curated by Dr. Laurel Kendall of the American Museum, an expert in Asian religious and cultural ritual and practices, was jointly developed with the Vietnam Museum of Ethnology, and its director, Nguyen Van Huy, was cocurator. The topic turned out to be as complex as it was fascinating. Vietnam has more than fifty ethnic groups, not all of which could be fairly represented in one exhibit. Nevertheless, the exhibition opened to wide acclaim and enthusiastic audiences in New York in the spring of 2004, and by spring 2006 it had been installed in Hanoi.

I was also in Vietnam for a second, biological purpose. Soon after arriving, I joined Dr. Eleanor Sterling, the talented director of the American Museum's Center for Biodiversity and Conservation (CBC), launched in 1995 to link the formidable effort being made by the museum's curators to discover diverse species all over the world with conservation needs and action. Vietnam, with its cultural diversity, its dynamic economy, and its unique but threatened natural habitats, was a logical target. In 1998 the CBC had joined forces with the Missouri Botanical Garden, the Institute of Ecology and Biological Resources (IEBR) in Hanoi, the World Wildlife Fund Indochina Programme, and BirdLife International's Vietnam Programme to make an exhaustive survey of important forestland and to apply the findings to help the government establish secure reserves. In addition to the research and conservation applications, the project was intended to enlighten and educate people about the tenuous natural wonders of Vietnam. To this end, Eleanor wrote, with coauthors Martha Maud Hurley and Le Duc Minh, *Vietnam: A Natural History*, a comprehensive and handsomely illustrated volume which was published in 2006.

Vietnam and Laos for decades had been responding to local and international demand for timber. During the 1990s, Vietnam ranked second only to Thailand in exports of wood from mainland Southeast Asia to the European Union and Japan. By the end of the century logging in Vietnam had reputedly declined; there had been a sharp decrease in economically retrievable timber, and the government acknowledged that runaway depletion of forestland would soon deprive the country of this resource entirely. Deforestation, a complex process not simply confined to logging, is not always easy to estimate, especially when reliable records of past forest cover and forest loss are nonexistent. The real rate at which deforestation in Vietnam slowed is controversial, and reports about it are conflicting. Of main concern to us were the primary forests, or those that show little or no evidence of past or present human exploitation. Vietnam purportedly lost a staggering 51 percent of its primary forests between 2000 and 2005.

The logging industry is the blunt edge of the wedge into the forest. Logging roads become lifelines for migrating people who slash and burn, grow crops, hunt for meat, establish villages, carry on commerce, build dams and irrigation systems, and develop towns and cities—in other words, do all the things people normally do when they colonize new land. We have seen this pattern of invasion throughout the world—in the Congo, in the Amazon, and, a few centuries back, in the virgin woodlands of North America. Whatever people gain from the forest in the way of goods and cropland, they may lose in the form of invaluable biodiversity. Vietnam is now a signature example of this problem. In its postindependence phase since the war with the United States came to an end in the 1970s, the Vietnamese government relocated almost 5 million people from the crowded lowlands along the coast and the Mekong Delta to the biologically diverse uplands. The exploitation of land was intentional; logging was succeeded by agriculture, including vast plantations of such cash crops as coffee. Unfortunately, the government officials who endorsed this transformation overlooked the fact that these fragile upland forests were incapable of supporting so huge an influx of humanity.

If these threats to Vietnam's natural habitats weighed heavily on Eleanor Sterling, she did not show it. Two mornings after my arrival in Hanoi she greeted me at my hotel with a broad smile. Eleanor knows what it is like to experience the profound and magical isolation of the forest. Her dissertation fieldwork as a doctoral candidate at Yale's School of Forestry required that she observe, over several years, the secretive comings and goings of lemurs in a thick stand of rainforest on an uninhabited island off the northeastern coast of Madagascar. But Eleanor is not simply a reclusive scientist; she is also a person comfortable with the world at large, an internationalist, who quickly learned the most appropriate dialect of Vietnamese for her work. Her sincerity and quiet optimism attract many fans and young team members, who believe in her and in what they are doing.

"Today we are going to the institute. They are a great bunch, good friends, great scientists," Eleanor announced.

"Will I get to see a saola specimen?" I asked. I wanted to skip the main course for the ice-cream cone.

"Yes, but first you will have to work."

In this case, work meant a meeting or two or three, an activity that fails to delight me. I too am a field person, and I had just escaped a New York full of meetings. Fortunately, the staff at the IEBR turned out to be delightful. Particularly memorable was Professor Nguyen Tien Hiep, who greeted Eleanor with a hug of sibling affection. Dr. Hiep loves the field too. With his wide, bright eyes and a round face always poised for a laugh, he rhapsodizes about plant life like a forest wise man, a youthful Yoda. We followed Dr. Hiep up a few flights of external stairs, like a rickety fire escape, to his office. There we sat and sipped tea as he proudly showed us his latest pressings of plants collected in the wild. His workplace, with its windows open to rustling trees that crowded the building, seemed more like a tree house than an academic office.

The forests of Vietnam are distinctive not just for their saolas and other rare mammals. Their flowers and other plants, many of them not found anywhere else in the world, are showy and diverse. A special gift to Vietnam is its cycad flora. These spiky, somewhat palmlike plants are not palms at all, but nonflowering plants that hold seeds in cones consisting of overlapping seed-bearing leaves. The blossomlike appearance of these seed-bearing structures has inspired some botanists to argue that they are closely related to the most primitive flowering plants. But cycads preceded the appearance of flowering plants by more than 100 million years. They heralded the Mesozoic Era, the age of the dinosaurs, beginning about 250 million years ago, while there is no evidence of a true flowering plant that predates about 130 million years before the present. Cycads grow in many habitats in Vietnam, including the steep forests in the Truong Son Mountains. Indeed, Vietnam is enriched with a diversity of twenty-four species, a number that exceeds that of any other country in Asia.

Cycads are slow-growing plants. Removal of a few of them can put a whole population at risk because they do not rapidly reproduce new plants. Many of Vietnam's cycad species are threatened by deforestation as well as by selective picking of the most handsome individuals for ornamentals. Gardens in China, Taiwan, and the Republic of Korea are often the destinations of these transplants. Dr. Hiep has led a national effort to save the cycads of Vietnam, as well he should. The 2006 Red List of Threatened Species, issued by the International Union for the Conservation of Nature and Natural Resources (IUCN), rates sixteen of Vietnam's cycad species as either endangered or vulnerable.

When it comes to flowering plants, the bounty of Vietnam's flora is impossible to summarize quickly. I'll mention just one group. The moist forests and marshes harbor a spectacular array of orchids, as many as 897 named species and perhaps more than 200 or 300 that are yet to be named. Orchids are marvelously fastidious. Many of them are pollinated or fed on by only one species of insect. Some orchids lure insects to pollinate by mimicking the shape and color of the opposite sex of a visiting insect. The deceived insect—a fly, wasp, bee, or something else—may indulge in unproductive copulation, but in the meantime, it will have deposited pollen it has carried from another plant. The seductive orchid has thus accomplished its mission to propagate. Some orchids have long, tubular flowers that require insects with bizarrely elongate mouthparts—primarily flies, moths, and butterflies—to feed on the nectar they contain deep within the base of the tube. As the insect probes, the foreign pollen grains on its head and antennae come off and stick to the orchid host. Other orchids trap insects in nectar-filled cavities, allowing their liberation only when the incarcerated insect, in its furious attempt to escape, has left its pollen behind. Charles Darwin was obsessed with orchids, both for their beauty and for their biology. Horticulturalists love them. Many orchids are threatened by the same deforestation and avid collecting that threaten cycads. In Vietnam, slipper orchids (genus *Paphiopedilum*) are in particular danger. Several of these slipper orchid species qualify for the Vietnam Red Data Book, although all species are globally threatened.

We climbed back out of Dr. Hiep's eyrie and followed him to the main building of the IEBR. Warmth and humidity filled a room where several scientists were gathered around a table. A colorful map with detail as intricate as the woven patterns in an oriental rug extended along a whole wall,

Vietnam slipper orchid,
*Paphiopedilum
micranthum*

showing in multiple colors the complex patchwork of Vietnam's natural and developed tracts of land. Dr. Hiep and his colleagues began talking very earnestly about the effort to survey the rich forests in Vietnam's border territories. The conversation segued to a discussion of the planned international reserve that would cross into Laos. Concerns were raised: there were too few scientists to do the work and too few being trained; the government was being less than generous with the support needed to carry out such ambitious scientific and conservation projects; collections were poorly housed, and buildings needed renovation. Eleanor was patiently encouraging, noting the gains that had been made and expressing hope for more support in the future.

I can't recall the next meeting in detail. Eleanor and I were in another office, an elongate one with a narrow window at one end and a large rug in the center that formed the landmass for a coffee table. The rug was surrounded by dull leather couches that reminded me of those I had seen in Mongolia in its Communist days. We were supposed to discuss the establishment of a major natural history museum in Vietnam, a splendid idea. I tried to engage, but my jet lag had not yet dissipated, and the day was long, the cups of tea notwithstanding.

Eleanor nudged me as if to say, "Now you get to see your saola." We followed our guides through a patchwork of buildings into an alleyway, across a street, then across a weed-choked plot of ground—no orchids or cycads here. More buildings. A door was forcefully opened into a building not much larger than a two-car garage, and we entered a very dark room with antlers, skeletons, skulls piled up on shelves, specimens in jars, and a few display cases with some stuffed mounts. This would have to do for Vietnam's temporary natural history museum. As we threaded our way among the cabinets, someone pointed to my left. There, barely distinguishable from other shadows in the display case, was a stuffed mount of *Pseudoryx nghetinhensis*.

The next day I arose at 4:30 a.m. as I had on the previous mornings in Hanoi. After years of travel to eastern Asia I have learned that the photoperiod shift that comes with flying west over the Pacific, while disastrous for late-afternoon meetings, is perfect for early-morning activity. I joined my museum colleague Gary Zarr for a jog around Hoan Kiem Lake in the middle of the city. Hanoi is magical at this hour. The French colonial buildings, mostly pale yellow with white trim, burn gold in the dawn. The purple of the morning sky conceals a pollution that soon rises with the buzz of thousands of small motorcycles recently imported from China. The surface of the lake itself ripples like waves of blue silk in the morning breeze. This is a view of the lake from afar, however. Up close, bits of trash float on a reddish brown aqueous film, a mixture reminiscent of those bowls of nuoc nam sauce served with delectable spring rolls. But here the resemblance ends; indeed, the lake itself offered the aroma of dead fish faintly laced with sewage.

Hoan Kiem Lake means "Lake of the Returned Sword," a name inspired by an elegant legend. In the fifteenth century King Le Loi finally repulsed Ming invaders from China after ten years of struggle. His victory relied on a magic sword provided by local fishermen. The war over, the king went for a recuperative boat ride on the lake only to be accosted by a giant talking turtle that requested the return of the magic sword. The king complied with gratitude, and the name for the lake has stuck. The turtle is not a fabrication. It does not talk, as far as we know, but it is spectacular, the largest and rarest of all softshell turtles, at 6 feet long and 550 pounds (250 kilograms). The turtle's identification is somewhat controversial, but most experts believe that it is a giant variant of the Shanghai softshell turtle, *Rafetus swinhoei*. It is not known how many of these monsters live in Hoan

Kiem Lake, perhaps as few as only two or three. Now and again one of them sticks its head out or even floats its body over the unctuous surface of the water. Such sightings are the source of great excitement and celebration, a credible Loch Ness monster making its rare appearance and, according to Vietnam tradition, a very good omen. The lake in its polluted condition, however, seems a highly unlikely sanctuary for even the few turtles that remain. Still, Hanoi has invested serious money in dredging centuries of accumulated sewage-infused mud and removing excess nitrogen, phosphates, and other nutrients that have spawned algal blooms that die off and poison the lake, killing fish and possibly the giant turtles in the process.

On this, our leisure day, Gary and I turned out to be the only tourists from the American Museum ready for migration. The others in the party decided to stay and explore Hanoi, a cultural nexus we barely knew. The two of us took the short flight to the port city of Da Nang, where we were greeted by our guide, Traung, who had already received the news that our ranks had thinned. He was gamely holding up a sign reading "Gary" and "Mike" with strikeouts through the names "Ellen," "Lisa," and "Linda." Our first stop was the Cham Art Museum, an elegant building in typical pale yellow filled with magnificent statues commemorating various gods, kingdoms, wars, and heroic acts. Da Nang of course is a name and a city of legend for another history. It was of profound importance to any American who remembered the progress of the so-called Vietnam War or followed it as it occurred with great interest and anxiety. The city had been used as an air base by U.S. forces, although the Vietcong concurrently had managed to burrow a complex network of tunnels underneath it. This prompted me to ask Traung about *the* war.

"Which war? Oh, you mean the American war. We have had so many wars, you see," he replied.

Before coming to Vietnam, I had imagined its postwar landscape as being the epitome of scorched Earth. But everywhere I looked there was green, a combination of richly wooded hills separated by rice paddies connected by huge networks of irrigation channels. This seemed a place of bounty even for the 80 million people who lived there. Where was the land laid waste by war and poisoned by Agent Orange? It turned out that my impression was too myopic to be accurate. U.S. armed forces had sprayed Vietnam's countryside with more than 20 million gallons (76 million liters) of Agents Orange, White, Blue, Green, Purple, and Pink, a kaleidoscope of toxins aimed

at defoliating both forests and agricultural fields and thus revealing the lurk-
ing enemy. Many of these agents contained the highly toxic substance
dioxin. Apparently the effects of this noxious cloud remain; satellite images
show the straight lines of scars through forestland extending for more than
twenty miles, the defoliated areas where vegetation has failed to grow back.
Some of the soil also contains a high level of toxins and thus a low potential
for renewal. As Eleanor has noted in her recent book, scientists are still striv-
ing to understand more precisely the effects of these defoliants.

The high point of our tour was a visit to Hoi An, a port of exquisitely preserved buildings, some dating back to the fifteenth century. Traung took us
to his favorite restaurant. Through years of travel I have learned to be wary
of such arrangements. A special deal made between a tour guide and restau-
rateur usually results in what is at best mediocre and at worst execrable food.
Such was my experience at the Great Wall of China, the Great Pyramids of
Egypt, and the Great Pyramids of Teotihuacán outside Mexico City. But
there was no reason for suspicion in this case. Gary, Traung, and I were
treated to nine courses of the most astounding food I had ever savored. I remember a single scallop floating in a swirl of vinegar, sea salt, ginger, lime
juice, and chives. The scallop seemed outrageously delicate and fresh, as if
I were scooping it out of some enchanting tide pool. The fish was just as
good, and the beer ample. The meal had a certain meaning to me beyond
its culinary powers. All its components — the icy crispness of soybean shoots,
the slightly steamed slivers of asparagus, the potato chip–thin slices of ginger, the delicate fish broth whitened with coconut milk, and the strategic
shocks of tiny green and orange peppers — seemed to flow together like an
ecosystem of food. As we stacked a pyramid of cans of 3,3,3 ("Ba, Ba, Ba")
beer on the table, Traung grabbed a guitar off the wall and started singing
Vietnamese folksongs and American rock oldies. I took the guitar and did a
slurred version of Dylan's "Don't Think Twice, It's All Right" and an even
less refined rendition of the Brazilian masterpiece "Felicidade." There was
a call for one more round of Ba, Ba, Ba.

We slowly disengaged from the restaurant, moved out to a blinding afternoon sun and headed to dockside and the shade-offering tin roofs over the
town market. The eruption of bounty at the market was at its daily peak; displayed was a range of fruits, vegetables, fish, and spices almost obscenely
sumptuous in their textures, smells, and colors. Although most of the world
subsists on a handful of crops — primarily wheat, rice, soybeans, and corn —

more than fifteen hundred species of plants serve as food for humans. I could swear they all were in the Hoi An market.

Near the edge of the market was a small stall where a rotund, elderly man sat like a Buddha on an ornately enameled stool, his scowling wife behind him. Surrounding the man were a great number of large glass jars, each with something long and helical suspended in a murky root beer–colored liquid. The submerged objects in question were snakes, and this was the famous snake wine, a strange concoction that is believed to provide acute virility, a sort of organic folk medicinal answer to Viagra. Before we could demur, the old man dipped cups in a very large jar and offered it to Gary and me. Not bad, a bit like a cognac that had failed some refinement. The appreciation on my face registered. The old lady squinted and laughed, exclaiming, "Good night!"

It was not our intention to indulge in this patronage. Snake wine is a problematic product. Many of the snakes used for the purpose, including several cobras, were endangered species. But customs are not easily abandoned here and elsewhere. Perhaps the most emblematic of victims in this regard is the tiger (*Panthera tigris*). The demand for traditional medicines in Asian countries, notably China, has led to what could eventually be the total extermination of tigers everywhere. It is estimated that more than 60 percent of China's 1.3 billion people still ascribe to the salubrious effects of tiger bone, bear gallbladder, rhinoceros horn, and dried gecko. What few Vietnamese tigers are left lurk in the Truong Son Mountains—perhaps on the lookout for that rare saola—and scattered other parcels of remote forest. The drastically reduced population of Vietnam's largest wild cat is, again, the result of deforestation, lack of prey, and overhunting by people to obtain putative medicines to sell in the booming Chinese market. By the early 1990s, procurement and sale of such remedies were a six-billion-dollar-a-year business. The Environmental Investigation Agency (EIA), an independent, international group, estimated that at least one tiger a day was being killed for Chinese medicine. Nearly half this trade came through Hong Kong; an estimated nineteen hundred kilograms of tiger bone, the equivalent of 400 to 500 tigers, were exported to Japan from Taiwanese waypoints in 1990. Other countries on the receiving end included the United States and Great Britain. One might suppose that a commodity this rare and tenuous is expensive, and it is. Only a few years ago tiger bone was fetching about $140 to $370 per kilogram (2.2 pounds), and a bowl of virility-inducing tiger penis soup could be slurped for $320. This seems an outrageously low value for the devastation of a priceless species in the wild, especially in a world where cultivated renewable product, such as a bottle of first-growth Bordeaux wine, can command hundreds of dollars per liter.

To be sure, efforts have been made to stop the slaughter and plug the commerce. In recent years organizations like the World Wildlife Fund have stimulated wider concern about endangered species and spurred investigations of medicinal alternatives. Surveys in 2003 revealed that endangered species products showed a marked decrease in traditional Chinese medicine shops in North American cities. But Asia itself still represents the primary challenge. China has been a member of the Convention on International Trade in Endangered Species (CITES) since 1981, but for years it has failed to enforce any of its regulations, especially those pertaining to the importation of tiger parts from India and other countries. In 1999 China did respond to international appeals by passing a law prohibiting the killing of tigers and other endangered species for traditional medicines. Hunting of tigers in the wild is now also at least officially prohibited in all countries where they live.

In March 2006 a workshop in Beijing brought together practitioners of traditional Chinese medicine (TCM) with researchers, wildlife experts, and government officials to discuss the practices that are destroying Asia's own natural resources. The government has also committed research funds to identify medicinal alternatives. However, the funding levels are comparatively modest, considering the vast amounts of new money China has poured into scientific research and technological development, and it remains to be seen how effective China will be in enforcing its new law. There were probably more than 100,000 tigers in the wild less than a century ago. The 2006 Red List still categorizes *Panthera tigris* as endangered; recent studies estimate that there are fewer than 2,500 mature tigers today and no localized population with more than 250. Persistent demand for traditional medicines, fueled by increasing consumer wealth, encourages active, albeit now officially illegal, tiger trade. The threat of eventual extinction of this species in the wild is still with us.

On our return to Da Nang we stopped at another famous spot, China Beach, where U.S. officers used to take their R and R. I could see why. This long, beautiful beach made a graceful curve against a siege of waves feathered by a cool onshore breeze. It was like the beach in my hometown, Santa

Monica, only with clearer water and an elegant grove of palm trees unbroken by hotels and other commercial buildings. Gary, Traung, and I joined some locals in downing a few more beers (this time Tigers). We then reluctantly joined our driver, appropriately deprived of alcoholic refreshment and waiting patiently in our van, for our return to the airport. By this time Traung was in not much shape to guide. He only said, "This has been one of the best days of my life." We agreed.

On the return flight to Hanoi, I thought of Traung's enthusiasm, fleeting tigers, snake wine, giant turtles, saola taxidermy, Dr. Hiep's cycads, the IEBR, and Vietnam's past and future. It seemed so odd to be immersed suddenly in a country that as a teenager I had sworn I would never go to, certainly not then for the purpose of killing and dying. But now Americans and Vietnamese were intimately connected in their work to initiate a healing process that celebrates Vietnam for its wondrous culture and art and to participate in a desperate effort to hold back the wave of another kind of destruction, euphemistically known as development, assaulting this beautiful country. And Vietnam was no different from, no more beautiful and no more vulnerable than, other places I had visited, no more than the green bamboo-clad hills of Sarawak, the thorn forests of Madagascar, or Lake Tana, with its waters of robin's-egg blue set against the scarred landscape of Ethiopia. Vietnam was another pitched battle in a campaign to stave off what, in moments of cynicism or despair, seemed inevitable.

Conservation efforts in the world today are broadly aimed. The targets are whole regions, landscapes, ecosystems, and numerous species. Yet people like to have symbols for motivation, touchstone species that seem to capture in one image both precious nature and the urgency to protect it. In this regard the saola of Vietnam and Laos is a very effective symbol. It is not only a cornered species, nearly gone before we even have gotten to know it, but a hanger-on from a lost world. This is not a sci-fi world of spewing volcanoes, tortuous vines, dinosaurs, and giant gorillas. It is instead a world whose forests extended uninterrupted from the sharp limestone mountains of the Indochina peninsula to the azure Pacific, a world without sprawling cities, motorbikes, rice fields, and scorched Earth, a world full of animals like the saola that we shall never have the pleasure and the honor of seeing alive. Twelve thousand years ago much of this part of Asia—indeed, much of the world—

was inhabited by dramatic, big, and strange beasts. Fossil deposits are the only evidence we have that these spectacular big mammals were spread all over the world. Today only in Africa do they persist in such great variety.

It is important to recognize as well that the saola is not the only symbol of this lost world. Vietnam has an arresting number of unique species, what in science we call *endemic* species, species found nowhere else but in a single region, sometimes a single valley or a single stream. Other Vietnam endemics include some of its orchids and cycads, the warty pig (now likely extinct), three species of primates, and several barking deer, known for their diverting vocalizations that emanate from the forest primeval. Why this plenitude of endemics? To try to answer, we must consider not only the impact of humans on wildlife but also the effects of climate change. One cascade effect from climate change is particularly profound in the tropics. A warming trend at the end of the last glaciation some twelve thousand years ago caused sea levels to rise 125 meters (410 feet). Coastlines and lowland areas all over the world were affected, but flooding of the land and disruption of the biota were most pronounced in the extensive lowland areas, notably in the tropics and more specifically the tropics of southeastern Asia. By eight thousand years ago more than half the land areas supporting lowland tropical forests during the earlier phase of glaciation, with its lower sea level, had been entirely flooded. On the mainland, animals and plants migrated to mountainous regions, such as the Truong Son, above the flooded plain. Former coastal areas, peninsulas, and land bridges became small islands, little emergent arks stuffed with surviving species. This is why many of these refugia have remarkable and unexpected concentrations of large animals, including orangutans, tigers, banteng cattle, gibbons, and hornbills, as well as saolas and barking deer.

What else once lived on the lands now submerged? The fossil record in the tropics is poor, and we do not have a very good idea about what was lost, but there is a basis for an educated guess. Simple application of some classic ecological principles offered by Robert MacArthur and E. O. Wilson suggest that this much land loss meant at least a 10 percent loss in species through extinction. Moreover, because of the confined land areas, surviving populations have probably declined from their former supersaturated levels, and many of them are in a precarious state. Tigers are threatened throughout their entire range, but perhaps most endangered on Indonesian and other islands that are the remnants of once extensive habitat.

Another aspect of the shifts in the tropical land biota twelve millennia ago warrants mention. Mammals bigger than fifty kilograms (110 pounds) living in tropical forests, especially in South and Central America, suddenly disappeared. Of the bulky, browsing herbivores, only the tapir remains, in these New World forests, as an important agent in spreading seeds of large trees and other plants through its feces. This disappearance of massive herbivores was likely due to a combination of climate and human hunting, and it unquestionably had a serious impact on the regrowth of plants in many areas. Recent studies of the Amazonian ecosystem show that the fluctuating abundance of even somewhat smaller mammals, such as the white-lipped peccary, is directly linked to the propagation of seedlings of common trees.

The big mammals of ten to twelve millennia ago were victims of extinction events ascribed to various causes: climate, competition from invading animals, disease transmission from early humans or their domesticated animals, and intensive hunting pressures from those same humans. The last of these, the overkill theory, is currently the explanation that scientists most widely accept. Nonetheless, some reject it and note that the extinction of large animals has not been as marked in historical times as in prehistoric times, despite the great surge in human populations, habitat destruction, hunting technology, need, and motivation.

But this argument is actually blind to the real situation. Extermination of large mammals in recent centuries has been much more intensive than for mammals in general. Five species of large hoofed mammals have gone extinct in just the last five hundred years: the Arabian gazelle, red gazelle, bluebuck, and two species of hippopotamus from Madagascar. Several other species have gone extinct in the wild, though in some cases they have been restored through captive breeding and reintroduction programs: Przewalski's wild horse from Central Asia, Père David's deer, the Saudi gazelle, the black wildebeest, and the Arabian oryx. One might also include on that list two species of bison, given their dismal history of virtual extinction in the wild. Moreover, several large species would surely have suffered extinction without the protective measures taken just in recent decades: all five barely surviving species of rhinoceros, the mountain zebra, several species of the ass, and many of the ruminants, cud-chewing artiodactyls, such as the camels.

The litany of extinct, nearly extinct, or seriously endangered large land mammals is hardly a full measure of the historical destruction of the biota, of course. In this book I focus on the traumatic evolution of life on terra firma, but it would be irresponsible not to mention that large animals of the oceans have been harvested to extinction, or nearly so, in recent centuries. One of the most arresting examples is the green turtle (*Chelonia mydas*), animals that glide from sea to land and back, swimming the warm waters of the Atlantic, Pacific, and the Caribbean and alighting onshore to lay their eggs. In past times these animals were so plentiful that they sometimes served as a shore themselves; lithographs from the seventeenth century show sailors disembarking from their ships and walking to the beach of a Caribbean island on the backs of turtles. The subsequent demise of the green turtle is a tale of spectacular devastation. When Europeans first arrived in the Caribbean, they encountered an astounding number of green turtles—historical records suggest an estimated 33 million adults. That number was reduced thirtyfold, to about 1 million, largely the result of overfishing in the eighteenth and nineteenth centuries. Exploitation continued worldwide—large populations of green turtles were exterminated in Australia's Moreton Bay in the early twentieth century—but these later events merely built on a long history of massive overfishing. The largest nest sites today are in Costa Rica (twenty-two thousand females) and on Raine Island in the Great Barrier Reef (eighteen thousand females). The IUCN's 2006 Red List notes that thirty-two index sites worldwide have shown a 48 to 65

The Loggerhead turtle, *Caretta caretta*, an endangered species of marine turtle

percent decline in mature nesting females over the past 100 to 150 years. Destruction of populations is not confined to this species of marine turtle. The once abundant Loggerhead turtle (*Caretta caretta*) is now also on the IUCN endangered list.

Moving farther offshore, we are familiar with the continuing decline of the great whales, a trend only arrested somewhat in recent years by international law and enforcement. Some whales, like the gray whale, are now on a population rebound, but others, like the North Atlantic right whale (*Eubalaena glacialis*), are hardly in a state of security. A recent sighting and photograph of a solitary right whale with her newborn calf were front-page news and a cause for celebration worldwide. The mixed success in saving the whales notwithstanding, we continue to devastate marine life for sustenance. In the process we are drastically denuding an impressive list of creatures: sharks, swordfish, Chilean sea bass, North Atlantic cod, several strains of salmon, lobster, and many others. At a conference at the American Museum of Natural History held in 2002 that dealt with this sober subject, Daniel Pauly, a well-known fisheries scientist, predicted, "We'll soon be eating jellyfish," a less than delectable marine creature with virtually no nutritive value. Some of Japan's fish markets have recently issued a demand for same.

Of course these are only the most recent waves of decimation in the history of life. Major biological catastrophes punctuate the history recorded in the abundant fossils we have extending back over 500 million years, and it is likely that extinction events also marked the shadowy history of life on Earth all the way back to its beginning 3.5 billion years ago. Across this vast expanse of time, multitudinous species have originated, persisted, died out, and been replaced by new species, a turnover that occurs at various rates whether or not mass extinction events occur. In this way, species, like individuals, are mortal: they have given life spans. Extinction is a way of life. The result? Most of life on Earth belongs to a lost world. There are 1.75 million living species already named and registered, and 10 to 20 million or more living species yet to be discovered, but 99.999 percent of all life that ever existed is extinct.

Fortunately, the extinct species representing those early lost worlds are not totally lost. The fossil record outlines at least the history of life, and we can appreciate its profound impact on the world today. In addition to the continual churning and overturning of elements in the biota, marked by the extinction of old species and their replacement by new ones, paleontologists have identified five major, or mass, extinction events over the past 500 million years. Yet most of these catastrophes of the distant past are mysterious, their causes ill understood. When we move closer to the present, things become clearer. As we shall see, we have good evidence that 65 million years ago many organisms, including all dinosaurs except birds, were wiped out when a piece of rock as big as Mount Everest collided with Earth near what is now the southeastern coast of Mexico. And then, as I have noted, we have vivid evidence—despite some skeptical reaction to it at first—that between 50,000 and 10,000 years ago colonizing, predatory humans were in many places the principal force behind the extermination of large animal species. The human factor becomes very clear in historical times. Thus the cause of the current biodiversity crisis—what many scientists are calling the sixth mass extinction—is disturbingly clear.

We are beginning to learn that all mass extinction events have similar effects regardless of their causes, whether human intervention or asteroid overkill. The recovery of resilient organisms and ecosystems shows a particular pattern and particular tempo, and we know that what eventually recovers is never exactly what once had been. Studying these patterns helps us assess the nature of our own current life crisis and its likely outcomes. As we look to the future of life on Earth, we must keep looking back. The fossil record—beyond all the theory, sophisticated ecological modeling, and prognostication about the anticipated state of the world and its biota—is the only direct evidence we have for what actually happened.

2

LUSH LIFE

As a ray of morning light catches a layer of frosted limestone that forms the 29,035-foot summit of Mount Everest, a spider barely twitches, hanging from a strand of web like a dying climber. The spider is not in fact dying, but it is, so to speak, barely alive. Spiders are accustomed to prolonged periods of starvation, yet dependence on food in the form of prey at this altitude seems nothing less than suicidal. Still, other things live here: insects of various kinds that are potential victims and food. Perhaps even several species of worms have burrowed into the substrate of decayed rock. We don't really know a good deal about what lives up here. Humans on their way to the world's loftiest point of land are hardly preoccupied with this investigation.

A few thousand feet lower in the Himalayas, life is easier to observe. Yaks range up to 20,000 feet in the summer season; ibex, argali, gazelle, and chiru ascend to 18,000 feet. These animals feed on the stunted vegetation that also supports several species of rodents, hares, pikas, and mice. Carnivorous mammals at these altitudes that prey upon the hoofed animals, rodents, and birds include the snow leopard, the lynx, the wolf, the fox, and the weasel. At this upper edge of the snow-clad pines, gray monkeys, or langurs, leap with uninhibited energy and grace. As for smaller creatures, expeditions have brought back beetles and grasshoppers from 18,500 feet and lizards from 17,000 feet.

But these altitude records for living species are anchored to the ground. Mountain climbers in bivouacs perched at twenty-five thousand feet have seen ravens, kites, and hoopoes flying overhead. Spiders are known to float on balloonlike webs, tracking jet streams across continents and oceans at even higher altitudes; these hardy travelers have been ensnared on airplanes! Doubtless tiny spores of bacteria and other microscopic organisms also survive at these atmospheric heights. The diverse life that swarms and floats on or above this planet may have even tipped into the upper atmosphere. Who knows? Some extremely resilient microscopic creatures may even thrive suspended at the very edges of space. Yet to be accomplished with any satisfaction is a census of life at these altitudes.

At the other extreme, in the abyss of the deepest oceans, we have also barely initiated our explorations for life. Feathery-tufted worms in tubes, pale crabs, and other species creep along or cling to the edges of the deepest trenches on the ocean floor. More than a mile below the sea surface, vents spew out sulfur-laden boiling water from deep inside Earth's crust, forming grotesque phallic chimneys, often hundreds of feet in height. Bacteria and other organisms, most of them unnamed or unaccounted for, thrive in these Hadean geysers. There is even a suggestion that some forms of life adapted to extremely high temperatures, like the bacteria living in deep-sea vents or boiling cauldrons in places like Yellowstone, occur thousands of feet below the surface of Earth's crust.

Between these extremes of height and depth is life as we think we know it: thousands of mammal species, tens of thousands of fishes, hundreds of thousands of plants, possibly millions of insects, and a hugely uncertain number of other invertebrates, fungi, and microscopic organisms inhabit-

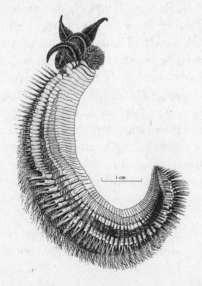

1 cm

Alvinella pompejana, a
tubeless worm from a
hydrothermal vent in the Pacific

ing oceans, rainforests, deserts, soils, lakes, ponds, and swamps. Fossils are, as I declared earlier, the only direct evidence of past life, but because the fossil record is incomplete, the true lushness and complexity of life can only be appreciated in real time. Here it is important to recognize a simple and humbling fact: the diversity of life — even life that exists today — is both staggering and, currently at least, incalculable. The 1.75 million named species is far short of the actual tally. A possible estimate of 10 to 20 million or more species is a testament to both the overwhelming richness of life and our humiliating ignorance of it.

Much of this vital diversity, this concentration of living, wriggling stuff, is of course not on our radar screen. In chapter 1 I considered the large mammals, both living and extinct, that attract enormous human interest. (People don't go to zoos to see bacteria.) But our preoccupation with the big beasts gives us a skewed sense of nature in its magnificent intricacy. Nature may come to mean largely what professional land managers refer to as wildlife, and even this latter term is vague and misleadingly broad in its connotations. However resplendent "wildlife" is to us, it represents only a small part of Earth's natural ecosystems and a still smaller part of life's diversity.

Consider those charismatic mammals. This branch of the animal kingdom with its wondrous variety of prancing, climbing, digging, flying, and swimming forms comprises more than 5,000 different living species, but this represents only a smattering of the known diversity of other life-forms. By contrast, the diversity of green plants, the main food source for most mammals (the taste for meat, shared by hyenas, lions, tigers, wolves, killer whales, and most humans, is a much less common mammalian predilection), far exceeds that of its hairy consumers. We have thus far recorded about 300,000 species of plants; about 240,000 of these are the angiosperms, flowering plants that for the most part sustain their species through reproduction that depends on their extravagant color, scent, and delicate beauty. We can appreciate this in a wholly dramatic way when we ponder the yellow and white incandescence of a field of daisies or the chromatic flurry of cherry blossoms that in the spring can paint a city like Washington, D.C., a dazzling pink. But there are angiosperms of subtler inflorescence — the grasses, for example.

One might think that plants, especially angiosperms, in their lushness, visibility, and abundance, are the most diverse of land species. But their

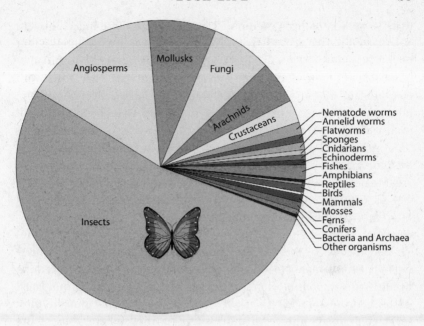

The relative diversity of different groups of organisms showing the dominance of insects and other arthropods

number is far surpassed by another major group of visibly abundant organisms integral to the functioning of modern ecosystems, the insects and other joint-legged arthropods, relatives such as arachnids (spiders) and crustaceans (crabs and kin). There are more than 940,000 named arthropod species (830,000 of them are insects, 350,000 of which are beetles!), representing an impressive 54 percent of the total named species on Earth; the actual number of living arthropod species is certainly much larger. It has taken and will take tremendous scientific effort to come up with a better arthropod census.

To appreciate that effort, we might review the statistics on our database of the natural world as represented in museum collections. The American Museum of Natural History, where I work, has a collection of 32 million specimens and artifacts—everything from pre-Columbian ceramics and Madagascar silk textiles to dinosaur bones, shells, bird and mammal skins, snakes, frogs, and fish preserved in jars of alcohol, rocks, gems, minerals, meteorites, frozen tissues for DNA sampling, and of course insects. More than 16 million of this museum's 32 million "things" are indeed insects. Scientists at the Museum sally forth in more than 120 expeditions a year to collect and survey the world. Their work yields a spectacular procurement rate of about ninety thousand specimens a year. Now gone, with good reason, are the days when explorers made large collections of rare, precious, and endangered mammals, birds, and other vertebrates. But even if we still were collecting that way, those vertebrates would represent only a small portion of our annual effort. In fact, a huge portion of the ninety thousand specimens we bring back each year are insects. The great majority of the more than a billion biological specimens in all the world's natural history collections are insects. Biologists estimate that there may be at least seven million more species of living insects yet to describe!

Why exactly are there all these insect species? We have some emerging indications, but not the full answer. Many insects—for example, dung beetles—are responsible for such vital services as recycling biological wastes and thus providing enriched nutrients in the soil. But the most spectacular function served by insects is pollination, whereby insects are not only the facilitators but the enablers of reproduction in flowering plants. The two go together in what is surely one of nature's most powerful acts of cooperation. One would think, then, that the diversity of plants and insects would be more or less equal; it would be logical to envision a system wherein one plant species was fed upon and pollinated by one insect species. And when we consider the ancient origins of the flowery ecosystem, we learn that certain orchids and other flower species have developed such exclusive partnerships. But many other flowers are not so fastidious. Studies have shown that 337 insect species representing 37 families of flies, beetles, bees, and wasps routinely visit carrot flowers, for example. So how many different insects interact with a given species of plant on average? If entomologists agree that the actual number of insect species is some multiple of a million, perhaps 8 million, then the average ratio of plant to insect diversity is one to twenty-seven. We cannot satisfactorily explain this stunning preponderance of insects. Whatever the underlying causes, the insects are perhaps the most riveting expression of evolution gone baroquely wild. With the possible exception of the myriad microscopic organisms we are just beginning to know, insects are the diversity regents, true lords of the terrestrial realm.

In every Animal there is a world of wonders; each is a Microcosme or a world
in it self.   —Edward Tyson, 1679

In my undergraduate years at UCLA I took what I thought would be a relatively stress-free elective course in entomology. I was wrong. The problem
was not the teacher. Professor Belkin was a kindly, soft-spoken man, not the
world's most riveting lecturer, but clear and to the point. The information
on insects that he provided was satisfying, and the exams were relatively predictable. The problem had to do with what I had anticipated as the fun part
of the course. Each student was required to make an insect collection for
virtually all the local insect orders and a representative number of their families. I was already trending toward a career in paleontology, with a macho
preference for a rock hammer or a mattock, and I took up an insect net reluctantly. Soon, however, the assignment became fun and productive.
While sometimes traveling with my fellow rock musicians to a band job, I
would deliberately command our equipment-filled van to pull over and
stop on the side of the road in Topanga Canyon. Ignoring, even enjoying
the taunts of my colleagues, I would make a few sweeps of my net over a creosote bush. With this brief stop, I might nab a couple more samples of families of flies or beetles.

Things went well until the term was nearing its end. Professor Belkin
warned us that a few groups would be elusive. I tried to capture the requisite
number of dragonflies, for example, but found that with their 270-degree
vision they could easily fake me out. Other noncooperative groups included certain wasps, flies, and beetles. With great desperation, I spent a
Saturday morning securing my pathetic collection on straight pins and writing the labels for each specimen in minute dots of India ink. I looked over
to see a fellow student with a magnificent collection of walking sticks, giant
green moths, iridescent beetles, and multihued grasshoppers. He was from
Hawaii, and his relatives had merely mailed him these exotics, captured in
their own gardens. Professor Belkin was delighted with my rival's work and
even stated that some of these would become incorporated in the university's entomology collections. Not fair.

Our tense Saturday in the entomology lab was suddenly interrupted. A
fellow student, a big, brutish dude with a scraggly beard, rushed in.

"People, listen to me!" he said. "I was accepted to UC Davis Medical
School last term. I only need to pass my courses from now on. But I am

flunking this course. I need two insects from each of you. If I don't get them, I'll wreck your collections."

This command, as if God were instructing Noah, naturally caused a ruckus. There were several shouts of "Shut up!" The noise attracted a professor across the hall, trying to do some undisturbed weekend research in developmental biology. He rushed in and grabbed our disrupter by the arm. By this time the antics of this undeserving future doctor had become legendary. He was the same person who had climbed naked into a tree on campus and had a friend take a picture of him there, although he at least had the restraint to cover his private parts with leaves, and had then sent this self-portrait to various teachers. One faculty member posted the photograph outside his office door with the caption "DO NOT TAKE CANDY FROM THIS MAN." The developmental biologist told our intruder to leave the room. This incited an emotional outburst of explanation and pleading mixed with great consternation, ending in complete despair, with the student issuing an absurd comment: "Sir, you're just jealous because you don't study things with six legs." To which the developmental biologist replied, "Get out!"

The premed student's attitude would not have gotten him very far in the study of biology during much of its history. Nor would our desultory efforts to assemble study collections. For centuries, specimen collecting had been a very big deal, the core of biology. After all, it is impossible to take full stock of nature unless one can somehow freeze it in time, capture a still life so that one can scrutinize and actually dissect complex, intricate shapes and contours. Expertise in collecting and identifying species has sadly diminished at most universities in recent years, however, as other priorities, such as molecular biology, neurobiology, and biophysics, have taken hold. Even so, the notion that one should collect specimens in order to understand nature has a much longer history than these newer disciplines have.

Aristotle not only was the fountainhead for modern thought in philosophy and science but loved to collect things. Between 347 and 343 B.C., he traveled to the island of Lesbos, married his first wife, and spent a good deal of time collecting and studying fish. His later great biological works draw heavily on information about the fishes and other organisms characteristic of the island and its surrounding seas. Aristotle had an audacious, and to this day unfulfilled, ambition: to catalog the entire biological diversity of the world. In the end he described "only" about five hundred species of organisms, but that was certainly a good start, and it required exhaustive collecting, careful planning, and social and political savvy. Legend has it that

Aristotle married up in order to procure the wealth necessary for amassing zoological collections. It is also commonly stated that he conscripted many slaves to collect for him and even coaxed his former pupil Alexander (as in "the Great") to send back specimens gathered along the campaign trail in Asia Minor and northern India. Modern scholars doubt these stories, but it is likely that Aristotle did recruit numerous students and allies for his massive task. His famous school the Lyceum is also known as the Peripatetic School, allegedly because of Aristotle's habit of strolling about while teaching, a fitting name for the place that launched the exploration of the natural world. Twenty-one centuries after the father of natural science laid his work to rest, Darwin remarked that the intellectual giants of his own time were "mere schoolboys compared to old Aristotle."

Charles Darwin himself came to his great synthetic insight into the evolution of life through a passion for collecting animals, plants, and rocks. His obsession drove some to distraction. His father, in reflecting on Darwin's indifference to medical studies at Edinburgh University, exclaimed, "You are good for nothing but shooting, dogs, and rat catching, and you will be a disgrace to yourself and all your family!" Others were not so negative. Charles transferred to Cambridge and took up studies for the clergy, a more leisurely vocational choice that not only pacified his father but offered Darwin, through his own design, free time to roam the fields. When he arrived, that august institution was infected with a kind of beetlemania. Darwin found himself in a furious race to collect beetle species against a well-known champion, "Beetles" Babington. Classmates spurred him on. One drew a cartoon of Darwin, complete with top hat and butterfly net, mounted on a giant beetle over a caption with the exhortation "Go it Charlie!"

Darwin's enthusiasm for collecting paid off. He was recommended by his role model and teacher, the Cambridge botanist John Stevens Henslow, to fill the slot for a naturalist on the Admiralty ship Beagle. After providing the references required to convince his father of his participation in this venture, Charles Darwin, a mere twenty-two, boarded the Beagle on December 27, 1831. Five years later Darwin returned to England after a circumglobal voyage that he later remembered as the most important experience of his life. This was the grand tour that set his thoughts in ways that eventually emerged as his theory of evolution by natural selection. The cruise was also a collecting extravaganza. Darwin had managed to collect, preserve, label, and organize thousands of beetles, tortoises, birds, plants, and fossils representing more than fifteen hundred species.

Collecting animals and plants does not of course alone qualify as science. The collected specimens have to be given labels. As I have noted, the fundamental category for identifying an organism is the species. But what is a species anyway? How can we say, for example, that there are twenty-seven species of insects for any one species of plant? Is an insect species actually some kind of equivalent to a plant species? To answer these questions, we must delve momentarily in the realm of biology known as taxonomy, the science of recognizing different kinds of organisms and putting them into meaningful classifications. Taxonomy is in turn part of a broader enterprise known as systematics, the science aimed at elucidating the great branching history that links Earth's species together and using diverse evidence—from the fossil record, genes, development, anatomy, and behavior—to recover this history. Without these sciences our understanding of life's diversity, both past and present, would be bankrupt. We would have no language to describe the fundamental units of nature, and we would be unable to group them in coherent ways that reflect their ancestry and descent. Without taxonomy and systematics we could not say that a tiny six-legged creature with antennae and three body segments is an insect and a hairy, warm-blooded, upright, two-legged, two-armed, toolmaking, talking creature is a human primate.

The same analytical approaches that are used to identify insects or humans and the groups to which they belong can be applied to a rare, newly discovered species like *Pseudoryx nghetinhensis*, the endangered saola of Vietnam and Laos. The binomial system and the formal classification scheme that I discussed with reference to the saola, which is used in virtually all scientific literature today, are a European development attributed to the work of the great eighteenth-century Swedish naturalist Carolus Linnaeus. But some form of "folk" taxonomy has emerged from virtually every region and culture. We know that many peoples have developed traditional skills and languages for identifying species, and their success at this practice is impressive. Certain tribes in the Amazon are familiar with thousands of species, many of them important as sources for foods and medicines; outsiders have learned much from these great native naturalists. A recent study by David Fleck, Robert Voss, and Nancy Simmons examined the taxonomy of bat species used by the Matses Indians in Amazonian Peru and found it surprisingly comprehensive. Although Matses have only one name for all bats, *cuesban*, they use descriptive terms to distinguish forty-three different bats. Of the fifty-seven species collected by scientists in the area over the years, thirty-four species had also been hunted by Matses, whose notebooks provided previously unknown information on the roosting habits of several of them. Moreover, at least fourteen Matses names closely correspond to standard Linnaean taxonomic names. Here are some examples:

Linnaean Name	English Common Name	Matses Name	Translation
Phyllostomus hastatus	Greater spear-nosed bat	cuesban maipue	red-headed bat
Thyroptera tricolor	Spix's disk-winged bat	cuesban tacsedëṁpi	little white-bellied bat
Saccopteryx	Sac-winged bats	cuesban cabëdi	variegated-backed bat
Desmodontinae	Vampire bats	cuesban intac chishquid	blood-sucking bats
Emballonuridae	Sheath-tailed bats	cuesban dëuishquedo	fleshy-nosed bats

Fleck and his coauthors noted in their paper that the Matses also share with other cultures an aesthetic delight in nature, even bats, and are expressive in their romanticism. As evidence they cited this winning entry in a Matses letter-writing contest, poetry intended both to humor and to enamor a sweetheart:

Cuesban-n inchësh-n chiuish bacuë sin-aid istuid-ash cuishonque-an-ac-bimbo-ec mibi ush-quin is –ash cuishonque-e-bi

"Just as bats start vocalizing joyfully when they find ripe fig fruits at night, I rejoice when I see you in my dreams."

The ubiquitous naming of living forms has ancient roots. The particular pathway to our Linnaean taxonomic system can be easily identified. In the fourth century B.C. the same Aristotle who loved to collect developed a rational classification that comprehensively covered some major groups of animals. Aristotle's student Theophrastus accomplished the same for plants. For centuries both classifications stood as a framework for understanding nature. But the concept of a species required refinement, and this did not occur until the seventeenth century in the revolutionary classifications of John Ray and somewhat later in the works of Linnaeus himself.

Until the nineteenth century, European scholars who cataloged species regarded them as myriad separate acts of God. Linnaeus proclaimed that his massive census of life was a testament to the "incredible resourcefulness of the Creator." But the father of modern taxonomy then had second thoughts. In his later years Linnaeus suspected that not all species he had named had been created. Some, he speculated, might have arisen from the union of two very different species to form a wholly new one, a thought doubtless influenced by his observations of frequent examples of hybrids in the plant world. But Linnaeus did not let these suspicions violate his basic belief that the matter of origins was known only to the Creator who had made it all possible.

It was left to other visionaries of the late eighteenth and early nineteenth centuries, including Comte de Buffon, Jean-Baptiste Lamarck, Charles Lyell, Étienne Geoffroy Saint-Hilaire, Robert Chambers, and ultimately Charles Darwin and Alfred Russel Wallace to propose that species had indeed arisen through natural processes as descendants of earlier ancestors. Darwin and his historically overlooked codiscoverer Wallace transformed taxonomists into systematists, into scientists interested not only in organizing diversity into meaningful groups but in explaining these in terms of their evolutionary history. The importance of this conversion became ever more apparent as paleontologists uncovered increasing fossil evidence of ancient life, of lost worlds where species were very different from those of today. The twentieth-century exploration of the genome lent further validation to the notion that we and all other species evolved together in the great branching tree of life.

Such progress may suggest that we have a clear general definition, a crystalline concept of *species*, that we may confidently say that one species is as recognizable as another, whether insect or plant or mammal or bacterium. Ah, but it is not so easy. Biologists have indulged in endless debate about the nature of species and how to recognize them. There is widespread agreement that species are discrete groupings of individual organisms that have evolved to become isolated from other very similar groups. But how do we decide at what level such groups of individuals actually are distinct enough to be called separate species?

One approach is to recognize species as aggregates of individuals *reproductively* isolated from one another; they do not interbreed even with closely similar groups, or at least they do not interbreed to produce fertile offspring. Species recognition based on reproductive isolation is sometimes called the biological species concept, and it is usually associated with the highly influential biologist Ernst Mayr, who died in 2005 while still a productive, outspoken scientist at the age of one hundred. Of course many "new" ideas have antecedents. In *Histoire naturelle*, his great series of forty-four volumes published between 1749 and 1788, Buffon recognized that the horse, *Equus caballus*, and the donkey, *Equus asinus*, do not normally interbreed, and when they do, they produce either mules or hinnies, which are infertile hybrids. To him, their status as separate species was therefore uncontroversial.

Unfortunately what works in distinguishing horses from asses may not help us distinguish species of bamboos, butterflies, beetles, or bacteria. In addition, many species, particularly microbial ones, lack sexual reproduction entirely. Even where sex is the norm, as in animals and most plants, its use as a criterion for recognizing species is limited. We have observed directly the breeding habits and the reproductive viability of only a paltry few of the 1.75 million named organisms. Indeed, for many biologists, reliance on the biological species concept to describe nature is hugely impractical.

Alternatively, we can sometimes infer that two very similar forms are separate species because of the small differences between them. This is especially convenient if those differences pertain to the reproductive apparatus. Many entomologists use such features in their taxonomy. They identify forms as separate species when the male sex organ of one species simply doesn't fit in the necessary lock and key way with the female organ of another, otherwise closely similar species; clearly the insect in question is reproductively isolated from all species except the one having the female with the right fit. By extrapolation we may find that other traits of a group of individuals—in shape, structures, or certain genes—suggest that it has parted company with its relatives somewhere back in its evolutionary history. This is commonly called an evolutionary species concept.

So how many traits does one have to take into consideration in order to know a group is different enough to be its own species? Sorry, we are back to square one. The answer is there is no easy answer. While the description and naming of species are not completely arbitrary, they are hypothetical. We might hypothesize that two populations of grasshoppers living in Ohio are two separate species on the basis of the differences in coloration of their wings or a few base pairs of their DNA. Later we might find that these two populations actually interbreed and produce fertile offspring, requiring us to question the original hypothesis. This can and does happen, as many sci-

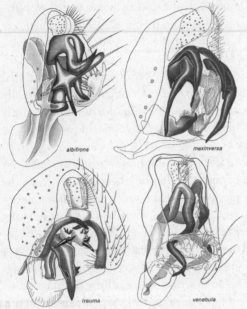

albifrons

mexinversa

Male genitalia in species of the
fly *Cladochaeta*

trauma

venebula

entific papers have documented. The converse situation can also occur: a
population originally thought to be a single species may be later recognized
as comprising a pair of, or several, species. Many gene studies have distinguished species, called sibling species, by differences in their DNA, even
though these species look virtually identical in body form (morphology).
And splitting up former species has occurred not just because of studies of
DNA. Closer scrutiny of outward features in more familiar groups has led
to this sifting. Most traditional classifications recognized about ten thousand species of birds, but newer classifications, based on a more precise
knowledge of the branching evolutionary history of bird lineages, recognize
many more. This is due, for the most part, not to new discoveries of birds in
the wild but simply to revisions in our recognition of species already known
to science. If anything, we've underrated even our own museum collections
as a testament to the awesome diversity of the living world.

When my adult hometown of Manhattan embarked in 2000 on its latest
census, the challenge of that enterprise was a matter of frequent news. The
census takers had to deal with all kinds of obstacles: sketchy records, chang-

ing addresses, peripatetic lodgers, people with no traceable names or identities, and simply secretive individuals who had no wish, for a variety of reasons, to become another statistic. The same was true on a national scale,
whether for denizens of a major metropolis or hermits in the hills. The
same is also true of the global census of living species, of our current biodiversity. There are as yet unknown species more secretive and certainly more
elusive than the rarely seen loner in apartment 3B. Moreover, only a few
people are qualified to carry out certain aspects of the biological census. I
am trained to know something about mammals, and I could no more distinguish and classify two species of midges than a midge specialist could
place the saola correctly among the artiodactyls. Taxonomists and systematists may spend their whole lives absorbed in the study of one small group or
even one or a few species.

Despite this impediment, there are encouraging signs of a new infusion
of energy in the quest for a decent global census of the biota. Edward O.
Wilson, Peter Raven, and others described both the audacious mission and
the daunting goal of the census takers. The call for action was issued in the
early 1990s. How far have we come since then? With no little embarrass-
ment we must admit that we lack a precise notion of the scope of the task.
In the 1980s provocative studies suggested staggering counts of true diversity. When certain trees in a tropical rainforest were fogged with a bug
bomb, what fell out was an alien world; sometimes five out of every six
species of insect were new to science. Extrapolations based on these spot
samples yielded big numbers, a planet with as many as 30 or maybe even 50
million species. Subsequently, some have questioned the size of this projection and have revised it drastically downward, but we are still in the dark.
Current estimates in the literature range from an improbably low 3.6 million to an equally improbably high 100 million, with most estimates coming in at around 10 million.

Why so much imprecision? Is it because of those frustratingly copious
insects? True, insects do represent a staggering amount of diversity, but we
actually know something about more than 800,000 species of them. Other
major groups are so poorly understood that we would hardly have the jump
on a group of alien taxonomists landing on Earth tomorrow. About 100,000
fungi have been described and named, but this group is thought to number more than 1.5 million species. There are other blanks on the biological map. When it comes to sheer biomass, one poorly known group, the

worms, may be the champions: not the familiar segmented Earthworms in the gardens or at the ends of fishhooks, but the more secretive pale, smooth worms, the nematodes, which perform many of the same functions in soil management as their more familiar segmented kin. Four out of every five animals on Earth may be nematodes. In fact, it has been suggested that if all the land were completely stripped of everything but nematodes, all the outlines of the continents and their textures would be perfectly replicated by a swarming global mass of nematodes. About 15,000 nematodes species have been named, and the existence of millions more is suspected.

Even more profound than these mysteries is another yawning void in our census of life. Bacteria and the even more primitive one-celled organisms called the archeans—what E. O. Wilson characterized as the black hole of systematic biology—are the oldest, most ubiquitous, most rapidly evolving, most physiologically diverse, and most poorly known of all organisms. There are about four hundred different species of microbes, many of them undescribed, living in every human body. Bacteria of the genus *Prochlorococcus*, possibly the most abundant organism on the planet and the ultimate source of food in the sea, were not known until 1988. About six thousand species of bacteria are recognized, but to assert that these fairly represent the microbial world would be as if we identified a few paintbrush strokes and claimed to visualize the *Mona Lisa*. A shortcut method to assessing the diversity of microbes relies on samples of soil or water for different kinds of genes and DNA, suggesting the presence of different kinds of bacteria. This approach yields a much bigger number, one that might be in the millions.

We shall look more closely at the wondrous substance called DNA and how it is studied in order to assess the diversity and evolution of life, but for

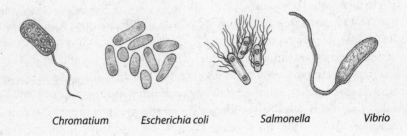

Chromatium   Escherichia coli   Salmonella   Vibrio

Four genera of bacteria

now we can say that DNA sampling is leading us to new and shocking revelations about what is out there. One enterprise concerns the domain constituting roughly 70 percent of Earth's surface, the oceans. In 2003, Craig Venter, the famous comapper of the human genome, cruised the world in the *Weatherbird II* as well as his sleek sailing yacht, the *Sorcerer II*, spot-sampling the ocean for microorganisms. Several hundred liters of water were collected from a half dozen sites near the island of Bermuda; DNA from the water samples was then extracted through microscopically fine filters and analyzed. Venter and his coworkers estimated at least eighteen hundred different microbial species in the infinitesimal aliquots of ocean represented by their samples. One could only imagine what other new species thrive in other waters at various depths. The global quest to answer this question continues.

But the microenrichment of the oceans pales in comparison with what might exist within solid ground. Estimates for soil bacterial diversity also focus on DNA. If DNA from a single organism is purified and heated, the strands of the double helix forming the DNA molecule "melt," separating into single strands. If you then cool the DNA, the strands will come back together, or reanneal, and the rate at which they do this indicates what you've got. Big and complex DNA will reanneal slowly. Studies in the early 1990s using this technique showed that the DNA extracted from a small soil sample reannealed very slowly indeed, about seven thousand times more slowly than the average rate for the smaller, simpler DNA of a single species of bacterium. But this does not necessarily indicate the presence of a big, complex organism; rather, investigators interpreted the DNA in the sample as an artificially pooled "genome," simply a collection of DNA from many bacterial species. The obvious conclusion was that about seven thousand different species of bacteria were residing in a few grams of soil.

That estimate has now been revised upward to an astounding new number. In a paper published in 2005, Jason Gans and coauthors from the Los Alamos National Laboratory looked again at the reannealing rates of DNA in these soil samples. They applied new analytical techniques that accounted not only for diversity but also for the relative abundance of bacterial species. Why is this important? Earlier estimates assumed an equal abundance for all species, and in nature this is virtually never the case. An acre of prairie land on the Great Plains may harbor hundreds of thousands of individuals of one grasshopper species and only one pair of great horned owls, yet each species counts as only one data point in the biodiversity census. Now imagine you were skimming over that same acreage in a hang glider, sweeping the high grass with an insect net. You'd be likely to get plenty of grasshoppers, but small, rare insects would not be nabbed so easily. In fact, odds are you would miss many of the furtive creatures. The same applies in the study of the bacterial genomes in soil samples. When Gans and his colleagues accounted for what might be easily missed—those rare and elusive species—they concluded that the clumped genomes represented far greater diversity. In a measly ten grams of pristine soil—that is, soil unpolluted by heavy metals—there lurked in reality as many as a million species of bacteria, exceeding early estimates by almost three orders of magnitude. So, on a global scale, how many bacteria and archeans are there actually? I won't even hazard a guess. We have barely crossed the border into this microscopic underworld.

Obviously, then, taking the census of the living planet is an almost ludicrously big job. But perhaps achieving it is not so remote a goal. There are about six thousand taxonomists worldwide. If, for the moment we relieve ourselves from describing *all* the incomprehensible numbers of bacteria species and stick to a more conservative goal of accounting for an incomplete census of these tiny creatures, we still have an estimated 8.3 million species left to be counted. If we divide this figure by the current workforce, we come up with a job order of about 1,383 species per taxonomist. That seems downright doable; after all, the late-eighteenth-century botanist Carl Peter Thunberg discovered and described nearly 2,000 new plant species in his travels to Japan, Ceylon, and southern Africa in a matter of a few years.

Lest we are lulled into complacency by the achievements of Thunberg and other indefatigable taxonomists, we need to consider a major impediment. The overwhelming majority of those six thousand current taxonomists are not trained to recognize or describe species in the "mystery" groups—the fungi, nematodes, and sundry protists, not to mention the unfathomably diverse bacteria and archeans. Even groups such as the insects that have attracted a respectable number of specialists represent a formidable challenge. Lee Herman, a colleague of mine at the American Museum, published in 2001 a catalog more than forty-two hundred pages long on staphylinids, or rove beetles. According to this catalog, which records taxa identified between 1758 and the end of the second millennium, 53,132 species of these small, slender beetles have been named. Earlier workers

named a few thousand species, but the task is getting harder because some
of the new species are rarer and more remotely located. Moreover, the specialist must take stock of the massive amount of previous work, decide
whether names are valid or invalid, and synthesize new classifications that
confer stability on the names. Herman himself is responsible for 232 new
species, and this represents a good chunk of a scientific career. His catalog
is a comprehensive accounting for roving rove beetle specialists, but unfortunately, similarly monumental catalogs are not available for many key
groups of beetles and other insects.

On the other hand, we live in the age of Wi-Fi and digital imagery. Why
could not this greatly improve our efficiency? This possibility has been explored and promoted by systematic biologists. Several websites are able to
network the efforts of bioexplorers, and they offer handy keys, maps, images,
and descriptions that freely crisscross among specialists on opposite sides of
the planet. These are meant to make taxonomy easier, and they do. As they
improve, they allow someone not so familiar with a group of animals or
plants to wander into a local forest, click a digital camera on a creature, pipe
that image to a remotely located expert through a website flashing on the
screen of a handheld computer, and perhaps get a species ID, or an indica-

Scanning electron
micrograph of a
staphylinid beetle,
body length 4.2
milimeters

tion that the image is of a species new to science, before dinner in camp that evening. Inspired by this progress, E. O. Wilson has promoted the idea of a new encyclopedia of life, with roughly one electronic page per species, and predicts that our technology, combined with sufficient planning and focus, could provide a reasonably complete census of life by 2020. What remains to be done is of course a prodigious task. Since the publication in the mid-eighteenth century of Linnaeus's *Systema naturae*, the great seminal work for modern taxonomy, we have accounted for only about 10 percent of an estimated 10-plus million species. It is proposed that we deal with the remaining 90 percent in one-tenth the time. Why not try? There are much less beneficial, even deleterious projects that require and have attracted even more massive human resource and effort. The great census of life has another poignant dimension. As I have said, an estimate that 30,000 species a year may be going extinct is not unreasonable. That means that by the time the census is completed under this audacious schedule the loss could account for one-quarter of the biota.

My interest in the biodiversity crisis surged in the early 1990s, when I attended a series of job talks presented by candidates for an ornithology curatorship in the American Museum. Some of these candidates reported that the very bird populations that drew their passionate interests had been disappearing during the few years it had taken them to do their dissertation research. Now those local populations are all but exterminated, and the species to which they belonged highly threatened or endangered. The understated style that is standard in a scientific talk belied the pain of loss felt by these bright young scientists. I was shocked by the wanton destruction and by the way it was colliding with the human need to discover what exists in the world around us. However serious it is to confront the sixth extinction, it is even more disturbing to contemplate that in a few years we might fail to learn even the fundamental quality and quantity of that loss. We would then have an unprecedented mark on history; we would be the generation that let life slip through our fingers.

3

EPHEMERAL LIFE

How old are humans? Flowers? Bees? Dinosaurs? When was the first tree? Coral reef? Jellyfish? Worm? Bacterium? These are questions relentlessly investigated by paleontologists. In this case, science seems to reflect a basic human curiosity about birth dates. We want to know when things became things. Why this obsession? Perhaps one, but not the only, reason is that it gives us some orientation about the division of time and the scope of our legacy, the life on Earth that came before us.

The 4.6-billion-year history of Earth has been sliced into a calendar with designated time intervals. The first two-thirds of this history are very fuzzy and ill defined. Thereafter, especially during the last 500 million years, represented by a much better fossil record, time becomes easier to tell. This is because of a particular quality of the history of life: it comes in waves. If we subdivide major groups of organisms, such as plants or vertebrate animals, down to their fundamental units, species, we get an appreciation for life's vicissitudes. Over time, waves of species are replaced by other species. Decimation and darkness give way to rejuvenation and flourish. Then come disaster and decimation again. No species we know of has spanned the entire history of life or even a large part of it. There are famous examples of endurance: cockroaches, horseshoe crabs, and other lineages. But despite the strong similarity between a cockroach from a 300-million-year-old coal forest and the unwanted creature skulking under the kitchen sink, they are certainly not the same species. Indeed, paleontologists have argued that even the most enduring species do not outlast about 20 million years (still an impressive span) and that on average, species probably endure for about 1 or 2 million years. These facts have a big effect on how we see life of the past, present, and future and clearly demonstrate that extinction is as much a part of life as birth and flourish.

If we simply applied an average duration—say, about 1.5 million years— to the life spans of species and combined that with the assumption that species split into two or more "daughter" species, the curve for life's diversity through time would be smooth and upwardly trending. The fact that the diversity curve does not look this way is one of the great discoveries of science, one that occurred in an exciting phase of enlightenment in the mid-nineteenth century, at nearly the same time that Darwin published his *Origin of Species*. In 1860 the paleontologist John Phillips documented the diversity of marine fossils, such as trilobites, crinoids, clams, snails, and ammonites of a rock series representing millions of years exposed on the British Isles and parts of Europe. He noticed that this record showed sharp dips in diversity where many species—even major groups like trilobites that contain such species—dwindled and disappeared. These dips were followed by a sharp rise in diversity and abundance of new species and, often, major new groups. Phillips recorded two very deep dips followed by two high peaks and at least two less precipitous peaks and dips. He cleverly formal-

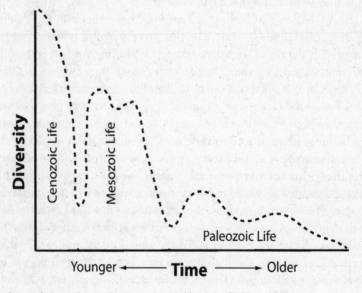

Time intervals based on extinction proposed by John Phillips, 1860

ized this pattern as divisions of a great calendar of Earth time over 500 million years, from the Cambrian Period to the present. These divisions were the Paleozoic Era (or ancient life), the Mesozoic Era (or middle life), and the Cenozoic Era (or recent life). Each era boundary was marked by a period of lowered diversity—in other words, by a period of accelerated extinction. The end of the Paleozoic was marked by the devastating Permian extinction event, while the boundary between the Mesozoic and the Cenozoic was marked by the great Cretaceous extinction event that wiped out sea creatures such as the ammonites. Later in the nineteenth century paleontologists also observed that the extinction event at the end of the Cretaceous eliminated the dinosaurs (but not completely; birds are a branch of dinosaurs and are thus the one lineage in the group that survived extinction).

Over the years the Phillips geological calendar has been much refined. In addition to the Permian and Cretaceous mass extinctions, we now recognize three other mass extinction events. Two occurred at the end of the Ordovician and Devonian Periods during the Paleozoic Era; an additional one occurred at the end of the Triassic during the Mesozoic Era. Also, the calendar has been recalibrated in million-year dates that differ notably from those used by Phillips. Nevertheless, the basic pattern Phillips described has only been recalibrated and made more elaborate; it has not been seriously challenged. Phillips gave us the data that allow us to recognize mass extinctions and their capacity to recur. He disarmed any presumption that the life we see around us is necessarily here to stay.

The notion of life ephemeral can be disquieting, especially when we consider the signs of our own times. That brings us to a consideration of mass extinction event number six. That tens of thousands of species every year may now be going extinct is emphatically not an expected rate. Paleontologists have come to recognize that the rate at which species on average naturally go extinct, known as the background rate, serves as the metronome that marks the timing of the appearance and disappearance of species over the last 500 million years. Studies published in the 1990s noted that current extinction rates were one hundred to one thousand times the background rate. They also predicted that the rate would increase another tenfold, and by the mid-twenty-first century we might lose from 30 to 50 percent of all species on Earth. Such was even the case for the mysterious microworld. According to the study by the Los Alamos team, the stupen-

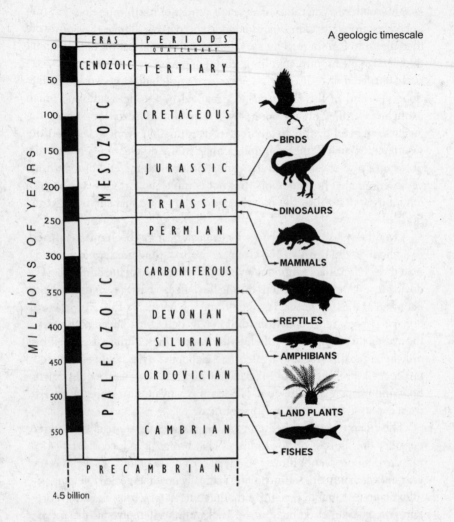

A geologic timescale

dous diversity of soil microbes precipitously declines in samples of soil polluted by heavy metals. This has disturbing implications for the health of the environment. Fewer soil microbial species mean fewer options for recycling important nutrients in the soil, the work of these minuscule biofactories. It is little wonder that the degradation of soils worldwide is one of our most serious environmental problems.

Human destruction of Earth's environments, particularly the conver-

sion of land for agriculture or human occupation and the resultant fragmentation of habitats, is ultimately the greatest threat to biodiversity,
though there are other major and contributing factors. In recent years scientists have also documented the serious impact of a synergy between habitat fragmentation and the apparent dramatic change in world climate.
Studies of the fossil record show that during past times of climate change,
species either migrated to new areas where their adaptations allowed them
to survive or went extinct. But today species marginalized by dramatically
changing climates may have nowhere to run, as they are boxed in by huge
and ever-expanding areas of human occupation.

Measuring the effect of all these forces is a complex business, since the
impacts occur at different intensities depending on the location, habitat, or
kinds of organism affected. Biodiversity is not distributed evenly over the
planet. Some confined areas, such as uplands in the tropics isolated by high
mountain ranges, contain a disproportionately high share of endemic
species, like many of the Vietnamese orchids, with small ranges and limited
capacity for migration. They cannot escape unfavorable climatic shifts,
overhunting pressures, invasions from exotic species, or slashing and burning of the habitat. We call such areas of high biodiversity and many endemic species "hotspots." There are many in the tropical land regions as
well as certain marine environments, such as coral reefs and shallow coastal
areas. Notable hotspots include the Atlantic coastal forests of Brazil, the
mountains of central China, the plant communities of California and
the Cape of South Africa, the west African rainforests, the tropical Andes,
the shallow reef areas in the Caribbean, and virtually the entire island of
Madagascar.

Different groups of organisms are variably affected by environmental
degradation. Some organisms have very narrow temperature tolerances or
limited powers of movement and migration. Many plant species, for example, live and die according to the timing of the first frost and the first thaw,
and the effect on their survival of a shift in those dates is clear. But we are
just beginning to get a sense of the mosaic response of biodiversity to these
pressures. We do not know the reasons why certain groups are more susceptible than others. For example, for some reason populations of frogs, a diverse group with more than five thousand species, are showing considerable
decreases even to the point of near extinction in many parts of the globe.
One cannot assume that species more capable of movement or migration

are more resilient. A now classic study in 1996 by Camille Parmesan showed that local populations of Edith's checkerspot butterfly (*Euphydryas editha*) went extinct in many southern and lowland areas of the range of the species in western North America, likely because of a warming trend that affected its habitats.

Birds and animals capable of long-distance dispersal and migration are among the greatest casualties of the invasions of humans, their domestic animals, and exotic species. The impact is, if anything, worse on less mobile birds that live in constricted and isolated habitats, including patchy forests on small islands. It is estimated that as many as 2,000 species of birds on many Pacific islands went extinct with the arrival of the first humans some three thousand years ago. Today the Pacific has 289, or 24 percent, of the world's globally threatened birds; 14 percent of these are critically endangered, giving the region the dubious distinction of registering the world's highest extinction rate for birds. Of the 129 species of birds that have gone extinct since A.D. 1500, 63 species are from the Pacific. If, as Emily Dickinson wrote, "Hope is the thing with feathers," we are in big trouble.

Let us again consider the research showing that species are going extinct at thousands of times the background extinction rate and that we are thus likely to lose 30 to 50 percent of all living species within this century. People questioned such predictions and started looking harder at many of the habitats where species were supposedly most seriously threatened, including the so-called hotspots in the upland tropics. How do these early predictions stand up? In 2006, of 40,177 species assessed for the International Union for the Conservation of Nature (IUCN)'s Red List of Threatened Species, 16,119 are now categorized as threatened with extinction: 1 in 3 amphibians, one-quarter of the world's pines and other coniferous trees, 1 in 8 birds, and 1 in 4 mammals. The basic conclusion is that biodiversity loss is accelerating, not slowing down.

Recent compilations also show a wide range of predicted casualties, depending on the species in question and where they live. Global warming will be particularly pronounced in the Arctic and will reduce sea ice habitat for polar bears, probably reducing their populations by 30 percent. Desert areas are now also at risk. Seventeen years ago, when I first led paleontological expeditions to Mongolia's Gobi Desert, the goitered gazelle (*Gazella subgutturosa*) numbered more than 140,000 and was widespread throughout Central Asia. It is now estimated that populations in Mongolia have been reduced by more than 50 percent because of illegal hunting. In the wide-open spaces of the Gobi Desert and the central steppes, little has been done to control the activities of poachers. Overall the rate of decline has exceeded 30 percent over the last ten years. This gazelle has been reclassified in the IUCN Red List from "Near Threatened" to "Vulnerable."

In the oceans of the world, new data also show a distressing decline in big fish, especially sharks and rays. The angel shark (*Squatina squatina*) has been reclassified from "Vulnerable" to "Critically Endangered" and is declared extinct in the North Sea. This culling of big marine species extends deep into the whole marine ecosystem. Centuries of overfishing have created a step-down pattern. In earlier times hunters and fishermen targeted the big animals at the top of the food chain in shallow seas near the shoreline: whales, sea cows, seals, sea lions, green turtles, and the biggest individuals of coastal fish. When these more visible, often more proximate species were thinned out, fishermen expanded their reach to the seemingly unlimited bounty of the open ocean. Far-ranging whales were an early casualty, and fish species, big and small, soon fell victim. Depletion of the biggest and thus most productive species was followed by catches of smaller fish and so on down the line. The modern saga of commercial fishing is a march down the food chain. Purse seining, dredging, longlines, and other methods allow for massive, indiscriminant catches that victimize threatened species like the eastern spinner dolphin (*Stenella longirostris orientalis*). They can also badly damage the sea bottom. In the North Atlantic, commercial fish populations of cod, haddock, hake, and flounder have fallen as much as 95 percent. Worldwide, other depleted casualties included a 90 percent loss of bluefin tuna, for which market demand fetches a price up to twenty thousand dollars for a fifteen-hundred-pound adult.

Freshwater fish may be confronting even greater stress. Freshwater ecosystems are especially vulnerable because of a combination of qualities. First, they can support an inordinate amount of diversity. Freshwater fish species, for example, represent roughly 41 percent of all twenty-four thousand living fish species, even though (since most freshwater is locked up as groundwater or glacial ice) all these species inhabit only 0.009 percent of the total water in which fish can live. Second, many of these diverse freshwater species don't move around much, and their entire life cycles may be confined to a single isolated stream or lake. They do not have the option, as various species on land do, of dealing with a rise in global temperature by

shifting their ranges to cooler habitats at higher latitudes and higher elevations that mimic their original environments. Meanwhile, the amount of
livable area for these assaulted species is shrinking.

The 2006 Red List shows that freshwater ecosystems are in far worse
shape than forests, grasslands, and coastal estuaries. More than 20 percent
of the world's 10,000 freshwater fish species have become extinct, endangered, or threatened in recent decades. They are at the top of the 2006 Red
List. Of the 252 endemic freshwater Mediterranean fish, 56 percent are
threatened with extinction, the highest proportion of any regional freshwater fish assessment so far; 7 species, including the carp relatives *Alburnus
akili* in Turkey and *Telestes ukliva* in Croatia, are now classified as extinct.
In East Africa human activity threatens nearly 28 percent of the freshwater
fish diversity. As for larger freshwater species, *Hippopotamus amphibius* is
now classified as "Vulnerable," primarily because of its marked decline in
the Democratic Republic of the Congo (DRC), the result of the high demand for meat and tusk ivory. Water may be the source of all life, but
humanity's overt dependency on precious and confined freshwater has become a death sentence for a significant portion of the world's biodiversity.

Many plants are also in danger. The Mediterranean region, one of the
world's thirty-four biodiversity hotspots, supports nearly twenty-five thousand species of plants, of which 60 percent are endemic, found nowhere
else in the world. These plant communities are being assaulted by urbaniza-
tion, mass tourism, and intensive agriculture. Several plant species are facing extinction, surviving in only a score of sites populated by a total of less
than twenty-three hundred mature plants.

Recent surveys of many other sites representing many different habitats,
including hotspots, show similar patterns of decline and potential extinction. A 2003 study by Taylor H. Ricketts and other authors homed in on
centers of imminent extinction, where highly threatened species were confined to one site. They found that within 5 taxa surveyed worldwide—
mammals, birds, selected reptiles, amphibians, and conifers—were 794
imminently extinct species, three times the number that have gone extinct
since 1500. These were restricted to 595 sites in tropical forests, on islands,
and in mountainous areas. Some alarming patterns emerged. High numbers of mammal and bird species became extinct after 1500, and others became trigger species—that is, species whose extinction was imminent,
while reptiles and amphibians showed comparatively few historical casual-

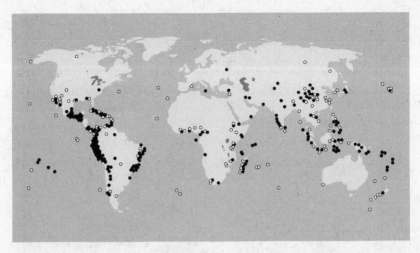

Centers of imminent extinction for both protected (light circles) and unprotected (dark circles) sites

ties but many more trigger species. Pine trees and other conifers with no recorded historical extinctions whatsoever also now showed a high projected number of trigger species. Patterns with regard to habitats were shifting as well. While much of the recorded historical extinctions had occurred on islands, threats now extended to the mainland. Disturbingly, only one-third of the sites surveyed were legally protected, and most were surrounded by areas densely populated by humans.

As we learn more about trends in climate change, the dual punch of global warming and habitat loss is seen in the scientists' projections for likely extinction. Many of the exquisite endemic plants of South Africa face extinction or drastic range reduction. Shifts in temperature and rainfall will affect biodiversity hotspots of extraordinary wealth in southern Africa. The Cape Floristic Region, or Fynbos biome, in South Africa has more than 8,000 species, many of them endemic and many of them spectacularly beautiful, including the large, radiating flowers of the *Proteaceae*. The flora here is a source for many of the world's favorite garden plants. Many of the Fynbos species depend on fire to germinate, but fires occur only about once every ten or fifteen years, severely limiting the resident plants' times for dispersal. The biologists Guy F. Midgley and Dinah Miller of the National Botanical Institute of South Africa used a computer simulation model to

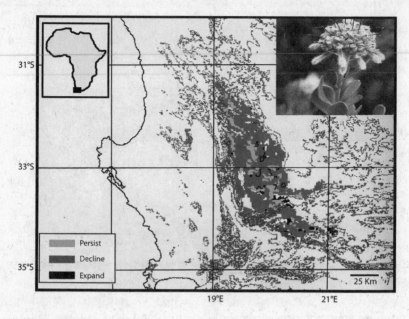

Predicted range shift of *Vexatorella amoena* (inset), a member of the protea family currently inhabiting the mountains above Cape Town, South Africa; shaded areas denote range projected to be lost by 2050, light shaded areas denote range projected to be remaining by 2050, dark shaded areas denote newly suitable climatic conditions for the species.

predict range shifts over the next fifty years in 343 protea species in the Fynbos flora. The model assumed a temperature increase during that time of 1.3° to 2.5°C for the southern Cape, with reductions of winter rainfall of up to 25 percent. According to the model, this climatic shift will have a shocking outcome. One-quarter of the protea species in the Fynbos hotspot will suffer *total* range loss, which effectively translates to extinction, by 2050; another 10 to 20 percent of the other species will shift so drastically to the south and upslope that their new ranges will have no overlap with the current ones. The ranges of some species will expand while others shrink, confining them to the very southwest edge of the Cape, an area already altered and fragmented by human activity. These latter species require intensive conservation management if they are to survive.

Another rich plant community in South Africa is represented by the five thousand species of succulent plants farther north, on the Karoo plateau.

The twenty succulent species in the Karoo, as modeled by Midgley and Miller, showed a trend toward southward migration with a median range loss of 60 percent by 2050; two species showed a stark 80 percent shrinkage in range. Unlike plants in the Fynbos community, these twenty species did not face near-term extinction. Why? Midgley and Miller reasoned that the Karoo succulents are more resilient to climate change; they do not depend on fire germination, as the protea species do, and although they need rainfall to release their seeds, the small seeds can be carried by wind, so they have options for reproduction and dispersal even with much less rain.

In a study published in the April 2006 issue of *Conservation Biology*, J. R. Malcolm and his coauthors used a computer model to predict extinction levels in twenty-five biodiversity hotspots; they found that at least 25 percent of the world's plant and vertebrate animal species would be extinct by 2050. They used a computer model that projected future plant and animal distributions under an assumption that CO_2 levels in the atmosphere would double. The model also explored different scenarios based on different assumptions about vegetation, biome scope and content, distribution, and migration capabilities. Extinctions were calculated using species-area and endemic-area relationships; in other words, the model worked with the rule that extinctions are likely to be more frequent when the area for habitation is decreased and even more frequent for endemic species. Hotspots destined to be unusually hard hit included the Cape Floristic Region, the Caribbean, Indo-Burma, the Mediterranean basin, southwestern Australia, and the tropical Andes, where the study predicted that plant extinctions per hotspot would sometimes exceed two thousand species. The model projections showed that global warming increased the expected extinction rate over earlier estimates, which had presumed deforestation as the major driver.

Of course we can't have a good sense of the state of most of the world's species, since we haven't even identified most of them. But some important data are emerging on the casualty levels we might expect. A 2004 study by J. A. Thomas and his collaborators found that butterflies on the British Isles were doing much worse than birds and plants. While 20 percent of the native plant species in the past forty years and 54 percent of the native bird species in the past twenty years had decreased in population, a whopping 71 percent of butterfly species had suffered population decline. In the ten-kilometer (thirty-three-foot) square plots surveyed, butterflies showed a decline in 49 percent of them, birds in 29 percent, and plants in 22 percent. These data suggest that something very fundamental is happening, though as the authors acknowledge, they concern only one region of the world. If other insects are as sensitive to such changes as these butterflies are, then a major part of both known and unknown diversity in the world is at great risk. We don't have many indications of this prospect for insects, but one comes from the 2006 Red List: of the 564 dragonfly and damselfly species so far assessed in it, nearly 1 in 3 (174) are threatened, including nearly 40 percent of endemic Sri Lankan dragonflies.

These declines—of butterflies from Britain, dragonflies from Sri Lanka, and familiar vertebrate and plant groups worldwide—confirm earlier predictions. The prediction of a 30 to 50 percent loss of species by 2050 is realistic. As the next Red List is under preparation alarming news is already emerging. In December 2006, a 3,500-kilometer survey along China's Yangtze River failed to turn up a single river dolphin (*Lipotes vexillifer*), one of the most specialized and wondrous of all aquatic mammals. The species is surely on the verge of extinction. In the next few years we shall have the answer to the following question: Is the Red List a guide for an unprecedented global conservation effort, or is it simply a census of death row?

4

ELEPHANTS, DUNG BEETLES, AND ECOSYSTEMS

On a brilliant February morning in the Serengeti I had the unpleasant shock of a wake-up call. This was not a buzzing alarm clock or the annoying jangle of the phone in my hotel room, but a thunderous crash, slash, and boom. It was 7:00 a.m. and a mere two hours' sleep after an exuberant party with my fellow travelers had been rudely truncated. "What the hell is that noise?" I yelled out loud to myself. I opened the sliding door of my room to a magnificent wood porch and stared into the gray eyes of a bull elephant turning an acacia tree into a compost pile. Behind him were three other harvesters doing their own bit for radical landscaping. To me and my sluggish brain it was all crash and chomp, a scene of drama and at the same time utter devastation. Those beasts were taking down whole trees! Trees that were perhaps more than half a century old!

The extraordinary amount of food required to keep an elephant alive poses a problem today for fragile African habitats. Although census data for the African elephant (*Loxodonta africana*) from earlier decades are uneven and somewhat unreliable, it is clear that three decades ago the ivory trade, along with trophy hunting and habitat loss, brought wild elephant populations to seriously low levels. At that time the Convention on International Trade in Endangered Species (CITES), signed in 1973 by more than 150 nations, included the African elephant in its depressingly long registry. A concomitant international ban on ivory was widely adapted, not without some controversy. Governments in southern Africa rejected the necessity of the restrictions, claiming that elephant populations in their region were stable or increasing. Nonetheless, the ban surely fostered the subsequent resurgence of elephant populations in many parts of Africa.

This has put unusual and very problematic pressure on the African elephant's food sources, however. An elephant clan of ten individuals foraging over twenty square kilometers (about eight square miles) of savanna woodland can remove 21,900 tons of vegetation a year. If most of that vegetation mowed by elephants was grass or herbs, there might be little concern, but elephants, as I witnessed from the porch of my hotel room, have a penchant for shrubs and trees; they break branches, strip bark, and uproot huge trees in the process. Included among the woody casualties are noble baobabs more than a hundred years old, often plowed by elephants for a mere mouthful of leaves.

Even this kind of concerted harvesting might once have been sustainable. In prehistoric times forests were widespread in East Africa, but those woodlands are now reduced and fragmented, many of them confined to the wildlife reserves that also protect elephants and other big herbivores. During the 1960s serious reductions of acacias and baobabs caused by elephant feeding were recorded in Ruaha National Park in Tanzania, where acacia stems were so severely damaged by elephants that 55 percent never survived to yield a mature plant. Over twenty-five years, high concentrations of elephants in the northern section of Murchison Falls National Park in Uganda destroyed 95 percent of *Terminalia glaucescens*, a normally abundant tree species that is not only an important element of the African flora but an important human commodity, used for housing material, firewood, and even dental sticks whose extracts kill off periodontal disease–causing bacteria. Thus stands of fragmented forests are no longer able to absorb the shock of large populations of elephants and big herbivores.

An elephant's prodigious consumption, as well as ours or that of any other species, is a way of promoting energy flow in an ecosystem. One species becomes food for another; the food is converted to energy to do work for the feeder; ultimately, the ecosystem is kept working as a whole. This consumption of one species by another is what biologists call trophic interaction (from the Greek *trophos*, feeder). Unfortunately, the system doesn't always work smoothly, as in the case of elephants and their dwindling habitat. The devastation of African forests by elephants is a riveting example of what happens when the delicate balance of interactions in a natural ecosystem goes awry.

When we speak of energy flow in an ecosystem, we need to make one thing very clear. That flow does not mean that energy is recovered. Energy is not recyclable. The ultimate source of energy is the sun. It is estimated that the sun on average bombards Earth with about 175 peta watts (a peta is 10^{15} or 1 followed by 15 zeros). Our biggest power stations have a capacity of about 1,000 megawatts so this amount of solar energy is equal to the output of about 175 million power stations. (A watt is a unit of power, or the amount of energy flow per second.) Solar energy is converted by organisms and transferred to other organisms. With each transfer some of the original energy is lost in the form of heat. Indeed, the ultimate destiny of energy in ecosystems is to be lost as heat. Plants take energy from sunlight and use it to convert carbon dioxide into glucose and other sugars in the process known as photosynthesis. Green algae and cyanobacteria (often misleadingly called blue-green algae) are also photosynthesizers. Bacteria in vents at the ocean floor and deep within the crust convert energy coming from chemicals, such as sulfur and methane, released from the Earth's volatile interior and use it to make sugars. These as well as the photosynthesizing organisms are called producers, or autotrophs, because they convert energy that is unusable by other organisms into a form—namely, into sugars—that can be transferred.

The organisms on the receiving end are the consumers. Primary among them are the plant eaters, or herbivores, which in turn are consumed by carnivores. Of course many organisms, like us, are both carnivores and herbivores, or omnivores. An additional category of energy users are the decomposers, prominently fungi and bacteria, that feed on dead organisms, scarfing up what remains of organic tissue and its stored energy, returning inorganic nutrients to the soil or water where they can again be utilized. Here is another important point about the ways organisms interact with one another and their environments: nutrients are recyclable; energy is not.

The trophic relationships between producers and consumers are obviously very complex. There can be several consumers at different levels. Grass can be consumed by grasshoppers, which are in turn consumed by spiders, consumed by toads, consumed by snakes, consumed by eagles. A food chain can have many steps. Moreover, the chain is rarely, if ever, linear. The trophic linkages usually take the form of a food web, with a variety of options for connection. Some food webs, such as the consumption of plankton by baleen whales, are simpler than others, such as the interconnections among plants, insects, mice, rabbits, spiders, toads, snakes, and predatory birds on a western North American prairie.

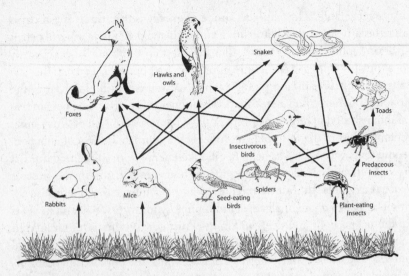

A simplified food web of a grassland ecosystem

Regardless of their complexity, all food webs share the same basic principle of efficiency when it comes to energy transfer. Each step up the food chain means significantly less energy available for transfer to the next level. Some of what is produced is never eaten, and as noted, some energy is lost as heat with each transfer. Greater amounts of energy are stored at lower levels than at higher levels, and the resultant energy diagram takes the form of a pyramid with the ultimate producers at the base and the top consumers at the peak. A very rough generalization is common in the ecological literature. Only 10 percent of the energy available at one trophic level will be transferred to the next level, and 90 percent of the original energy available is lost in the transfer. Since the amount of energy available directly relates to the amount of life it can support, the organisms at a higher trophic level will have only 10 percent the total weight, or total biomass, of the trophic level immediately below it.

This principle is hardly one of precision or constancy; energy transfer takes meticulous measurements and varies from one food web to another. The amount of plant material consumed relative to the amount available in a northern evergreen forest may be even less than 10 percent, whereas it may be as high as 60 percent in an African grassland. But there is, in general, a high inefficiency factor from producer to consumer. This is instructive in practical terms. Every time we eat meat we are depriving roughly nine other people of food that would be available if all ten of us ate the original plant material that was consumed by the animal we are eating. The energy pyramid also clarifies the problem with elephants and overharvesting. Elephant biomass in a prescribed area must not exceed about 10 percent of that of the acacias and other trees and plants they feed on. If it does—when elephants overharvest—the proportion in biomass is upset, and the result is defoliation without regeneration and eventually mass starvation of the consumers. As we shall see, elephants are not the most gluttonous beasts that have ever walked the Earth: the trophic roles of giant sauropod dinosaurs raise some fascinating questions.

The food pyramid might suggest that ecosystems have been built around a substandard, inefficient scheme of energy transfer, one requiring too much investment at one level to feed the next level. After all, why not prefer a system in which ten people can be sustained with rice or beans instead of only one with meat? Since we humans are omnivores and flexible in our food preferences, habits, and fads, this option might seem readily attainable. But many organisms are not built that way. Even some human vegans have found they must focus on certain natural foods or supplement their diets with vitamins in order to get minerals—for example, iron—that are conveniently available in red meat. Bona fide carnivores are more specialized still. The limitation begins right up front. The teeth of cats, lions, and tigers are built for cutting and slicing flesh. These animals can no more grind grass into pulp than a cow can devour a gazelle. And cats lack the lengthy gut and digestive enzymes as well as coinhabiting bacteria necessary to process plant matter efficiently. Carnivores not only have a feeding apparatus specialized for meat but are built to hunt for it. They apply bursts of energy, speed, agility, and intelligence to the task.

Predation is a way of life and an adaptation for many species, including humans. Yet in our tendency to personify nature, we are sometimes shocked by its violence. Even to a seasoned field biologist, the sight of a pride of lions bringing down a full-grown wildebeest is not easily forgotten. There are less visible but comparably skilled and dedicated predators. The arachnids, or spiders and scorpions, have a key function in ecosystems because they are often the primary consumers of insects. What seems to be a broad cultural aversion to spiders may indeed relate to some knowledge of their habits. The predaceous spider—in most species, the female—builds a web to snag her prey. The ensnared victim is often kept alive, but just barely. Over a period of days or weeks the captor takes little chunks out of its captive. So compelling is this symbol of prolonged voraciousness that soldiers of the Moche civilization of ancient Peru displayed the symbol of a spider in its web on their helmets and shields and practiced a ritual of capture, torture, and human sacrifice that mimicked aspects of arachnid behavior and struck terror in their enemies. Not that predation itself is always the easy life. Spiders so rarely catch prey in their webs that experts surmise that they live in a prolonged state of semistarvation.

A variation on the consumer lifestyle involves a parasite and its host. Often the latter becomes so debilitated that it survives solely to provide food for the parasite. This extreme form of exploitation is commonplace in nature, but why? According to the evolutionary principles first proposed by Darwin and Wallace, the pressure that predators and parasites exert on populations promotes the selection for stronger and more adaptable prey. Predation may also keep prey populations from becoming too large, in which case the species would in the end face a much greater catastrophe because it had exhausted its resources.

There are interactions other than predation that keep a particular species from exhausting resources. Individuals and the species to which they belong must compete for food and other resources. Central to evolutionary theory is the notion that the amount of a resource, whether chestnuts or bananas, is always exceeded by the demand for same. Competition is a "negative interaction," in a way, but one that determines the "weave," the innumerable interconnections, of an ecosystem at any given time. Ecosystems are equilibrated by the multitude of different species that must share the same food and space. These species must also, in the ecologist's parlance, "split the niche"—they must nudge others and give themselves elbowroom in a crowded habitat so they can feed, lodge themselves, and produce viable offspring. Some species, at a given time and under certain conditions, will be better than others at this game of survival and reproduction. When conditions change, so does the relative competitiveness of the species affected by those changes. Fifty million years ago in North America and Eurasia widespread warm, humid climates gave trees and tree-living animals the competitive edge. Some 15 million years later, with the onset of a global-scale cooling and drying trend, grasses that thrived in open, windblown habitats and animals that moved rapidly over great distances and fed on grasses dominated. Forests and forest-dwelling animals were marginalized under these drier conditions. After some mass extinction events, species, such as corals, that required highly diverse habitats disappeared, and other species, such as algae and other small organisms that thrive on a barren sea bottom, bounced back. The ability of some species to outcompete others under the new conditions changes the weave of the ecosystem.

In contrast, some interactions are spectacularly collaborative. Some species actually cooperate to survive, with the activity of one species benefiting the other. Flowers provide nectar for insects, and insects in turn carry the dusty particles of pollen, the plant's sexual enablers, on the fine bristles of their legs and antennae to another receptive flower. The process whereby insects facilitate flower sex ensures a broad spread of different genes among individual flowers of the same species. A robust variation in genes for sexual reproduction is critically important for virtually all sexually reproducing organisms, probably because it gives the species resiliency to deal with change or stress when conditions change. Such a reciprocally beneficial system between flowers and insect pollinators is known as an example of mutualism in biology. Another ubiquitous and extremely critical form of mutualism is that between nitrogen-fixing bacteria and plants.

Plant consumption, predation, parasitism, competition, and mutualism are important because of a general rule in biology. An ecosystem is a unit of nature made of diverse species representing all kinds of organisms, from bacteria to fungi to plants to animals. But an ecosystem is not simply the sum of all its species parts. For an ecosystem to work, its parts must interact. And the interaction—that which promotes energy flow—is invariably complex, with feedback loops and side eddies, as if the currents in an infinite number of intertwining streams were flowing in both directions. These interactions extend farther still, actually taking root in the soils, converting the chemicals from the atmosphere and the crust into useful nutrients, and infusing the atmosphere itself with life-sustaining and climate-changing gases. *Ecosystem* is another word for the living planet itself.

For all this interaction to sustain life on the planet, it has to be reliable. In other words, the function of ecosystems must remain stable over time. The decline in function of a component of the ecosystem, a species, must be compensated for elsewhere. An ecosystem built like a house of cards will collapse if only one component is removed. But how stable are ecosystems in the real world? This remains a profoundly important area of inquiry in ecology. Much of this research centers on what is commonly known as the diversity-stability debate. One would think that more diverse ecosystems are more stable because there are more opportunities to compensate for the decline in function of any one component. If a pollinating wasp declines in numbers or disappears altogether, any one of a number of other pollinating wasp species could likely take its place. Most ecologists had this view before the 1970s. But Robert May in 1973 published provocative research based on mathematical analysis and statistics that diversity tends to destabilize community dynamics and thus ecosystem function.

When I was a graduate student in the 1970s, I eagerly devoured May's book, and it became emblematic for me and a few of my peers as the new wave in biology. During subsequent decades ecologists have returned to the idea that indeed, more diversity on average can be associated with greater stability of the ecosystem. May's results were not necessarily refuted; he based his conclusions on statistically randomized communities. If the components of an ecosystem in the real world were indeed associated in this randomized way, one interaction would not complement another. More diversity would not contribute to stability and the dynamics of an ecosystem would oscillate wildly over time. In contrast, natural food webs are intricately constructed with both strong and weak interactions. The most stable ecosystems seem to be ones, like those in the tropics, that are hugely diverse and where many weak interactions mute the potentially destabilizing force of a few strong producer-consumer interactions. Elephants may be a major drain on forest resources, but if the trees are diverse enough, showing a great range of mechanisms for pollination, seeding, and regrowth, the forest ecosystem may be sufficiently resilient. Unfortunately, as in the case of African woodlands, these diverse ecosystems now occupy small and scattered land areas, so they cannot support such heavy consumers.

This brings us to another fundamental area of current agreement among ecologists. Even though high diversity may confer stability to an ecosystem, it does not guarantee that an ecosystem can withstand environmental disruption, such as the loss of effective land area, the loss of species, or the addition of new, alien species. Current theory and experiments show that drastic changes to a biological community can occur with even the removal or addition of a single species. In a classic study R. T. Paine experimentally removed the predatory starfish, *Pisaster ochraceus*, from an area of the Pacific tidal shore and found that the food web of the community became greatly simplified because the mussel, *Mytilus californianus*, suddenly liberated from the pressures of its main predator, competitively dominated all other species attached to the substrate. Unfortunately, the house of cards analogy seems to apply to nature after all. Our capacity to remove resident species and introduce alien species is not good for ecosystems and ultimately for us.

If one had to identify a single metaphorical image for the energy transferring and nutrient recycling ecosystems of the earth, it might well be a dung beetle. Ancient Egyptians took to this image; they revered dung beetles because they believed them responsible for the revolution of a dung ball–shaped Earth. Dung beetles, which belong to the Scarabeidae, the diverse family of scarab beetles, amaze all who have witnessed their arduous effort. They can roll a piece of dung the size of a tennis ball and bury it in the soil. It has been observed that one dung beetle can bury 250 times its own weight in a night (one appropriate generic name of a group of dung beetles is *Sisyphus*). Such perseverance pays off on a grand scale simply because there are so many dung beetles rolling and burying dung. In fact, they are a major means of biodegrading dung and dispersing seeds from plants eaten by elephants, rhinos, cattle, and other large animals, including humans, in many regions of the world. The cumulative recycling output is staggering. It has been estimated that scarab beetles transport to the soil forty thousand tons of human excrement each day in India. Their responsibility for other dung providers, such as cattle, triples this amount. At certain times of the year two-thirds of the excrement in India are buried by scarabs. That's an ecosystem at work.

It turns out that excrement disposal confers many advantages to many clients—to humans, to cattle, to the pasturelands, to the soil, to plants, and to the beetles themselves. Adult dung beetles are drawn to manure by odor; particular beetle species prefer a particular kind of animal dung. Incidentally, among the strong odors emanating from cattle manure and other excrement is the gas methane. Scientists have actually measured the amount of methane released to the atmosphere in this manner. The average cow releases, through a combination of defecation and flatulence, about three hundred liters of methane a day. Humans are far less productive; human

Dung beetles

males release an average of twelve times a day and females about seven times, contributing between four hundred and twenty-four hundred cubic centimeters of methane to the atmosphere per day. Dung beetles are apparently cued by a particular bouquet of odors; they will fly up to ten miles in search of the right dung. Some species even attach themselves to the tail of the dung producer in anticipation of the big drop. The liquid in the dung, to which biologists have applied the term *dung slurpie*, is their nourishment.

The real significance of this behavior comes with nest building and brooding. In many species a nesting pair of beetles shape the dung into a rolling brood ball. They then bury the ball; the female with her thicker legs does the main digging while the male hauls out soil from the tunnel. Beetles are spectacularly productive in recycling manure because the female typically lays only one egg in the brood ball. Then she seals the ball and the entrance to the tunnel and moves on, conscripting her male to help roll more balls and dig more tunnels for more eggs. After the hatching of the egg, the ball becomes a larval feast and is typically reduced by about 40 to 50 percent. Nestled in its dung ball at the end of a tunnel capped with soil, the larva is protected from predators and is not compelled to share its food with others. It thrives and matures as long as the brood ball is not destroyed. In about three weeks it transforms to adulthood, eats its way out of the brood ball, forms a new tunnel, emerges from the soil, and becomes the next generation of dung recycler; it will breed two weeks later. The generational handoff from one dung processor to another takes only six weeks.

Just as nature is complex, so are dung beetles. There are variations on the dung ball theme. Many species actually let the manure pile stand on the soil surface, where it becomes an odiferous island forming the lid of a great network of brood tunnels. Each branch of the tunnel system contains one or more dung balls sculpted directly from the lump of feces above. There can be scores of these dung balls in the filamentous community under the dung pile. When the larvae have matured, they eat their way through their brood balls, dig through the upper reaches of the tunnels

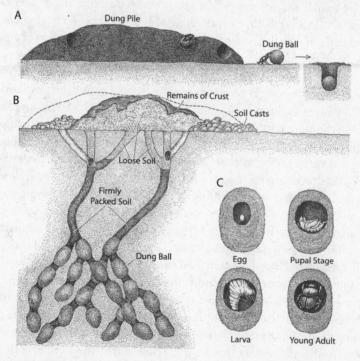

Different modes of dung beetle reproduction: (a) dung pile with beetle removing ball of dung and burying it; (b) dung pile in section with subsurface brood masses in tunnels; (c) brood mass containing different stages of developing beetle

backfilled with soil, and continue to feast on the remains of the dung pile above. The timing here is similar to that of the dung rollers—about six weeks—if the soil is sufficiently wet and the weather sufficiently warm. Otherwise the cycle is longer.

The impressive power of the industrious dung beetles transcends their mere success in processing dung. Not only is the dung beetle's redistribution activity a mechanism for dispersing the seeds of plants, but it also functions importantly in pest management. A single manure pat can produce a breeding ground for flies, including horn flies (*Haematobia irritans*) and face flies (*Musca autumnalis*), and be a place for scores of fly eggs. The beetles, competing with fly larvae for food, significantly reduce fly populations in livestock pastures. Even more significantly, their tunneling increases the soil's capacity to hold water, which sustains the plants rooted in it.

The tunnel diggers also profoundly improve the recycling of nutrients; they are extremely effective in enriching soils with nitrogen, an element essential to plant growth—indeed, as a key element in the formation of amino acids, the building blocks of proteins, essential to all life. That is why nitrogen is a basic ingredient of commercial fertilizers. Nitrogen is the most common element of Earth's atmosphere, but plants cannot process it in its natural form as the inert gas N_2. They need it to be "fixed," or converted into a usable compound in the soil before they can utilize it. Lightning can heat up the atmosphere enough to fix some nitrogen, but the preponderance of fixing is done by the bounteous, ubiquitous soil bacteria. These bacteria are particularly prevalent on the nodules of the roots of peas, beans, and other legumes, where they convert nitrogen compounds into a form the plants can use, and receive in return carbohydrates from the nurturing environment of the host plants. The fixed nitrogen is taken up in plant roots, and animals get it by eating plants or other animals that have consumed plants.

If a dung pile stayed unaltered on the surface, there would be little opportunity for this microbial industry. Instead, 80 percent of the nitrogen in manure would be lost through volatilization, converted to gas and released to the atmosphere. By efficiently transferring manure to the soil, dung beetles make significantly more nitrogen available for fixation by bacteria. With commercial fertilizers we have simply substituted this well-proved natural recycling system for one in which nitrogen is artificially infused into the soil. The dung beetles' aid in preventing volatilization of nitrogen, burying waste, and controlling parasites and pest flies represents an enormous cost savings for the world economy. In a remarkable analysis published in 2006, John Losey and Mace Vaughan calculated that the United States alone averts $380 million in annual losses because of the labors of these insects.

Dung beetles have much company. Other insects as well as worms, centipedes, spiders, fungi, and bacteria all play their part in rebuilding soils and cycling nutrients. Likewise, in the seas and lakes cyanobacteria supply useful nutrients like nitrogen to algae and other plants. Together these species provide perhaps the most fundamental intersection between living forms and the physical Earth, between the biosphere and the geosphere. Humans have both relied upon and exploited this interaction. The result is a high-stakes gamble.

I grew up five miles from the Pacific Ocean in a suburb of Los Angeles appropriately called Mar Vista. From the upstairs bedroom of our house I could make out the line of palm trees that marked the hazy shoreline. Some summers I was there in the waves, much to my parents' alarm, nearly every day. I sometimes wondered how I came to be so lucky to grow up at the edge of that vast, entertaining body of water, Earth's greatest single physical feature, rather than in those dreary housing tracts farther inland, in the middle of the stifling L.A. basin. Yet there was trouble in paradise. In various phases of the beach season a putrefying stench would assault us as my brother and I approached the beach. This was the evil vapor of the red tide, blooms of plankton that had died and decayed.

Plankton is actually a grab bag term for many different organisms that are passively carried over open water by winds and currents. Phytoplankton are microscopic plants, bacteria, algae, and single-celled protists that photosynthesize. Zooplankton range from microscopic protozoans and rotifers to shrimp and fish eggs and larvae to large, passive floaters like jellyfish. Massive decay and stench occur when plankton are too voluminous for other sea creatures to cull them efficiently. As the plankton die and drift to the sea bottom, their decaying remains deplete the dissolved oxygen in the bottom waters, killing fish, crabs, and other consumers that require this oxygen. All that death and decay release gases, such as methane and sulfur compounds, that contribute further to the stench.

As time went on, it was disturbing to see that the bouts of red tide were increasing. In recent years, returning to the playgrounds of my youth, I have been dismayed by the frequency and intensity of the red tides. Beaches are sometimes even closed, quarantined for long stretches of days or weeks because the toxins in the water that killed plants and fish are also harmful to surfers and swimmers. This hazmat environment is not simply the outcome of a natural correction. Red tides can be caused by an excessive influx into the sea of a variety of chemicals, primarily nitrogen, from unnatural sources.

Here we have an unfortunate outcome caused by too much of a good thing. We must return to the nitrogen cycle, so spectacularly facilitated by those indefatigable dung beetles, to show why. The nitrogen cycle is rather complex because it involves a number of intermediate compounds and requires a number of microbial methods of conversion. Nitrogen can be fixed expeditiously, and we have learned that some bacteria convert it directly to its usable form, nitrate, a nitrogen atom with three oxygen atoms attached (NO_3). But usually nitrogen fixation is more circuitous. When the decayed and waste materials of animals and plants are broken down by decomposers, the nitrogen from this process is usually bound up in ammonia (NH_4^+). Some plants can utilize ammonia directly to get their nitrogen, but

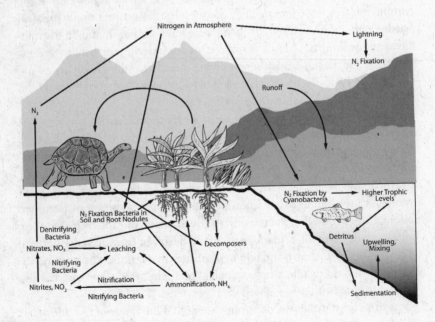

The nitrogen cycle

it is toxic for most plants and requires further conversion—first to nitrite (NO_2) by "nitrite bacteria" and ultimately to useful nitrate by "nitrate bacteria." Nitrogen in the air is also replenished by the decay of organic materials. This happens when the fixed nitrogen in the form of NO_3 or NO_2 is converted to gases, a process called denitrification.

Denitrification once again underscores the amazing versatility of bacteria. It is made possible by the work of denitrifying bacteria, which do not require oxygen for their metabolism. Those steaming cow pies on a cold morning are indeed giving off water vapor, but they are also releasing nitrogen to the atmosphere. The end product of denitrification is N_2, an innocuous gas that is plentiful in our atmosphere, but the intermediate products, nitric oxide (NO) and nitrous oxide (N_2O), are anything but innocuous. The former is a major component of smog, the latter a greenhouse gas that is contributing to global climate change.

Under natural conditions the whole nitrogen cycle works amazingly well, so well that it inspired humans to model nature in order to improve our lives. In the early twentieth century a German scientist, Fritz Haber, devised a scheme to short-circuit the nitrogen cycle by fixing nitrogen at high temperatures and pressures. The synthetic nitrogen fertilizers that this work made possible led to the food miracle of modern times, a tremendous increase in agricultural productivity. All this progress had unforeseen and unfortunate consequences because not all of the nitrogen fertilizer we apply to soils does its job. Some of it is washed out by rain or irrigation water and then leached into the soil or groundwater. Excessive nitrogen in drinking water causes certain kinds of cancer and also respiratory diseases in infants. Runoff enriched with nitrogen badly affects the health of coastal waters. As the platitudinous proverb goes, "All rivers runneth to the sea"; so doth fixed nitrogen. This nutrient overload, known technically as eutrophication, is the root cause of overexuberant blooms of plankton. All the unwanted nitrogen that exacerbated the red tide at my hometown beach had been washed into the sea from the nitrogen-rich products that humans had generated: soils infused with artificial fertilizers, the cuttings of nearly a million Los Angeles lawns, garbage, and excrement both human and animal.

The statistics on red tide are arresting. There are now more than 146 oxygen-deficient dead zones of coastal waters worldwide. These extreme depletion areas do not even include the beaches of Los Angeles, and some of them, such as those in the Gulf of Mexico, East China Sea, and the

Baltic Sea, are enormous, as large as twenty thousand square kilometers in area. More and more a growing source of the pollution is the runoff from irrigated fields where artificial fertilizers have been used or where large livestock herds are established. An example of such a problem spot is the heavily fertilized Yaqui Valley in Mexico, which feeds into the Gulf of California. This body of water in its natural state is nitrogen-poor, and its resident biota—shellfish, fish, and other sources of human food—are accordingly adapted to nitrogen-deprived conditions. They are then highly traumatized by the unwanted influx of nitrogen. Another example is the massive dead zone appearing every summer at the mouth of the Mississippi River, which has seriously threatened the shellfish and shrimp industries in the Gulf of Mexico. The red tide is now spreading off the coasts of many regions in Asia, Africa, and Latin America. A recent scientific paper predicts that by 2050 between 27 and 59 percent of all the nitrogen fertilizer that ends up polluting nitrogen-deficient marine ecosystems will come from upstream areas in the developing countries. Globally expanding industrialized agriculture is therefore severely debilitating an equally important source of food in the marine realm.

This nutrient overload is, if anything, even more widespread and more serious in the atmosphere and on land. The excess nitrates and ammonia in surface waters and soils, when denitrified, contribute mightily to smog and greenhouse gases. To make matters worse, another man-made short circuit of the natural cycle adds to these gases: factories burning fossil fuels like coal and oil release nitric oxide, nitrous oxide, and other gases to the atmosphere. These nitrogen gases are also responsible for the acid rain that is thinning forests and killing freshwater organisms in parts of Europe and northeastern North America. Finally, increases in nitrogen in both air and soils favor weedy plants that thrive on excess nitrogen. Like too many elephants in an acacia grove, this can throw ecosystems off kilter. In northern California, diverse plants that have historically thrived on nitrogen-poor soils are being crowded out by more opportunistic species. The result is the degradation and eventual loss of a treasure of biodiversity, leaving many habitats less productive than they once were.

The nitrogen cycle is one of the more complex systems, but not the only one, responsible for bringing the biological and physical world together.

Living organisms require other elements besides nitrogen after all. These are phosphorus (P), potassium (K), calcium (Ca), magnesium (Mg), sulfur (S), carbon (C), oxygen (O), and trace metals. Obviously light (namely, sunlight) and water are also required. All these items have their own cycles. Over the great expanse of evolutionary time, the rate at which they work has fluctuated, and they are thus an essential part of the history of the modern ecosystem. Understanding the ways the cycles have changed in intensity, especially as a result of human activity, can help us look forward as well as back in time. Recent perturbations of these cycles can, like weather vanes, warn us of the approaching storm.

Some of these elements cycle in ways that offer both good news and bad, as the nitrogen cycle does. When plants take up sulfur compounds such as sulfur dioxide (SO_2) they remove the oxygen atoms, leaving sulfur in a reduced or deoxidized state. Reduced sulfur is useful; it is an important component of proteins. In its oxidized state, however, sulfur becomes a sulfate and contributes to the acidity of rainwater; like certain nitrogen compounds such as N_2O, it is a key ingredient for harmful acid rain. Once again, humans have all too effectively contributed to worsening the situation; industrial emissions contain harmful sulfur compounds that we have simply released into the air, including sulfuric acid particles, a major contributor to smog and an agent for respiratory diseases. It is estimated that this human industrial activity is about as productive of sulfur compounds as all natural processes, such as outgassing geysers and erupting volcanoes.

Oil, coal, and natural gas are of course constituents of another fundamental cycle. They all are derived from life itself, from the accumulating dead and decaying organic matter, fossils, in sediment. The dominant element of all these substances is carbon, also expectedly the basic element in the biological compounds of all life-forms. An alien arriving on our planet might first enter into its recorder the primary diagnosis that all life on Earth is "carbon-based." It is possible that some life elsewhere in the universe, if it exists, could be built on another basic element with somewhat similar properties—silica, for instance—but we have not discovered any such alternative on Earth.

The carbon cycle, like the nitrogen cycle, can involve several intermediate steps. It too plays a leading role in the saga of the modern ecosystem, and its fluctuations through time have had cascading effects on climate, plant growth, organisms, us. Carbon is stored in four main reservoirs: carbon as CO_2 in the atmosphere; as organic compounds in living or recently dead organisms; as dissolved dioxide in oceans and other bodies of water; as calcium carbonate in limestone and buried organic matter—coal, natural gas, peat, and petroleum. During photosynthesis, plants in the presence of sunlight and water are stimulated to absorb CO_2 from the atmosphere to form carbohydrates and release oxygen. At night the system shuts down because sunlight, its ultimate driver, is not present. (But get-well flowers are taken out of hospital rooms at night not because they release CO_2 but because their moisture can cultivate bacteria.)

This fluctuation in plant activity occurs over longer time frames than the daily one. Outside of tropical regions, plant metabolism is closely tied to the changing seasons; plants photosynthesize like mad in the spring and summer but are desultory about it in the fall and winter. The result is a flux in the amount of CO_2 in the atmosphere—low during the winter and high during the summer. Thus the natural atmospheric CO_2 curve over several years for say, a forest in upstate New York, looks like a roller-coaster ride, with seasonal highs and lows. For the great majority of Earth's ecosystems,

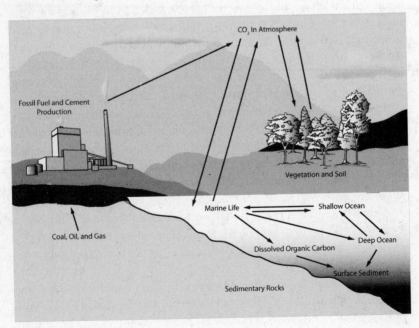

The carbon cycle

photosynthesis is like a heartbeat, its rhythmic manufacture of carbohydrates propelling life along at many levels. The producers and consumers all are transfer points in the carbon cycle.

When organisms eventually die, carbon is still part of the picture. Decaying and dead organisms are systems in which organic compounds are oxidized, combining with oxygen to form CO_2. But not all the compounds are oxidized. A fraction is redeposited as sediment. Trapped and compressed sediment enriched with carbon is what has formed Earth's great underground reservoirs of coal and petroleum. Carbon dioxide from the atmosphere also dissolves in oceans and other bodies of water. Aquatic plants use it for photosynthesis, and many aquatic animals use it to make shells of calcium carbonate ($CaCO_3$). The shells of dead organisms—plankton, coral, clams, snails, and other creatures—accumulate on the ocean bottom, eventually forming marine limestones. This is part of the sedimentary cycle, the reworking of rocks and their organic material through deposition, uplift, and erosion.

When we measure the uptake and burial of carbon and other elements as recorded in fossil shells and in limestone sediments, we gain once again important new clues about the changes in the proportion of gases, including CO_2 and oxygen, in the atmosphere over time; these mark the major climate changes that influenced the evolution of organisms. It may have taken millions of years for carbon to be transferred from one step to another in the rock cycle. By contrast, it takes only a day or even only seconds for a plant to convert CO_2 into useful carbohydrates. So CO_2 can increase or decrease over short or over long times, depending on the intensity of activity of its various sources. The rocks show us that 75 million years ago, for example, intense volcanic activity contributed an extra slug of CO_2 to the atmosphere.

In recent years it has been apparent that our own species is making major contributions to the carbon cycle. Human industry is transforming Earth's atmosphere by loading it with CO_2: burning forests to clear them for agriculture, logging, or land development; emissions from factories, cars, and other human-generated production. The trend began with the Industrial Revolution in the nineteenth century and has rapidly accelerated in the last few decades. This was first confirmed by graphing over time atmospheric CO_2 concentration based on careful measurements taken by Charles Keeling at a survey station on the volcano Mauna Loa on the island of Hawaii. The resultant Keeling curve has a steeply rising slope, showing that atmospheric CO_2 has increased by nearly 20 percent since about 1960. Using historical records and measurements in ice cores, we can extend the graph farther back in time. Extrapolation shows that a curve extending from about 1700 to the present day looks like the right half of a U—a striking image of the spectacular rise of CO_2.

Cores drilled in and extracted from glacial ice allow us to determine the CO_2 levels during even earlier times. There were relatively high concentrations of CO_2 in warm interglacial periods, when glaciers were receding, and low CO_2 levels in colder ice ages, when they were spreading. These core samples also clearly show that the current levels are significantly higher than those for the last 140,000 years. In fact, it is now estimated that the current levels of atmospheric CO_2 are higher than in any period since about 10 million years ago.

The CO_2 curve is one of the most important scientific discoveries in modern times, replete with meaning for the environmental present and future of the planet. CO_2, like water vapor, CH_4, and N_2O, significantly influences Earth's heat balance. We give greenhouse gases their name because

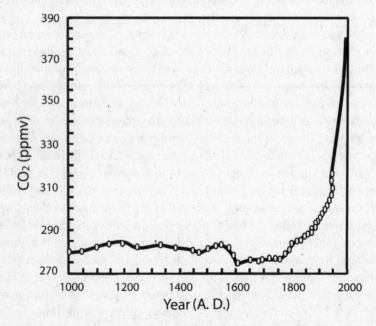

Increase in atmospheric carbon dioxide concentration since 1000 A.D. Ellipsoidal points represent measurements taken from ice cores.

they make the atmosphere seem like that warm, dense, moist air we feel when we walk on a cold wintry day in the Bronx, say, into the enclosed glass conservatory full of tropical plants at the New York Botanical Garden. As these gases accumulate, they form a "roof" over the air we breathe because they impede the rate at which atmospheric gases dissipate and escape into space, acting like a blanket of insulation around the entire planet.

There is a strong correlation between the increase in greenhouse gases, notably CO_2, and the increase in Earth's average atmospheric temperature. Such temperatures have risen $0.6° +/- 0.2°C$ since 1900, which is roughly correspondent to the period that experienced the rise of human-produced CO_2. It is predicted that by 2030–50 mean global temperatures will increase between $1.5°$ and $4.5°C$ depending on the region and latitude. The distant future is harder to predict, but different projections agree that by 2080–2100, mean annual temperatures will be $2°$ to $8°C$ higher, depending on location, than they are today.

That may not seem like much. But the key figure here is not the mean or average global temperature, which takes into account all the extremes of hot and cold from all regions in any given year, but the drastic upsurges in temperature that most models predict are likely at higher latitudes, including the polar regions. If those extremes vary just one or two degrees, the change in climate is bound to be significant, even traumatic. Sea-level rise from melting polar ice alone would flood many coastal cities around the world. Moreover, there would be new areas of climate and weather instability, perhaps a proliferation of hurricanes even in the South Atlantic, where they have not yet occurred. A shift upward of two to six degrees would be hugely disruptive, perhaps even catastrophic, and on a global scale.

It is all the more worthwhile, then, to consider other phases of Earth's history when the composition of atmospheric gases changed and when global average temperatures were as high as or higher than that which we are experiencing today. Animals and plants survived those periods of global warming, so what's the big deal? The big deal comes in the way climatic shifts changed ecosystems because it often led to the extinction of species and their replacement by new species. Moreover, those prehistoric changes are recorded on the scale of hundreds of thousands, even millions of years. What we see in the fossil record are the long-term readjustments to the changes; it does not disclose the trauma on a shorter timescale, say, a hundred years, for which the fossil record is too fuzzy to show us a pattern. But we know these effects could include the elimination of local ill-adapted populations, shrinking geographic ranges, and loss of food and other resources.

How would such events affect human survival and quality of life in our own time? It is not completely certain, but uncertainty is no cause for comfort. It is safe to assume that we consumers, at the top of the food chain, dependent as we are on so many parts of ecosystems for nutrition, clothing, habitat, and energy, may have little protection from large-scale disruptions caused by drastically warmer climates.

Given our sense of self-importance, it is no surprise that you hear claims that never before has one species so radically altered the environment—so much that the very physical workings of the planet and the composition of gases that form its atmosphere have been fundamentally disrupted. That is not strictly true. When photosynthesizing bacteria emerged and proliferated some 2 billion years ago, their labor produced so much oxygen that this gas could no longer be absorbed by water and the sediments at the bottom of the seas. Oxygen escaped to the atmosphere and accumulated to levels that spawned new species and new ways of life. Some of the species that exploited this oxygen-laden atmosphere included the lines leading to multicellular organisms—plants, animals, and us. But there were also casualties, species that were completely exterminated. Who will be the winners and losers in the warm, CO_2-enriched world of the future?

5

EVOLUTION—
LIFE THROUGH A NEW LENS

More appealing than knowledge itself is the feeling of knowing.
—Daniel J. Boorstin, *The Discoverers*

The amazing knowledge we have about our planet—its diversity, its history, its ecosystems, and its human-induced traumas—is not just the result of our highly charged, curiosity-driven intellect. We acquired this knowledge in a particular way. It is a product of science. And science itself is a peculiar and even limited form of enlightenment. A cartoon says it all. Two scientists are standing in front of a blackboard full of complicated equations interrupted by a big gap. Within the gap are the words "and something incredible happens here." One scientist admonishes the other: "I think your theory needs some refinement." Scientists do not have the option to make things up. They can generate new theories or hypotheses, but these must be testable; they must be compared with our scrutiny of the world, what we can sense, what we call empirical observations.

This important distinction often gets muddled. If the difference between science and other forms of knowledge were clearer to more people, we might not have the social mess we now are enduring over such issues as evolution versus "creationism." Contrary to the opinions of its advocates, the idea of intelligent design is no more scientific than the ancient Hindu myth that the world is carried on a stack of four elephants perched on the shell of an enormous turtle. Intelligent design requires the presence and actions of an unseen creator, a god. One's beliefs are one's own business, but any idea invoking God cannot be a matter for science, since God is not susceptible to empirical observation. Moreover, the realities that credible science insists on may be brutal, but they cannot be easily dismissed. Experts in materials science, engineering, and hydrodynamics predicted for decades that the levees around New Orleans could not withstand a direct hit from a major hurricane. In a wealthy country whose citizens are informed by some of the world's best scientists and aware of the flooding catastrophes and subsequent remediation in lowland countries like the Netherlands, the devastation of New Orleans by Hurricane Katrina in 2005 was one of history's most avoidable disasters.

In this book I am taking a long look back at numerous ancient species, many of them extinct. But the science of paleontology shows us that the processes effected by those organisms, as well as the forces acting upon them, were much the same as those at work today. Paleontology is closely linked with biology, indeed, in many ways is part of biology. And there is no better way to understand biology than to examine its core, the theory of evolution, which is the bridge between what is happening now and what happened before. It applies equally well to the cultured bacteria in the petri dish of a modern laboratory as to a swarm of giant dragonflies in a 300-million-year-old swamp. Its principles are critical to our understanding of the ancient origins of pollination in flowering plants by insects, a subject to which Darwin himself contributed seminal insights. It helps us understand the patterns of recovery in ecosystems following mass extinction events. More than any other scientific theory, evolution is critical to our understanding former traumas, the current crisis, and the prognosis for the living planet.

One of history's great ironies is that Darwin's theory of evolution, recognized since its inception as one of the most profound achievements of the human mind, did not belong to him alone. It was issued simultaneously and independently by a much less familiar figure, another Englishman, explorer, and collector. Alfred Russel Wallace, like Darwin, had an early passion for collecting beetles. Perhaps this shared obsession with the most evolutionarily prolific group of organisms on Earth was necessary to arrive at the theory that explained the stupendous diversity of all life. Both Wallace and Darwin credited the insight of the mathematician Thomas Robert Malthus with sparking their notions of competition and the struggle for survival. But in other ways the men were not similar. Darwin came from landed gentry. He had a passive disposition, and at least for a time he harbored scruples concerning the collision between the science he was doing and his religious beliefs. Wallace came from an impoverished background, was attracted to debate and controversy, and early in life drifted away from the church and other established institutions. Darwin arrived at his theory in a series of steps based on years of accumulating observation. Although Wallace also collected data in extensive fieldwork, he claimed to remember the precise moment when the inspiration came to him, in the midst of a raging fever contracted during a collecting trip to the Moluccas. But Darwin, the timid gentleman, and Wallace, the renegade, came to precisely the same revolutionary theory, what some have called their "dangerous idea."

Both Darwin and Wallace grew up in a world where almost everyone, at least in Western cultures, embraced a single explanation for the diversity of life. The Bible and other mythic works proclaimed that the Creator had imbued life with wondrous variety. Each different kind of organism, each species, was the individual act of the Creator. Species were therefore immutable and thus certainly resistant to transformation into another species. The most immutable and distinctive of all species was man himself, for man was a species of intelligence and mastery, a species uniquely created by God in his own image. We have no evidence that when Darwin entered Cambridge to study for the clergy, he had the pious devotion generally associated with that vocation, but he shared the general view of life and its origins. When he stepped on the decks of the *Beagle* to begin his voyage, he was, in effect, a creationist.

Yet already in 1831 there were inklings that not all was right with the creationist view. The French biologist and philosopher Comte de Buffon had almost a century earlier risked censure in claiming that Earth was extraordinarily more ancient than a literal reading of the Bible suggested and that the duration of Earth's history had been sufficiently long for animals and plants to change, producing new varieties. Jean-Baptiste Lamarck, another eighteenth-century scientist, who arrived at an erroneous mechanism for evolution, nevertheless correctly argued that over time organisms might change to adapt to new conditions and that these changes would be preserved in future generations. The word *evolution* was first used by the early-nineteenth-century zoologist Étienne Geoffroy Saint-Hilaire to describe the process of development in embryos. The British scientist Charles Lyell, who had contributed in 1830 the pivotal work *Principles of Geology*, claimed that species had appeared and disappeared during the vast period of time indicated by the slow and stately reshaping of Earth through mountain building, erosion, and other processes. Even Charles Darwin's own grandfather, the famous freethinking and provocative Erasmus Darwin, turned a notion of evolution into poetry, asserting metaphorically that the "urges" of plants and animals were driven by "lust, hunger, and danger" to develop new forms. It remained for Darwin and his overlooked contemporary Alfred Russel Wallace to describe evolution more precisely and to explain, for the first time, how it worked.

When he boarded the *Beagle*, Darwin the creationist was nevertheless clearly influenced by the musings of Buffon, Lamarck, and others; he even brought along a copy of Lyell's important book. But it is clear that the greatest influence on his thinking on the journey were his own observations. His *Beagle* notebooks suggest that something on the voyage changed his mind about species. When Darwin returned to London, he went furiously to work, exhaustively studying the variation in animals and plants and the domestic breeding of dogs and pigeons. He read "for amusement" Malthus's *Essay on the Principle of Population* (1798), the central theme of which was that populations, including human populations, tend to increase geometrically, whereas subsistence tends to increase only arithmetically. In other words, at any given time the amount of resource available in the form of food or other subsistence is always exceeded by demand. Accepting this assumption, one would have to conclude that unchecked population increase could have hugely catastrophic effects.

This dark Malthusian view sparked Darwin's famous series of premises. First, he noted that nature shows enormous variation, different organisms having markedly variable traits that represent many different approaches to survival. Second, these traits are passed on from one generation to the next. Third, with limited resources in nature only certain individuals, those inheriting the most advantageous traits, can survive to compete for those resources. These individuals are favored, or "selected," because they have traits that allow them to adapt well to a given situation. A change in conditions, which would influence the availability of resources, could "select" for individuals with certain other traits. In this way species could change through time, as their populations were dominated by organisms with new, better adapted traits, which would eventually produce millions of new varieties simply because, as Lyell and others had shown, there had been so much time. Evolution took its time, perhaps millions of years, instead of the few thousand years suggested by a literal reading of Genesis.

Darwin had set down these thoughts by 1842, but he kept them a secret for sixteen more years. An obsession with scientific detail, a need for scholarly validation, and persistent illness (largely, it is now thought, psychosomatic) were factors here, but the primary reason for the delay had deeper roots. Darwin, a man of weak constitution and retiring demeanor, was strongly reluctant to promote publicly an idea that he knew would be engulfed in huge controversy. At last he was prompted to go public through the strong urgings of his colleagues and, most important, by the threat of being scooped by Wallace, who amazingly had arrived at the same insight and conclusion. The story has an obvious lesson. My colleague at the American Museum Niles Eldredge, who was also the curator of our Darwin exhibit in 2005, is fond of repeating that if you have a truly good idea, you had better publish it fast because someone else is likely to have the same good idea sooner rather than later.

On July 1, 1858, three papers, two by Darwin and one by Wallace, that set forth in brief the theory of evolution by natural selection were read by the eminent Charles Lyell and J. D. Hooker to the Linnean Society of London. Neither author attended, and the readings failed to elicit any discussion by the thirty society fellows present. This subdued event was part of a year that, Thomas Bell, the society president, declared with little prescience, "has not, indeed, been marked by any of those striking discoveries which at once revolutionize . . ." Yet the unexpected appearance on the scene of Wallace's work and the occasion of its and Darwin's first public outing was momentous enough for Darwin. He was ready to launch. By the next year Darwin had published his 512-page book *On the Origin of Species by Means of Natural Selection, or the Preservation of Favoured Races in the Struggle for Life*. The book was scientifically exhaustive and rather technical for the general reader, but its grand ultimate vision and its graceful passages engaged a literate and intellectually hungry English society. It became a huge seller for those times, requiring a printing of more than four thousand copies. It also became, as Darwin ruefully had predicted, enormously controversial. Nonetheless, its impact was profound and inevitable. By 1876 Darwin's *Origin of Species* had sold sixteen thousand copies and been translated into virtually every European language, as well as Japanese. Ironically, Alfred Wallace drifted away from some of his earlier convictions and increasingly entertained a belief in a "Higher Intelligence." His fame was soon hugely eclipsed by Darwin's, and history has further skewed this rather unjust imbalance in acclamation. Darwin's name has become a metaphor for a new vision of life itself. By the time he was buried in Westminster Abbey in 1882 alongside Sir Isaac Newton, Darwin's explanation for what he called "endless forms most beautiful and most wonderful" had already defined the coming century of biological research. Science had fundamentally changed, and so had the world.

How did such profound insights come about from the thoughts of men with limited information and tools at their disposal? Darwin had an obsessive curiosity and a deep knowledge about beetles, barnacles, Galápagos tortoises, pigeons, dogs, and orchids, but neither he nor Wallace nor anyone else at the time understood the fundamental mechanisms, let alone the biochemistry, of inheritance. Nonetheless, the Darwin-Wallace theory of evolution was a powerful predictor. When eventually such inheritance mechanisms were actually observed and tested, they fell right into place; they vindicated rather than defied the theory.

Seven years after the publication of Darwin's *Origin of Species*, the mystery of exactly what gets transferred from generation to generation had the beginning of an explanation. An Austrian monk, Gregor Mendel, conducted experiments cultivating twenty-eight thousand garden peas in his monastery in the Moravian town of Brno. When Mendel's results were eventually published in 1866, the science of genetics was born. The work was largely ignored, however, then rediscovered thirty-five years later, long after Mendel's death. By the early twentieth century laborious experiments by Hugo de Vries and others recording the traits of rapidly reproducing fruit flies had illustrated both the complexity and the predictability of inheritance. The flourish of experimental genetics in the next decades culminated in the late 1950s with the disclosure of the very chemical nature of the genetic material—DNA.

As this century begins, we have the complete map of all 25,000 genes and 3.2 billion nucleotides, or base pairs, in the DNA of those genes in humans and complete genomes of several other species. Darwin and Wallace knew nothing about nucleotides or genes, and had no idea of how genes mutated to build new variants of organisms in populations. Nevertheless, their theory endures because it predicted the existence of these traits and mechanisms.

Darwin also predicted the riveting evidence for the close affinity among species based on their common descent. He spent some time at the London Zoo observing Jenny the orangutan, comparing its behavior with that of his own children. This and other observations led Darwin to claim provocatively that our membership within the primate family tree was even deeper than Linnaeus's classification had suggested. Numerous traits and behaviors were shared by humans and their closest kin, the chimpanzee, orangutan, and gorilla, and this kinship defined their shared evolutionary heritage. If he were alive today, Darwin would doubtless take some satisfaction from knowing that modern genomics emphatically supports his claim. The human and the chimp genome each have about 3 billion DNA nucleotides, of which 96 percent are exactly alike. The Darwin-Wallace explanation for the variation in tortoiseshells, the seemingly gratuitous abundance and diversity of beetles, and the humanlike qualities of Jenny the orangutan not only is profoundly validated by modern biology but *is* modern biology.

In his notebooks, Darwin drew sticklike trees with branches showing how species were connected by descent with modification. Both he and Wallace forcefully argued that species were kin and that understanding the pattern of their kinship was a fundamental goal in biology. But in pre-Victorian times, when every ship came back to Europe with new and yet stranger creatures from afar, this notion of kinship was hard to embrace. Konrad von Gesner's monumental *Historiae animalium*, issued in four volumes between 1551 and 1558 (and one additional posthumously in 1587), had profuse illustrations, including Albrecht Dürer's famous woodcut of the Indian rhinoceros, one of the most compelling visions of nature's beasts ever rendered (but an actual example of which Dürer had never seen). Gesner, despite his authoritative coverage, had no trouble blurring the line between myth and reality. Alongside rhinoceroses, elephants, and moles, he confidently inserted unicorns and three-hundred-foot sea serpents. And legends like this endure. The north woods of Wisconsin still harbor real wildlife— deer, black bear, water moccasin, wolverine, and gray fox. But the local beast of greatest reputation is the hodag, *Bovis spiritualis* (drunken cow), a creature putatively native to Wisconsin, with "the head of a bull, the back of a dinosaur, and the leering features of a giant man. Its legs are short, its claws are long, and its tail is spear-tipped." An example of this most formidable distinctive species had not actually been seen since one was allegedly captured in the late nineteenth century, though papier-mâché versions propped on floats in the Rhinelander Hodag Festival look rather child-friendly, like a creature in a Maurice Sendak book.

Dürer's woodcut of the Indian rhinoceros

One can see why the blurring between myth and reality is so tempting. An egg-laying duck-billed platypus, an armadillo, a walking stick, a Venus flytrap, not to mention the bones of a long-necked sauropod dinosaur, are all arresting at first encounter, a challenge to our wildest fantasies of the bestiary. Life not only is diverse but can be very weird. The pycnogonids, or sea spiders (not closely related to true spiders), could well stand in for invaders from Mars in H. G. Wells's *War of the Worlds*. These creatures have eight spindly legs sprouting from a ludicrously tiny body, so tiny that their digestive organs extend into the legs!

When we reflect on the weirdness and variety of life-forms, it is hard not to focus on the differences that set species apart. Yet the profound biological insight is that all those differences have emerged out of a basic sameness. At some basic level all life-forms are alike. Millions of animals share the same body plan, for example. If you split them down the body lengthwise, the two halves are mirror images of each other, an architecture biologists call bilateral symmetry. Those same animals have appendages, fins or limbs, that also exhibit bilateral symmetry. As we extend out over the galaxy of life, we find this basic attribute breaking down; starfish and sea urchins, for example, have a starburst architecture of the body and appendages—in other words, they are radially symmetrical. Organisms like plants or fungi have of

The pycnogonid *Anoplodactylus lentus*; head and body length about 4 centimeters (1.6 inches). Head, proboscis, and head claws are oriented toward the bottom of the photograph.

course grossly different body plans. Yet all these organisms as well as others, including amoebas, bacteria, and algae, have many traits in common: similar proteins, metabolic pathways, cells, parts of cells, and DNA. That all life, which is built on twenty-two amino acids and reproduced from a code written in DNA, came from one source, one ancestor, and one event is not a controversial notion.

Appreciating the reality of the common ancestry of all life requires us to look beyond the superficial qualities of a species to their meaningful, often masked similarities that indicate true kinship. We need a framework, a scaffolding for this, and it takes the form of a branching "tree" with a "trunk," "limbs," "branches," and "twigs." This branching structure is a phylogeny (from Greek: *phylon*, tribe, race, and *genetikos*, relative to birth, from *genesis*, birth). All the millions of species that constitute life, whether living or extinct, make up the myriad branches of a single phylogeny, a single tree, the tree of life.

How do we precisely determine the correct pattern of the branches within the tree of life? One important breakthrough of modern biology is that we have gotten better at this, in a large way because of the methodical

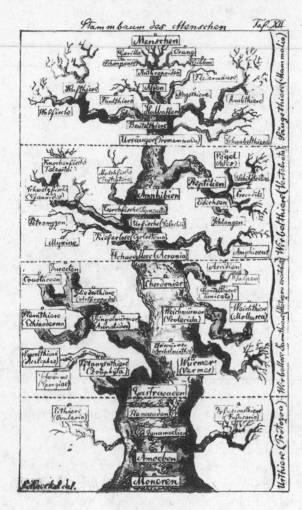

A tree of life as proposed by Ernst Haeckel, 1874, in his famous book *Anthropogenie* (*The Evolution of Man*)

advances of an area known as cladistics. We have also discovered evidence for the branching of the tree of life at several levels: we can identify similarities in social systems, behavior, superficial traits, anatomy, tissue structure, cell structure, protein composition, chromosome makeup, not to mention the structure of DNA, the basic chemical of the genetic code and inheritance. As we amass these data, the patterns of relationships, the branches of the tree, become clearer.

DNA has revolutionized these investigations of diversity and the tree of life. DNA, or deoxyribonucleic acid, takes the form of a double helix, where two strands of sugar-phosphate molecules are twisted around each other and linked by myriad chemical "ladders." The ladder rungs are very simple molecules made up of carbon, oxygen, nitrogen, and hydrogen bonded together in a way that forms a ring; these are called nucleotides, or bases. There are just four kinds of these nucleotides, and they only pair up in particular ways. For example, the base adenine (A) links only with thymine (T), whereas guanine (G) links only with cytosine (C). Hence one strand of DNA with its attached nucleotides is very like a negative of the other. If you know the section of one strand that attaches nucleotides ACGTAG you can determine the sequence on the other strand, in this case TGCATC. These sequences of pairs form a code that is the fundamental property of all genes. Genes are simply segments of DNA of varying length. When DNA unzips during cell division, it can duplicate itself because of the complementary pairing between the base of one strand and the base of another. Because DNA can self-replicate, an organism can manufacture more DNA when needed—for example, during reproduction or cell growth.

DNA has another very critical function, at which it works in tandem with a related nucleic acid called RNA, or ribonucleic acid. RNA, like DNA, has a sugar-phosphate strand or backbone, but unlike DNA, is

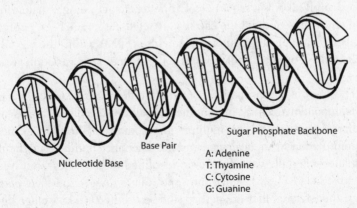

Sugar Phosphate Backbone

Base Pair

Nucleotide Base

A: Adenine
T: Thyamine
C: Cytosine
G: Guanine

A schematic of DNA

only a single-stranded molecule. In addition, RNA does not contain the base thymine; instead, it has uracil, another molecule with a ringlike structure that always attaches to adenine. RNA can assemble on a template a single strand of DNA, pick up a genetic code in the form of a base sequence, and move out of the cell nucleus to other parts of the cell. In this way it acts as a messenger, bringing a code that has been read from DNA to parts of the cells, such as the ribosomes, where proteins are manufactured.

The four different nucleotides in DNA and RNA can be thought of as a code consisting of 4 different letters. Such a depauperate alphabet—exactly 22 letters short of our own—does not seem to offer many options for reproducing genetically variable organisms or for making the blueprints for the hundreds of thousands of proteins that make up an organism. The DNA of each gene is made up of hundreds or thousands of nucleotides, however, which are arranged in sequences that vary tremendously. The human genome, as noted, consists of about 3.2 billion nucleotides in about 25,000 genes. Now if we use a gene probe, a short segment of DNA with a common base sequence of 3 letters consisting of AGC, we can find 40,049,089 matches—that is, other sequences of AGC—in the human genome. This number is less (it can be more) than the 44,767,202 matches expected in random 3.2 billion letters because the genome contains information in its code, and this information is not randomly distributed. The 6-letter code GGGTCG has only 79,358 matches when 699,434 matches would be expected if DNA were randomly distributed. Indeed, all it takes is a sequence of 16 letters to get a single match in the whole human genome. One can readily see that there may be enormous numbers of unique 16-sequence segments of DNA within the 3.2 billion nucleotides of our genome. The possibilities for genetic variation are staggering.

The amount of information contained in the genomic universe is even more stupendous. Think of genomes as galaxies, each one representing a different species. We can multiply the estimated 10 million species by the information in the average genome. The latter number is difficult to calculate—after all, we haven't even identified most species—but we can take an educated guess. Since most species are insects, soil invertebrates, fungi, and bacteria, they are perhaps the best source for an average figure. There is a small but growing number of species for which the genome is now completely sequenced. About 100 of the comparatively small genomes of the Archea and bacteria have been completely sequenced, but only a sprinkling of other kinds of organisms have complete genome sequences. The fruit fly, *Drosophila melanogaster*, has 13,600 genes; the sea urchin, *Strongylocentrotus purpuratus*, has 23,500 genes; the yeast fungus, *Saccharomyces cerevisiae*, has 6,275 genes; the roundworm, *Caenorhabdites elegans*, has 19,000 genes; and a common bacterium that inhabits our intestines, *Escherichia coli*, has 4,800 genes. These aren't many data points, but a ballpark figure for an average would be about 10,000 genes. This is about a third the size of the human genome, or an estimated 1 billion nucleotides per average genome. Ten million species multiplied by 1 billion nucleotides equals a very large number:

$$10,000,000,000,000,000$$

But this number is still far short of the actual amount of genetic information in the biological universe. What makes DNA fingerprinting useful in solving crimes or in exonerating unjustly convicted felons on death row is the fact that every individual (except identical twins) of every species is distinctive in at least a few nucleotides of a few genes. And the resultant number of *that* individuality is incalculable.

There is a caveat here. DNA has one severe limitation when it comes to capturing the pattern of evolution: with some rare and often controversial exceptions it cannot be isolated and sequenced in fossil organisms. I have argued that 99+ percent of all life is extinct, and fossils are all we have to capture much of that history, though the fossil record is woefully incomplete. Fortunately, in recent years we have made some spectacular improvements in it, and at the very least, it is in some critical cases good enough to help us figure out some of the major branches of the tree of life, cases where DNA is all but powerless.

I would argue that much of the wonder of discovery in the future will lie in our looking longer not only inward into our biology but outward to our myriad biological relatives that inhabit or once inhabited this planet. We must link all the emerging genomic information with information on other levels of organization, the protein products of gene expression, chromosomes, cells, tissues, organs, and even elusive biological traits concerning behavior and social systems. Life is not merely a collection of genes. Life is an effusion of diverse entities made up of genes and everything else. Phylogenies should draw on information about all these levels of or-

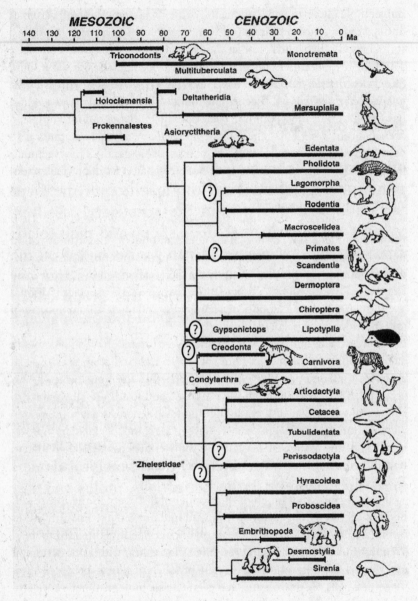

A phylogeny for mammals based on anatomy and fossils; horizontal black bars represent the age range of the group; question marks indicate controversial branching points in the phylogeny where gene studies have shown different relationships.

ganization. Then we shall better understand the architecture of the tree of life and the process, evolution, that built it.

Somewhere, most likely in southeastern Asia or China, a tiny new kind of microorganism emerged several years ago in populations of both wild and domesticated birds. The cohabitation between a microorganism and its host is not in itself a cause célèbre. As I have said, it is estimated that we humans are the walking ecosystems for more than four hundred species of microorganisms living in our digestive tracts, our mouths, and our skins—everywhere in and on us. Many of these are necessary for our survival, but others are unwelcome lodgers. Such is the case for this new organism that invaded birds, an influenza type A virus known as (HPAI)H5N1. HPAI stands for "highly pathogenic avian influenza," and H5N1 is code for the properties of the virus itself. Viruses are the organisms of utter simplicity in the biological world. (HPAI)H5N1 is a microscopic intracellular parasite containing a genome of only eight very rudimentary genes, eight single strands of RNA. This limits both the structural complexity and the functional range of the H5N1. This virus essentially does only two things: it parasitizes host cells, and it reproduces explosively. Importantly, it does those things better than virtually any other organism on the face of the Earth.

The *H5* in H5N1 stands for a subtype of the protein hemagglutinin coded by one of its genes. This protein binds the virus to the cell it infects. The *N1* stands for neuraminidase, an enzyme on the surface of the virus that speeds up the release of progeny viruses from the infected cell. Hemagglutinin and neuraminidase are of obvious relevance to medicine, and a variety of influenza A viruses are named for the subtypes of these proteins. Other strains include, for example, H1N1 (Spanish flu), H2N2 (Asian flu), and H3N2 (Hong Kong flu). All these strains are bad, but (HPAI)H5N1 seems particularly bad. It is transmitted rapidly from one bird to another, killing its erstwhile hosts in the process. Some strains of H5N1 have been known since 1959, but the extravirulent "Asian" strain of H5N1 is thought to have evolved only between 1999 and 2002. Since then it has killed tens of millions of birds and prompted the culling of hundreds of millions more. The outbreak first occurred in South Korea, Vietnam, Japan, and Cambodia; H5N1 infection has now spread to most Asian countries, many European countries, and a few countries in central and northern Africa. Many

North American birds are infected by a weakly pathogenic strain of H5N1, but it is safe to predict, with the discovered infection of thousands of migratory birds, that highly pathogenic avian flu is on the path of global conquest.

The likely spread of avian flu in birds is not the only problem. In 1997 it was discovered that humans who had ingested infected birds or been in prolonged close contact with them could contract the disease themselves. That same year 18 cases of avian flu in humans, 6 of them fatal, were reported in Hong Kong. Human infection subsequently appeared in Vietnam, Cambodia, Indonesia, Turkey, Azerbaijan, Iraq, Djibouti, and Egypt and reemerged in China in 2003. As of July 11, 2007, there were 318 reported cases, including 192 human deaths.

Contemplation of the future of avian flu is very disturbing. Viruses reproduce at a staggering rate, allowing many generations of them to experience gene changes brought about by mutation. While many mutations are bad news—scientists use the restrained term *deleterious*—some may confer advantages to the organism, especially if a new mode of life is adopted to deal with changing environments. To us, the latter mean droughts, global warming, more hurricanes, less freshwater—big-scale items. To a virus, a changed environment might mean simply a subtle shift in the pH level—from, for example, a more acidic to a more alkaline medium—of the cells of its host. What happens if humans infected by diseased birds offer the virus an environment that is even more nurturing to it than the original avian host? Why not simply short-circuit the pathway requiring an intermediate avian host and hit humans with a full volley, moving directly from one human to another? Avian flu virus could very well shift to this new pathway. The result would be a horrendous pandemic in Earth's human population, with mortality rates that could be in the hundreds of millions.

Our anxiety about the future of avian flu is well founded because it is rooted in our understanding of evolution. Viruses and their devastating effects are a particularly relevant example of evolution in action. As Darwin and Wallace recognized, organisms show much variation in nature, and natural selection works on this variation to separate victims from survivors. Adaptations are not ironclad; the very conditions that favor an adaptation may change. Thus some kind of avian flu virus might evolve that flourishes in the environment offered by human hosts rather than bird hosts. The probability of this occurrence is being debated. The sharp focus of this debate, and the energy and money invested in it, show that we recognize the deadly potentials of viruses and the seriousness of our "arms race" with infectious organisms.

We knew this when the American Museum of Natural History opened a special exhibit on Darwin in November 2005. We also knew that the exhibit, like the publication of *Origin of Species* more than 150 years earlier, was sure to engender controversy. Sentiment against Darwin and his theory, at least in the United States, still runs strong and negative. Recent polls show that roughly half the American public rejects the theory; they subscribe to the belief that humans, at least, were specially created by God. The antievolutionist perspective has always been distinctively popular because of the large number of Christian fundamentalists in the United States, and organizations like the Discovery Institute have effectively marketed the argument for what is called intelligent design and given it wide exposure in the media.

However, as Judge John E. Jones ruled in December 2005 at the end of a controversial trial in Dover, Pennsylvania, intelligent design is not science at all. It does not offer any explanation we can test scientifically in the measured observation of real things. It is instead another expression of "creationism," deceptively wrapped in scientific-sounding jargon and convoluted arguments about the unlikelihood of certain transitions. For 150 years the Darwin-Wallace theory of evolution by natural selection has resisted all challenges from any kind of scientific alternative.

The persistent rejection of evolution for religious or social reasons has a real, practical downside. Evolution not only is the framework for modern biology but inspires the vigilance required to sustain humanity's health and survival. As our exhibit at the museum opened—to a wave of favorable reaction from the press and public, thankfully—we were pleased to see our visitors spending plenty of time in the section it devoted to viruses and the study of infectious diseases. It is ironic that many who claim to reject evolutionary theory completely do not realize that they accept it when they worry about the spread of AIDS or the present and prospective dangers of avian flu since only evolutionary theory explains why we should worry at all. Indeed, it inspires not only vigilance about our health but hope for a healthy future. A scientist recorded in a video shown at the exhibit described the situation well: he could not treat AIDS patients—giving mixtures of drugs and withdrawing them when the virus adapted and became resistant to them—without evolutionary theory to guide his strategy.

As Darwin and Wallace told us, adaptation of organisms, whether human or bacterial, depends on how well they fit with changing environments. We know Earth's environments are changing on a huge, nay, global,
scale. Our survival, we are fairly certain, depends on the stability of the
global environment. One breakthrough of modern biology is its disclosure
that disrupted natural environments—slashed and burned tropical forests,
expanding deserts, and polluted lakes and seas—create evolutionary laboratories the results of whose experiments are not always good for us. Degraded
habitats seem to select for novel organisms, some of which carry new diseases and new destructive agents. It would be disturbing to reject these
warning signals, not to mention the possibility of species extinction as predicted by evolution, by rejecting the theory of evolution itself. A blind faith
in the superiority of the human species and its divinely ordained and brilliant destiny could be fatal. Humans have not yet seen a reversal in the erosion of the biosphere through the intercession of an intelligent designer. On
the other hand, evolution shows us how to live with the complex, coevolving organisms that share the modern ecosystem with us, both the beneficial
ones and the dangerous ones. Understanding evolution at least gives us a
fighting chance for survival.

Part Two

THE WORLD
BECOMES MODERN

We live in the flicker—may it last as long as the old earth keeps rolling! But darkness was here yesterday.

—Joseph Conrad, *Heart of Darkness*

6

ANCIENT GROUND

Let us replay time at fast forward.

About 4.2 billion years ago—about 400 million years after Earth became Earth—the planet was enveloped in the blue of voluminous oceans. The seas were first constituted from the hot spit of Earth's interior—stinking, sulfurous vapors and steam, condensation of acid rain clouds, and water vapor ejecta of volcanoes. Then, about 3.5 billion years ago, life emerged in the form of a cellular sac of water, DNA, and a few proteins. The seas, with this new microscopic life, rolled on. Then, about 900 million years ago, more complex life emerged. We have the evidence for this: wispy shapes of organisms made up of many cells are imprinted in rock slabs. A few hundred million years later more fossils show us patterns that are evocative of, well, creatures—jellyfish-like forms and curly wormlike animals with long antennae and many legs.

The oceans confined life for very long, but not forever. Nearly 500 million years ago sea life eventually floated, oozed, or crept out onto land. The first lowly land plants—scum life—later became swamp life, whose decay formed the great coal deposits of the world. Swamps dried up, ferns and pines sprang forth, and land-dwelling animals included the lineages that foreshadowed birds and other dinosaurs as well as mammals. Flowers bloomed and insects pollinated, giving rise a bit more than 100 million years ago to a familiar ecosystem, a thoroughly modern ecosystem. In this section I want to highlight the events that set the stage for the actual blossoming of the modern ecosystem in the late age of the dinosaurs and make the case that it is the legacy of the world of today, the world as we know it.

If the first known life is 3.5 billion years old, then the first life adapted to land—i.e., terrestrial life—is young in comparison, only about 475 million years old. Before then life and marine life were synonymous. What kinds of organism succeeded in this remarkable evolutionary transition? Land's first colonizers were very different from many forms of terrestrial life today. Indeed, the "plantlike" organisms that were the first life on land are not easy to place in any of the present-day groups of plants or their relatives. Nor are they easy to identify as some kind of moss, lichen, or nonplant fungus. However, plants eventually evolved within this alien, primitive assemblage of land's first colonizers that foreshadow the familiar. On this ancient terrain there were eventually some animals too, and it is no surprise that those animal species were the primitive members of that gaudy, absurdly diverse, and hugely successful group, the insects.

This sequence of events spans an almost incomprehensible stretch of time, which we try to make comprehensible by subdividing it into shorter intervals. Geologists have formally named these subdivisions, but this practice is not completely arbitrary. The science behind it warrants our brief consideration. I have cited dates in terms of millions or billions of years, but scientists also refer to Earth's history in terms of the subdivisions of time they have delineated. I have said that the calendar of Earth's time stretches back 4.6 billion years. Almost half that time, about 2 billion years, is taken up by the Archean Eon, and another vast stretch, another 2 billion years, is taken up by the Proterozoic Eon. These intervals of primeval time—roughly the first eight-tenths of Earth's history—are largely a blank; we know about them from rocks and a smattering of microscopic fossils so ancient that they have been identified only in isolated places on continents and islands. Even where such old rocks are extensive, as they are in the massive surface rock in central Canada and parts of the central United States, they represent only small slices of the temporal wilderness of the Archean and Proterozoic. These are the eons when oceans first formed, when life was born, when the atmosphere was infused with oxygen, and when one-celled organisms diversified into more complex creatures.

The other major subdivision of the timescale, the Phanerozoic Eon, extends back "only" 500 million years, but here the fossil record is notably enriched, depicting the rise and fall of biological empires as waves of new species replaced more archaic ones. This has allowed us to make a more refined and more precise division of time into eras and periods. (The Archean is subdivided into eras as well, and the Proterozoic into eras and periods, but these are not so concretely or precisely identifiable as those of the Phanerozoic.) The eras and periods within the Phanerozoic correspond to times when the fossils show us that certain forms of life were dominant. Their boundaries in several cases correspond to the mass extinction events first recognized by John Phillips during the mid-nineteenth century. The Paleozoic Era was largely a matter of diversification of marine fishes and invertebrates, but some of the late periods of the Paleozoic saw animals and plants invade land, dense swamp forests arise, producing the coal deposits of the Carboniferous Period, and fully land-dwelling vertebrates, such as reptiles, evolve. The last Paleozoic period was the Permian, which ended with the greatest mass extinction event we know of in the fossil record.

The succeeding Mesozoic Era saw the rise, the dominance, and, at the end of the Cretaceous Period, the fall of the nonavian dinosaurs. This was succeeded by the Cenozoic Era, often called the age of mammals, but also the time when many important groups such as snakes and certain kinds of insects and grasses took hold. Most of the Cenozoic is taken up by what is called the Tertiary Period and its last 1.8 million years, which is called the Quaternary Period. *Tertiary* comes from the Latin *tertiarius*, "of or pertaining to a third." This was originally meant to stand for the "era after the Mesozoic" (which formerly was called the Secondary). Thus Tertiary has now been downgraded to a period, one of the subdivisions (the other being the Quaternary Period) of the Cenozoic Era. Humanlike primates arose at some point late in the Tertiary; our earliest forerunners appeared about 7 million years ago. From a paleontologist's perspective, the Cenozoic goes on, and we are part of its history.

So many names for time! It's enough to make readers weary and students in Geology 101 stage a storming of the academic Bastille. Moreover, the names themselves don't help much—what does *Permian* refer to?—or are downright misleading. Mesozoic, middle life, is hardly appropriate, for example, because the Mesozoic Era comes very late on Earth's calendar; it's rather like an octogenarian referring to himself as middle-aged. Some of the names merely refer to places where rocks and fossils of a certain age were first described. The Permian Period denotes a time represented by rocks and fossils deposited in hills near the town of Perm in Russia. Cretaceous comes from the Latin *Cretaceus*, which means "chalky" and alludes to the chalk cliffs that align both sides of the English Channel. (*K* is used as

a symbol for Cretaceous to avoid confusion with the C symbol for Carboniferous.)

Still, the names are important in communicating notions about the history of Earth because they are strongly embedded in the language scientists
have used to communicate their ideas for more than two centuries. Boundaries for these named intervals are sometimes shifted upward or downward
as new evidence comes to the fore, but the continually scrutinized time intervals and the names for them have stuck.

At first, eras and periods merely indicated the succession of ages: Jurassic
before Cretaceous, Cretaceous before Tertiary, and so forth. A more precise
calibration of the timescale in terms of millions of years was a great breakthrough in the twentieth century, when it was learned how to measure the
radioactivity in rocks. Certain elements—uranium, potassium, argon, and
carbon—come in varieties, or isotopes, that have varying numbers of subatomic particles. Some of their isotopes are unstable, meaning that their
atoms lose either electrons or alpha particles. The degradation of the original
isotope releases radioactivity, and the whole process is called radioactive decay. It is profoundly important that the rate of radioactive decay is constant.
If a rock sample has two different forms or isotopes of an element, say, uranium, the relative abundance of these two isotopes in the rock will indicate
its age: an older rock has less of the original, or parent, isotope and more of
the derivative, or daughter, isotope, whereas a younger rock has a higher ratio of parent to daughter isotope. This is because radioactive decay in the one
has taken place over a much longer time than in the other. Uranium radioactive decay is very slow, and measuring it indicates the age of ancient
rocks—old enough, that is, to allow for a noticeable change in abundance
from a parent to a daughter isotope. In contrast, potassium decays more
quickly and is thus useful for dating somewhat younger rocks. An isotope of
carbon, Carbon 14, decays too rapidly to be of use in dating rocks or fossils
that are millions of years old, but it is effective in dating objects less than sixty
thousand years old. Conveniently, we have an assortment of radioactive elements with which we can date rocks and fossils of different ages.

A second, more recently discovered way to measure time comes from an
amazing, almost magical property of Earth's magnetic field. Let us start
with the less fantastic aspects of this phenomenon. Some grains of minerals
in rocks have been magnetically aligned, like iron filings drifting toward the
end of a magnet. This phenomenon is known as paleomagnetism, because

the alignment occurred at a remote time, when the rock with the magnetically responsive minerals first cooled and hardened, essentially freezing the alignments. The science of paleomagnetics, looking for these signals in rocks, has taken researchers to all continents, all oceans, and rocks of all ages.

What is responsible for the alignment in the first place? There is a big magnet involved here; it is Earth itself. Electric currents in its liquid outer core work like a huge magnetic dynamo that generates invisible magnetic lines of force around Earth that are positively charged at one end and negatively charged at the other. These charged ends are the magnetic North Pole and magnetic South Pole, which are located only 11.3 degrees off the "true," or rotational, North and South Poles. One would think that such a large-scale system would have a constant behavior, but it most emphatically does not. Clues to the complex behavior of Earth's magnetic field come from the rocks themselves. Rocks of different ages in different places have magnetic grains aligned in different directions, and they provide important information on two dimensions, time and space.

Now for the fantastic part. A shocking discovery, made in the 1960s, about the magnetic orientation of grains in ancient rocks is that they show that Earth's magnetic field has done flip-flops: during certain intervals the magnetic North Pole became the magnetic South Pole and vice versa. This schizophrenia in Earth's magnetic field seems bizarre, dependent on some form of alchemy rather than science, but the evidence for it is strikingly simple. Today the magnetic lines of force are oriented toward the magnetic North Pole. When lava cools, for example, the sensitive grains in it align themselves in a northward compass direction and stay that way, as if you had broken (or frozen) a compass and its needle subsequently refused to wobble. Magnetic grains preserved in certain rocks, however, show a southward orientation. After studying many rocks in many places, geophysicists concluded that the magnetic pole has reversed itself at certain times in the past. The mechanism for this polar reversal is not entirely understood, but it is widely thought that shifts in the currents of molten material inside Earth are sometimes big enough to flip the magnetic field.

Whatever the mysteries of its cause, recognition of magnetic reversal has had mighty consequences. A pattern of "reversed" and "normal" phases offers exciting elucidation because it can be correlated with the dates of rocks as established by radiometrics and fossil study, and then a more accurate timescale can be formed. The geologic record suggests that the field reverses about once every 1 to 5 million years, but not regularly or evenly. During the Cretaceous the field was stable for tens of millions of years or more, yet in other instances it flipped after only about 50,000 years.

In the 1960s the magnetic reversal timescales helped catalyze the revolution in geology that brought plate tectonic theory, the modern version of the theory of continental drift. Scientists had proposed that the seafloor is generated along vast linear features, such as the Mid-Atlantic Ridge, the immense serpentlike mountain range running north and south on the floor of that ocean. It was hypothesized that the ocean floor spread outward, eventually disappearing in trenches at the edge of ocean basins. In striking agreement, magnetic reversal timescales showed that the youngest seafloor rocks were indeed near the ridges and the oldest near the trenches, or subduction zones, at the edge of the oceanic plates. Moreover, the "striping" of normal and reversed rocks revealed in cores drilled from the ocean bottom were symmetrical on either side of a spreading ridge; on one side, the ocean floor showed a pattern that was a mirror image of the other side. It is hard to conceive of a more brilliant convergence of data and theory in the Earth sciences.

Paleomagnetics also allows us to explore the dimension of space. Rocks from divergent locales all of which have grains aligned along the normal

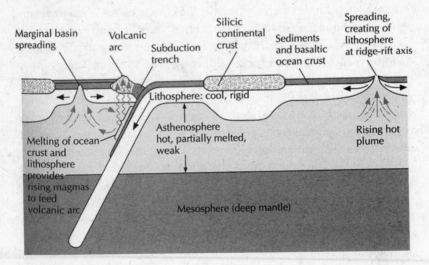

Subduction of oceanic crust

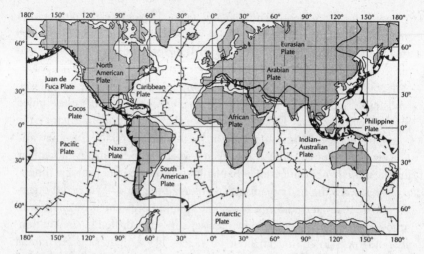

The major crustal plates

pattern toward magnetic north may show variations, with the grains oriented at slightly different angles. This shows us that the magnetic poles actually shifted slightly or "wandered" through time, but the basic cause of
this variation in alignment is the wandering of the rocks themselves. We
now know that continents and tectonic plates move through time, and so do
their rocks. Two different rock sequences of different ages in one region may
actually show different orientations, even if there was no magnetic reversal
event involved. One set of rocks cooled and preserved grains when the landmass that carried them was in one place; the other set of rocks cooled at a
later time after the landmass had shifted to a new location relative to the
magnetic pole. This phenomenon is a powerful component of evidence for
continental drift and plate tectonics. It is possible to determine the motion
of a crustal plate by how much it has "wandered " relative to the magnetic
pole through time—an important testament to the true nature and dynamics of a restless planet.

These time-space considerations show us an unavoidable fact: the evolution
of Earth's biota took place for the most part in the oceans. The first life on
land emerged only after nearly 90 percent of Earth's history had elapsed. In
plotting the emergence and history of the modern terrestrial ecosystem, we

cannot forget the aqueous world it came from. The oceans not only are the source of life but continue to harbor it in spectacular and diverse numbers and amounts. Moreover, the vicissitudes of life on land are clearly linked to physical and biological events in the world's oceans. The dry environments of the coast of California, for example, contrast strongly with the humid wetlands and forests of the same latitude on the eastern coast of the United States because of many factors, but perhaps the most influential factor is the cold current off California and the warm Gulf Stream off the Carolinas and Florida. The interplay of marine and terrestrial realms has made for many of the most dramatic changes in land plants and animals over geologic time.

What were the first organisms to colonize the land? When did organisms first strike land and make a go at hanging on? The answer is, as in much of science: We don't know precisely. In cores drilled in 475-million-year-old rocks in the ocean floor off places like Oman, on the Arabian Peninsula, we find virtually microscopic objects that look like parts of land plants. They have four reproductive parts, or spores, arranged as a quartet within a hard sac that is a protective coating preventing the precious spores from drying out when not enveloped by water.

If there is one thing that is common to all life, whether in oceans or on land, it is water. This fact explains the excitement of NASA scientists when they encountered evidence of the presence of water in the rocks on Mars. Anything that deprives an organism of its water is an enemy to its survival. And for all the benefits that air confers to organisms that breathe it, air has a nasty tendency to suck the water right out of an organism, which must consistently replenish its water and control its loss through evaporation. The odd little spore sacs from the Arabian Sea suggest that something was living on land nearly half a billion years ago, but what it was is another matter. The spore sacs with their quartet arrangement resemble those of liverworts, not very imposing plants that today are nearly ubiquitous. Their eighty-five hundred described species—doubtless many more are unknown—occur everywhere from high latitudes to the tropics. They thrive mostly in moist environments but also grow in very dry areas, extremely rainy areas, and even under water. Liverworts are small, leafy organisms, often with bright green, glistening, liver-shaped lobes. They remind me of something I might find in one of those exquisitely exotic broths in a restaurant in Vietnam. Like moss, they sometimes form carpets and squish underfoot, and they often grow in or near moss.

Visually, liverworts make good candidates for what we might imagine was the primal coating of fragile life on land. But we don't know whether these ancient spore sacs were from liverworts, from another kind of plant, or from something else entirely. Likely, in the early stages of invasion, the land regions of Earth were coated with a drab green film—a scum—clinging to soils, fringing hot springs, or fanning out over riverbanks and floodplains.

Fortunately fossils can be highly enlightening. In rocks from Ireland about 425 million years old, there have been found fossils of plants that we can actually see without a microscope. They are less than spectacular, to be honest—just some small bifurcating structures at most a few inches long. Similarly aged strata from Australia show club moss–like forms. Club mosses are very simple branching plants with scalelike leaves, and they are thought to be allied with ferns. Many other fossils from roughly this time interval have been found in Great Britain and Belgium. Some of these are flat disks a couple of inches in diameter with huge numbers of spores; one scientist had the perseverance to count thirty-five thousand on one disk! Another of these "plants" are little dark, shiny spherical globs less than an inch in diameter with small filamentous tubes radiating out in all directions. These creatures seem more like alien stowaways on interstellar voyages than anything growing on Earth today.

Eventually, however, plants came to look like plants. Some developed spines, probably to increase their surface areas for breathing in carbon dioxide. The early dramatic emergence of many species of land plants—what paleontologists call an adaptive radiation or simply a radiation—is best recorded in a lucky strike, a small unspectacular layer of rock underlying a moor near the town of Rhynie, twenty-four miles northeast of Aberdeen, Scotland. In many cases in paleontology, our vivid picture of life on Earth during certain periods of remote time can be due to one isolated but extraordinarily enriched locality. About 400 million years ago, in the beginning of what is known as the Devonian Period, the landscape of the region that now cradles Rhynie looked like the geological terrain of Yellowstone Park, with volcanoes and geysers. Some of the geysers spouted boiling water with a high silica content, which caused the vegetation, including the underlying peat layers, to become impregnated with silica, or silicified. The fossils in this Rhynie Chert are so perfect that they reveal their cell structure under a microscope.

So with all those plants exquisitely preserved in places like the Rhynie

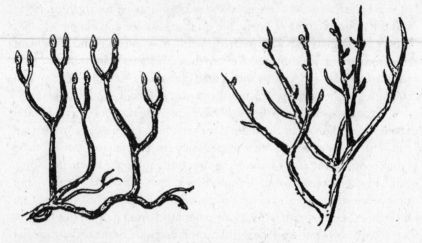

The Devonian land plants *Aglaophyton* (left) and *Rhynia*

Chert, where are all the animals? Particularly where are the insects? After all, a group so successful and diverse as insects should be of great antiquity. But scientists have been forced to accept the fact that there is no really good fossil evidence of well-developed insects until tens of millions of years after land plants first appeared. The exquisite preservational ambiance of the Rhynie Chert itself has for nearly a century yielded only some tiny springtails and mites, animals less than a millimeter in length. These forms are part of a group called the Hexapoda, which also included the diverse insects. But springtails and mites represent hexapods that branched off earlier than the great insect radiation. For decades these tiny inconspicuous creatures were all we knew about early animal life on land.

But we still live in an exciting age of paleontological discovery. In February 2004, in the international science journal *Nature*, two scientists reported on the discovery of the oldest-known insect in of all things the Rhynie Chert, which had been so resistant to providing such creatures. The specimen was discovered not in the field but in a forgotten drawer within the collections storage of the Natural History Museum of London. It is an insect head and little more, but it has very distinctive jaws attached to the head in two joints. This two-jointed jaw is a diagnostic feature of all insects, a feature that in part separates them from other hexapods. The jaw also has cutting parts, or teeth, differentiated into piercing structures ("incisors") and

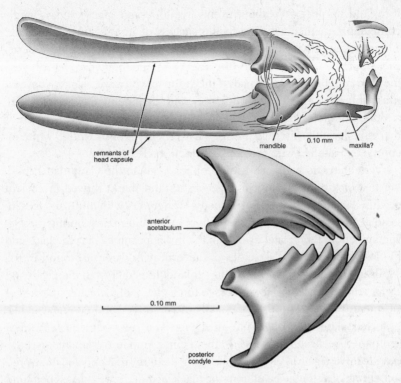

The preserved head and jaws of the *Rhynie* insect *Rhyniognatha hirsti*

crushing structures ("molars"). These features are rather advanced for insects and are almost always present in insects with well-developed wings. This led the authors, including one of my colleagues at the American Museum, David Grimaldi, to conclude that the Rhynie insect represents the first sign of an emergence of advanced flying insects, which underscores the antiquity of this most diverse group. Other scientists aren't ready to accept so easily the notion that this earliest fossil insect had wings. Nonetheless, all laud the discovery as a very important one in the study of insect evolution.

What about the lineage leading to the backboned, walking animals? In other words, what about the line leading to us? In our anthropomorphic bias we tend to be most interested in this very minor branch of the tree of life. The known vertebrates today comprise only about forty-five thousand species, and nearly half of these are fish. The vertebrates that actually made it to land number only about twenty-five thousand species today, hardly comparable in their diversity to the success of insects, fungi, or plants. There are fewer living land vertebrates, for example, than there are living cicadas, aphids, and leafhoppers (the insect order Homoptera). Nevertheless, the land vertebrates, or tetrapods—the group to which we belong—have an irresistibly intriguing history. Moreover, with their comparatively large size and their durable skeletons, vertebrates are more easily preserved as fossils than smaller and softer organisms. That vertebrate fossil record, while incomplete, is stuffed with eye-popping disclosures.

Vertebrates go back nearly 500 million years to the Cambrian Period, the first of the Paleozoic Era, when fishes and sharks thrived in ancient oceans. The debut of the first tetrapods—vertebrates with limbs and toes instead of fins—did not occur for another 140 million years, during the Devonian Period, much later even than the first appearance of land plants and insects. Well-preserved skeletons of these animals were first found in the 1930s in ancient rocks of eastern Greenland. Expeditions in the 1980s and 1990s added important new pieces to the skeleton of a creature of particular interest. The fossils suggest a squat amphibious creature, somewhat like a bloated salamander, about three feet long; it has been named *Ichthyostega* (from the Greek *ichthy*, fish; *stega*, roof). Unlike modern salamanders, however, *Ichthyostega* was a creature bridging the terrestrial and aquatic world, something of a strange beast from the black lagoon. Its well-developed hind limb does indeed have toes, but seven rather than the typical five or four seen in full-blown tetrapods. Moreover, the hind limb and foot are oriented and shaped like a paddle; they strikingly resemble the paddlelike forelimb of a modern river dolphin (which, of course, evolved its five-toed forelimb paddle independently from an ancestral walking mammal). The paleontologist Jennifer Clack and her colleagues suggest that *Ichthyostega* may have moved as a seal does, using its forelimbs to haul itself up on land and its hind limbs to propel itself through water. Other features, such as an ear structure finely adapted to hearing underwater, suggest that it spent a lot of time in the aquatic mode, though it was capable of crawling up on land.

Ichthyostega reveals some, but clearly not all, the steps toward vertebrate land life. What other creatures are helpful here? At the time *Ichthyostega* was discovered in the 1930s an animal called *Eusthenopteron* was also known. This form, occurring some 10 million years earlier than *Ichthyostega*, is easily recognizable as a fish. It has an elongate, streamlined body, like a torpedo, and several fins, both paired and unpaired, including a well-developed tail fin. None of these appendages have jointed toes; instead, the fins end in a series of numerous unjointed, elongate bones called fin rays, structures typical of fish. Paleontologists observed, however, that the bones at the base of the fin rays, especially in the anterior or pectoral fin, are shaped and arranged much as one would expect in aquatic forms related to the most primitive tetrapods like *Ichthyostega*.

There remained a big anatomical gap between the fishy *Eusthenopteron* and the amphibious *Ichthyostega*. Great hopes in this regard were placed on another fossil animal from eastern Greenland called *Acanthostega*, first discovered in the 1940s. *Acanthostega* was a contemporary of *Ichthyostega*. Features of its skull and skeleton seem prototypical of those in *Ichthyostega* and other tetrapods. While many of the typical fish fins, especially the unpaired fins, are absent, and the limbs terminate in jointed toes rather than fin rays, *Acanthostega* was nonetheless fully aquatic. All its limbs, each with eight toes, form paddles meant for swimming, and its tail has an oarlike fin supported by fin rays. *Acanthostega* has features that are clearly transitional between fish and tetrapods.

Ever since their discovery, scientists have debated about what these fossils reveal concerning the great leap from water to land. Resolution of this issue, which like all scientific resolutions requires a convergence in assumptions, theory, and observation, is difficult to achieve. That is why new discoveries of good fossils are so important. By providing new evidence, they can answer many questions in the ongoing debate. In 2005 Neil Shubin and his team at the University of Chicago found a remarkable set of fossils in Devonian-aged sediments on Ellesmere Island, a barren landscape of rock and ice in the Canadian Arctic. The animal, named *Tiktaalik*, lived between 385 million and 383 million years ago, so it was older than *Eusthenopteran*, *Acanthostega*, and *Ichthyostega*. Though *Tiktaalik* shows a transitional skeletal anatomy, it is, as expected, less specialized toward a tetrapod way of life than *Ichthyostega* but clearly less fishy than the younger *Eusthenopteran* and another important taxon, *Panderichthys*.

This can happen in paleontology. The age of a fossil can be deceiving. Many fossils show up that are either much older or younger than we might expect given their anatomical structure. Even living forms with very primitive features sometimes occur much later than we might expect, while conversely, some conservative lineages have persisted for very long times.

For example, the duck-billed platypus and the echidna belong to the monotremes, mammals that retain many primitive traits, including egg laying, that were lost or modified in the two other major groups of living mammals, the marsupials and placentals, that evolved and diverged more than 125 million years ago. Thus monotremes must be older than they, yet monotremes are still around. In this sense they are "living fossils" that tell us more about early mammalian evolution than do some 100-million-year-old fossil precursors of marsupials and placentals. The opposite is also possible: fossils with more advanced features sometimes occur in older rocks than fossils with more primitive features. The fossil record isn't perfect. Just the same, the anatomical features preserved in key fossils tell us a great deal about the transitions that occurred during real evolutionary time.

Why is *Tiktaalik* such a perfect example of this point? To begin with, the animal is so beautifully preserved that many details of its anatomy are readily observable. These features certainly indicate a form that is fishlike in many ways: a fusiform body, fishlike scales, and an elongate, vertically flattened tail with a fin and fin rays. It clearly had gill structures for breathing in water. But *Tiktaalik* has some striking terrestrial tendencies. It lacks several bones that in fishes link the head with the shoulder region, bones that protect the gills and facilitate the movement necessary for gill breathing. Instead, *Tiktaalik*, like a tetrapod, actually has a neck. The head propped on this neck could then emerge easily from the water and allow the animal to gulp air. In correspondence, the eyes are set high on the head, allowing the animal to see as well as breathe above the water's surface.

Most remarkable of all are the limbs, notably the forelimb. The limb bones are large, and their joints are flexible. So they were capable of a rotating movement in the "elbow" area between the upper and lower "arm." The toes are splayed out in a way that could have propped up the distal end of the limb, a sort of prototypical "hand," as it walked about on the bottom.

One can only speculate on the situation that promoted the evolution in *Tiktaalik* of this strange mosaic of terrestrial and aquatic traits. *Tiktaalik* was found in sediments representing former river channels and estuaries along an ancient coast engulfed in year-round warmth. Possibly the animal was multitasking, both breathing through gills and raising its head out of the water to take in air. During the Late Devonian Period, oxygen levels in the air were rising. Conversely, there may have been times when accumulations of bacteria in the water formed their own red tides, depleting much of the oxy-

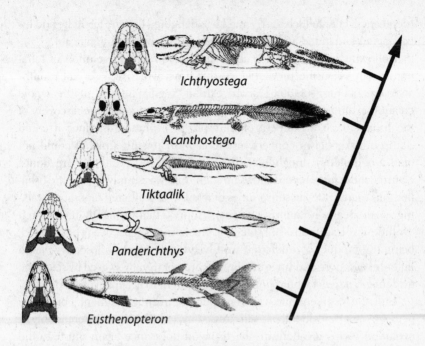

Ichthyostega

Acanthostega

Tiktaalik

Panderichthys

Eusthenopteron

The fish-tetrapod evolutionary transition, showing several key taxa

gen in the water, and during such times the air-breathing capability of *Tiktaalik* would have been particularly adaptive. Yet it is important to point out that this animal is not a full-blown tetrapod. Its fin bones are more like those of a fish than those of a tetrapod. *Tiktaalik* should be more properly regarded, as Neil Shubin and his colleagues wrote, as a "fishapod." The importance of the fossil is, in any case, profound. Jennifer Clack, an expert on tetrapod origins, remarked, "Although no single fossil can fully explain a complex evolutionary event, *Tiktaalik* is a true intermediate form, and it provides vital clues to the when, where, and how of the transition from water to land."

The discovery of *Tiktaalik* by Shubin and his team first came to my attention by a rather circuitous route. John Noble Wilford, a journalist at *The New York Times* who has covered much of our paleontological research, called me up in early April 2006. He had access to a prepublication release of the fishapod discovery that was to appear the next day in *Nature*. John was looking for a few comments on the discovery. I knew Neil well—he is a good friend and a talented scientist—and I was delighted that the fossils he and his colleagues had found and described were so dazzling and significant.

Following some questions about the anatomy and evolution of tetrapods, John asked the inevitable. At this time, which was several months after the opening of our Darwin exhibit, media interest in the whole creation-evolution debate was at high pitch. John asked if the discovery of the fishapod was a moment of triumph for evolutionists, since it could stamp out skepticism concerning the lack of "missing links." I could not offer a completely straightforward answer. First off, I said, the creationist claim that the fossil record shows no such links was simply wrong; the fishapod was merely the latest in a series of fossils—of feathered dinosaurs, walking whales, early hominins, and many more—that brilliantly reveal such evolutionary transitions. But since the creationist perspective is firmly held, being the product of a different worldview, a different philosophy, even a different emotion, and since creationists are not usually swayed by scientific evidence, *Tiktaalik* would not likely move them to accept evolution.

Still, the fossil serves an important purpose in enlightenment. For people who are uncertain about, or even stressed by, conflicts over the credibility of evolution versus creationism, the fishapod does once again illustrate the power of scientific evidence to reveal vividly the transitions that Darwin and Wallace expected for their theory. Darwin himself was troubled by the depauperate fossil record, so troubled that he devoted a whole chapter in *Origin of Species* to this problem, which he regarded as a weakness in the evidence for his theory. But as paleontology revved up in the decades after his death and on into our own time, subsequent discoveries, including that of the fishapod, have improved the prospect. And Darwin's theory shows us that such discoveries would be inevitable, given sufficient preservation of the animals in ancient rocks. He would have loved the fishapod.

Tiktaalik may be an icon for the grandeur of evolutionary transitions, but in its own place and time it was just another species in the ecosystem. That ecosystem had changed dramatically over millions of years, and it would change again. Insects, plants, and, later, vertebrates became major stakeholders in the great evolutionary game on land. So too were other species: microscopic bacteria, fungi, one-celled protozoans, as well as the inconspicuous and ubiquitous tiny animals such as mites and springtails. These first dominant land colonizers were hardly dramatic, hardly more than "scum lords." But soon the land world changed, became more vernal, woody, tree-choked, insect-ridden, and even populated with amphibious, backboned creatures with limbs instead of fins. The warm tropical climes that spread over Earth fostered a great hothouse of plants that made for swamps of continental dimensions. The decay of these swamps and the accumulation of their organic material within thick layers of sediment became the source of the world's great coal deposits. This was the first major transition of land ecosystems, from the microscopic to the visible, from the scum world to the swamp world.

The mighty transition from water to land and the subsequent revolutions in terrestrial life convey a general message about evolution then and now. *Tiktaalik* and other organisms explored new places and new ways of life not just because they had the opportunity to find unexploited sources of food and habitat. These species were likely under stress, attempting to escape unfavorable conditions such as toxic, oxygen-depleted water for clean, oxygen-enriched air. In much the same way, many organisms today are on the move, shifting their range to habitats that are less precarious than their original ones, which may be undergoing drastic change because of climate warming, deforestation, or other forces. The arctic fox is moving northward. The fragile Fynbos plants of South Africa are moving southward. Organisms with the capacity for movement and the flexibility to adapt to new conditions are likely to be favored in evolution over those that are not so mobile or flexible. The current shocks to the planet are extremely powerful sources of natural and unnatural selection.

We are a species too, and our habitat is a whole planet. Wholesale migration means space travel and colonization, a prospect that seems inevitable if our species survives long enough. It is fair, then, to acknowledge some momentous events of the past that foreshadow such an ambitious migration. The shift of a few vanguard species from water to land a few hundred million years ago was every bit as dramatic.

7

IMPERIAL COLLAPSE

From the rippling Allegheny Mountains in Pennsylvania to the river valleys, hills, and plains of Ohio and Illinois stretches coal country. That black, dirty, oily, shiny substance is exposed in seams, cliffs, holes, gullies, and man-made mines throughout the heart of this industrial and agricultural corridor in the eastern United States as it is in many other regions of the world. On a global scale coal is abundant. For years the United States claimed to produce the most coal in the world per year, about 1 billion tons, but the coal crown has passed to China, where prodigious mining has been stepped up with the boom in the world's most booming economy. Other great coal regions include parts of Russia and countries of the former Soviet Union. In all, it is estimated that there is enough coal in the world to last about three hundred years at its current level of usage—with no increase in world population or significant increase in consumption (two highly unrealistic expectations). Because it is the primary fuel for the world's electricity and serves many other industrial uses, coal is burned rampantly. Because it is a major source of pollution, it has been a mixed blessing since its very first use. China, with its unmatched bounty of coal reserves, is trying to develop programs to wean itself off a dependency on coal, promoting cleaner energy sources, yet its continued prodigious use there and elsewhere fosters an outpouring of pollution that is hard to beat back.

This magical and problematic substance, this fuel of the hearth and furnace, seems an unlikely example of a fossil. But coal is just that. It is simply the mass of compressed carbon settled out of the decaying bodies of millions upon millions of long-dead organisms. Carbon is that wondrous universal substance of life, the core of proteins and DNA found in every cell that seems so malleable and incredibly transformed under different conditions. Push a heap of carbon-rich fossils under thick layers of rock, and you have coal. Compress them further and push them deeper, and you have oil. Exert more pressure on carbon and push it deeper still, and you have on very rare occasions diamonds. The products of carbonized extinct life—the coal and oil locked up in seams and pools within Earth's crust—force us to recognize the importance of fossils, whether or not we find fossils intrinsically interesting.

Most of the coal in the vast reserves of the eastern United States, Europe, and many other regions is the product of a very ancient time. Some 350 million years ago, after vegetation and some primitive forms of animal life had begun to live on land, a spectacular transformation occurred. Plants got bigger, commonly exceeding eighty feet in height. They developed such familiar parts as leaves, wood, and seeds. The seeds of these early plants were exposed; they were not fully embedded in a protective structure or ovary. These were the seed ferns, groups notable for carrying their exposed seeds in cones, a trait we can see today in living pine trees and other conifers. Cone-bearing plants are known as gymnosperms (from the Greek *gymnos*; naked, and *sperma*, seed). In contrast, flowering plants, or angiosperms, which evolved later, carry their seeds in a protective chamber, or ovule (thus the prefix *angio*, receptacle), or enclose them in fruit.

The two major groups of plants are therefore identified according to their reproductive parts, and the working of these parts tell us a great deal about the reproduction and life cycles of plants. To explain how, it might be best first to consider some more familiar organisms. Humans and other animals develop through several stages: embryo, juvenile, and sexually reproducing adult. Insects and many other organisms have stages like these too, though often with more abrupt changes; a fertilized egg yields a pale little larva, then a caterpillar, then a pupa, then a reproductively mature butterfly. In any case, all animals have a fundamental division in the life cycle related to reproduction: as reproductively viable adults, they produce either sperm or eggs. The cells in sperm and eggs are haploid cells, also called gametes: they carry only half the number of chromosomes in a fertilized egg, or zygote. A fertilized egg and the cells it generates to form the developing organism have both sets of chromosomes derived from the union of the sperm and the egg and are thus called diploid cells.

However complex the life cycles of plants, their reproduction breaks down into two broad stages analogous to these. The reproductively viable stage of the plant is the sporophyte, which produces spores that develop into a gametophyte. The gametophyte can produce either male or female gametes or even both. The subsequent union of a male and female gamete leads, once again, to the sporophyte.

The alteration and emphasis between the sporophyte and gametophyte stages vary sharply from one group of plants to another. In mosses, for example, life is dominated by the gametophyte stage, gametophytes being the green, spongy carpets we associate with mosses. In other plants, the gametophyte stage is markedly reduced. In gymnosperms, the gametophyte stage involves tiny pollen and a much larger female gamete retained in the ovule. The ovules, or macrospores, develop in cones, as do the pollen, or microspores, which are carried on much smaller cones on the lower branches of often the same tree. Wind carries the tiny pollen to the cones containing the female ovules. The union of the pollen and the ovule during fertilization produces a developing embryo along with cells of the ovule to form a seed. The diploid seed is the beginning of the sporophyte stage, which will develop into the plant, even a towering tree, and it is exposed, being carried on a cone. A reduced gametophyte stage is even more marked in angiosperms, whose gametophytes are usually micro-size, or microgametophytes. Angiosperms during the gametophyte stage exist as tiny female ovules or as male pollen. As we shall see, the way gametophytes in flowering plants carry out their reproductive business is a key to their evolutionary success and is a primary explanation for their eventual domination of the world's modern land ecosystems.

The coal forests took root at a time when warm, humid conditions enveloped all the major continents and landmasses. This was due to a curious configuration of land, a concentration in one mega landmass, which scientists have called Pangea, near the equator. The lands of what are now northern Europe and North America were then in these southern latitudes. Huge masses of plants—seed ferns, ferns, calamites (a plant related to the living horsetail rush that grows near streams and ponds), and gymnosperm relatives of the cone-bearing pines—grew in the swamps of Pangea. As the plants died, their remains accumulated in wet peat, a global-scale compost heap, which through years of compaction from accumulating sediments above them turned to coal. It is no surprise that the time of great accumulation of coal swamps, occurring between about 354 million and 290 million years ago, is called the Carboniferous Period.

The Carboniferous environment had other striking qualities. At the beginning of the Carboniferous, atmospheric CO_2 levels were very high, about 1,500 parts per million per volume (ppmv). This might seem like a very tiny fraction, but it is not when we compare it with other times. Today the level of CO_2 is about 380 ppmv and rising. A rise by the middle of this century is projected to higher than 600 ppmv CO_2, which will create much warmer climates and have drastic effects on ecosystems and organisms. In the Early Carboniferous, global average temperatures as well as CO_2 levels were very high, about 22°C (72°F) (compared with the global average temperature of 12°C [54°F] today). Climates remained warm for millions of years, through the middle of the Carboniferous, as the swamp forests grew thicker, even though CO_2 levels began to drop. Why this disjunction? Some have suggested that it was due to the very proliferation of land plants, which actually created a situation in which the available carbon in decaying organic material was buried ever more rapidly in sediments. And the climate stayed warm because so much of the Pangean megacontinent drifted into the hot zone, the low-latitude tropics. But the temperature eventually dropped. By the Late Carboniferous both CO_2 levels and average global temperatures had become much lower. During this time a massive ice cap advanced from the south polar regions of the megacontinent, contributing significantly to a drastic fall in global temperature. The Middle Carboniferous, the heyday of the swamp forests, was also the interval during which the concentration of oxygen in the atmosphere was at a nearly all-time high, about 35 percent (compared with the 21 percent concentration of that life-sustaining gas in the atmosphere today).

These disclosures seem magically precise. How do we actually know the concentration of atmospheric gases 300 million years ago? The determination comes from an analysis of the isotopes of carbon in both fossils—that is, buried organic materials—and sediments. The key factor here is photosynthesis, the fundamental energy cycle in green plants and phytoplankton. As we already know, photosynthesizing organisms absorb CO_2 from the atmosphere in the presence of sunlight and water to form carbohydrates and release oxygen; they do this by extracting hydrogen from water and combining it with carbon to form the sugars needed for metabolism and for the production of carbon-based organic compounds. Naturally, the uptake of

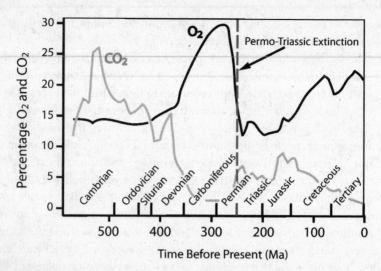

Atmospheric carbon dioxide and oxygen levels through time

carbon will be greater in times when more plants are photosynthesizing, in turn releasing more oxygen to the atmosphere. The situation is actually more complex because oxygen and organic material can recombine, with the effect of decreasing free oxygen in the atmosphere. If, however, organic material is buried rapidly under piles of sediment, it is prevented from this annealing, and then oxygen levels in the atmosphere remain high.

Thus the level of oxygen in the atmosphere depends on two factors: the amount of photosynthesizing plants and phytoplankton and the rate at which the organic material they produce is buried by sediments. This sedimentation rate is not dependent on biological phenomena, but is the result of Earth processes. Erosion eats at the edges of continents, leading to the deposition of thick sediments at the continental margins. The faster the deposition and the thicker the sediments, the more carbon is buried. This is a superb example of the profound relationship between the workings of ecosystems, the biosphere, and the dynamics of the physical Earth, the geosphere.

It is now possible to trace the concentrations of certain carbon isotopes both in organic material and in sediments over time. Particularly important here are changes in the amounts of the isotopes 12C and 13C; another useful isotope is the 34S form of sulfur. Concentrations of these isotopes are important indicators of the amount of atmospheric oxygen, since differing concentrations in organic material and sediments may indicate differing conditions. Sediments with high levels of 13C result from rapid burial, which also increases atmospheric oxygen; high levels of organic 13C may also indicate such conditions, but the two trends do not always coincide. In the last 15 million years, when oxygen in the atmosphere decreased, the percentage of 13C in sediments decreased too, but the percentage of 13C in organic material increased, caused, it is thought, by the evolution of new photosynthesizing pathways in plants and marine organisms. The calculations derived from these measures are complicated, and they depend on certain models and assumptions concerning the carbon cycle and its bearing on atmospheric gases.

Oxygen enrichment is of course not the only profile of a climate. Climates on Earth now and in the past have been controlled by a multitude of internal interactive components and external "forcings." The internal components include ones very familiar to us: atmosphere, oceans, sea ice, land and its properties (reflectivity, or albedo, biomass, ecosystems), snow cover, glaciers and other forms of land ice, rivers, lakes, and surface and underground water. The most important external forcings are the sun, Earth's rotation, changes in Earth's orbit around the sun, and the distribution of the Earth's physical features, whether ocean, landform, or profile of the ocean bottom. The whole complex system, a system that we know provides a stupendous amount of energy, is driven by the radiant heat from the sun, of course. About 30 percent of this energy is reflected back into space by clouds in the atmosphere or by Earth's surface. The rest is balanced by Earth's return of the same energy back into space through its emission of long-wave thermal radiation. But the natural greenhouse effect formed by water vapor, CO_2, and other gases, which keeps the heat from escaping entirely, blocks most of this thermal radiation. Without the greenhouse effect and the gases that cause it, Earth's atmosphere would be considerably cooler than it is.

Climates of the past can be reconstructed using more traditional means than models of the changing levels of carbon isotopes in organic material and buried carbon in sediment. The spacing of tree rings, for example, indicate slow-growth and rapid-growth stages for different seasons. The growth layers in fossil clamshells and corals offer similar clues. Ice that is tens of thousands of years old actually freezes in its air bubbles the composition of atmospheric gases at the time it was formed. Core samples taken from such ice show, for example, that there were significant amounts of air pollution, probably from mining, during Roman times. The shape of fossil leaf impressions, the distribution of plant fossils, and the isotope composition of fossil seashells also give indications of past climate. Finally, the abundant and diverse fossils of organisms belonging to groups that today are adapted to either warm, cold, wet, or dry climates give us further evidence, albeit indirect and imperfect, of past climates. In the case of the Middle Carboniferous Period, the accumulated evidence suggests a warm, sultry, oxygen-rich climate. Oxygen, abundant nitrogen, and other gases were being pumped out in massive amounts by the great plant factory, which produced an atmosphere much heavier with gases than the one we breathe today. This climate fostered an extravagance of carbon-based life that simply reinforced the system. It was the age of carbon: carbon in the living plants; carbon dioxide being absorbed by them; carbon dioxide being released to the atmosphere by volcanoes or, in the absence of sunlight, by plants themselves; carbon concentrated in coal from dead and compacted organic matter. The world has not since experienced such a lush and expansive greenhouse.

The fossil record of the Carboniferous coal forests has preserved more than plants. This was the abode of a diversity of tetrapods, early experiments in terrestrial lifestyles, like frogs and salamanders today, amphibians still tied to the water because their eggs lacked shells to keep the gelatinous ooze of yolk and surrounding membranes from drying out and killing the growing embryo. Many of these Carboniferous amphibians looked like fat, swollen salamanders. Others, like aistopods, had slender salamander bodies with boomerang-shaped heads. But when it comes to pure dramatic excess these vertebrates were literally overshadowed by others. Forests of the Carboniferous and the succeeding Early Permian Period teemed with giant mayflies, scorpions, spiders as big as horseshoe crabs, and five-foot millipedes. Most spectacular were the Permian-aged giant dragonflies, such as *Meganeuropsis permiana* and its relatives, like *Namurotypus*. *Meganeuropisis*, the record holder, had a wingspan of two and a half feet. The forest floor was dappled with shadows cast by dragonflies larger, albeit more svelte, than seagulls.

What accounts for all this gigantism? One hypothesis has to do with a surfeit of oxygen during the Permian Period. It has been argued that its oxygen-rich atmosphere may have favored giant insects whose respiratory systems were advantageously served by this abundance. Some have also proposed that greater oxygen concentration tends to spur more growth and larger size because it promotes molting, the intermittent process wherein an animal sheds its old skin and bursts out in a new larger body. But these correlations are not clearly demonstrated by experiments. Living insects, including dragonflies, subjected to greater concentrations of oxygen do not show significant increases in body size. Perhaps the mysterious growth stimulator was simply the thick air of the Carboniferous-Permian Earth—"thick" because of the increased pressure of the atmosphere caused by the accumulated gases, which would facilitate oxygen uptake by those big insects and supposedly make them even bigger. Some recent experiments with fruit flies suggest that it may be simply the pressure of the atmosphere, not the concentration of oxygen, that made the difference to these giants.

But are these explanations for giant coal forest insects even necessary? Some paleontologists, including the fossil insect expert David Grimaldi, don't think so. Carboniferous–Early Permian forests are preserved in rocks

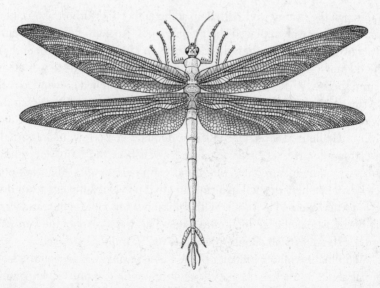

A large dragonfly (*Namurotypus sippeli*) from the Carboniferous with a wing span of 32 centimeters (12.5 inches)

that represent a time span of more than 75 million years. Fossils of large insects or other organisms are easier to find than small ones, so we are bound
to recover a few examples of giants that thrived during that vast time span.
Looked at this way, the need to explain the gigantism of early arthropod life
is simply a function of the bias in the fossil record, which has tended to preserve bigger organisms better. On the other hand, even this astute reflection
fails to explain why no insect as big as those swamp giants in the 290 million
years since has been discovered.

The stupendous swamp forests eventually diminished in expanse and
were replaced by other plant communities. Their demise was likely due to
a drift of the major landmasses and a change in climate. During this transition Pangea drifted northward, to cooler climatic belts, and the coal swamp
forests declined. Colder, more seasonal climates of the great landmasses
could not sustain the orgy of biological life we still associate with the tropics. The cold and the dry brought about a new order in the biota on terra
firma, a reshuffling of plant and animal species that yielded something
more enriched, varied, and seemingly but deceptively enduring.

As much of the swamp world dried up at the end of the Carboniferous, new
organisms decisively replaced the archaic ones. Ancient club mosses and
horsetails were confined in isolated swampland in the tropics, where their
dependence on water to disperse their seeds restricted them to wet areas.
Gymnosperms, whose seed dispersal advantageously relied on wind as well
as water, supplanted them in many regions. For the first time beetles and a
few other insect groups emerged, on their way to attaining a staggering and
as yet unchallenged diversity. Within vertebrates, a group with an entirely
novel reproductive system evolved and diversified: the amniotes. These
forms, which include reptiles, birds, and mammals, developed an egg protected by a leathery or shell-like covering that prevented the egg from drying out when exposed to the air. With this innovation some vertebrates were
totally emancipated from life in the water. This was a mark of the Permian
Period, extending from about 290 million to 250 million years ago.

A dominant amniote predator of the Late Permian was *Dimetrodon*, an
animal about 3½ meters (11 feet) long and weighing about 150 kilograms
(330 pounds). It had a huge head and jaws with knife blades for teeth and a
set of tall spines extending upward from its vertebrae. The spines possibly

supported a fanlike web of skin, something like the biblike skin flaps under the chins of some lizards or the webs between the fingers of tree frogs. *Dimetrodon* was a dramatic animal, but it wasn't a dinosaur. The Permian hosted many interesting creatures like it, but they died out at least 50 million years before dinosaurs appeared in the next time interval, the Triassic Period. *Dimetrodon* and other "finbacks," like the plant eater *Edaphosaurus*, are more closely related to us than to dinosaurs. They are an early group of synapsids, a long chain of relatives that include species leading up to mammals. Indeed, mammals and their ancient synapsid relatives are on an isolated branch outside a group that contains turtles, lizards, snakes, crocodiles, dinosaurs (including birds), pterosaurs, and the extinct marine plesiosaurs and ichthyosaurs.

These animals, such as forms like *Dimetrodon*, and the new robust plants, such as gymnosperms, built a biological empire that was diverse, global, and seemingly impregnable. Eventually however this empire collapsed in the greatest mass extinction event in the history of Earth. That catastrophe occurred about 250 million years ago, and it marks the end of the Permian Period. The devastation swept both land and sea. Marine creatures in Permian rock sequences from Alaska, the Italian Alps, and many other places around the world record the destruction of reefs and other marine fauna; 85 to 90 percent of all Permian marine species did not live to see the succeeding Triassic Period. On land the Permian extinction event was less pronounced, but still catastrophic, affecting roughly 70 percent of species, a traumatic change especially well recorded in South Africa, where the boundary between the Permian and Triassic is exposed in road cuts in places like Lootsberg Pass. There Permian rocks preserve an extraordinary array of synapsid species of a more specialized variety than *Dimetrodon*. Mixed among the Permian synapsids are a few specimens of an ugly, rather stumpy beast with a flat face and large protruding tusks called *Lystrosaurus*. Just above the Permian rock layers, in the lower Triassic sequence, the diversity of synapsid species decreases drastically. But *Lystrosaurus* and a few others had their day; they greatly increased in abundance, even though the diversity of different synapsid species was impoverished.

The record for land plants also shows marked devastation at the end of the Permian. (The floral record relies exclusively on the record of tiny microscopic fossil pollen.) Although many plant species became extinct, the larger groups of plants mostly persisted. Many species of gymnosperms, which were dominant in the Late Permian, disappeared but were replaced by others.

An extinction event in the remote past is like an unsolved murder. We know that death, in this case the deaths of huge numbers of species, occurred. We even have the bodies to prove it. But how was the crime committed? Who committed the crime? And why were the victims so easy to kill? To answer, paleontologists must find clues and piece together the evidence to come up with a convincing explanation. Then the trial begins. The opposition—likely other scientists—vault their spears and arrows at the prosecutors, questioning their arguments and their evidence. Debates, often highly charged, ensue. We hope for an answer or at least a consensus from them; perhaps it will not come. The process is not an easy one, and in the case of Earth's greatest disaster, the Permian mass extinction event, it is particularly difficult.

To get some idea of the dimensions of the problem we can consider various views of experts on the subject. A paper by Samual Bowring, Douglas Erwin, and Yukio Isozaki addresses the following questions:

How rapid was the Permian extinction event? Was it a gradual, stepwise phenomenon or an overnight disaster? Determining the actual time over which the extinction event occurred is difficult, because sediments and the fossils they entomb do not always mark time in a uniform way. Rocks can be deposited very rapidly or very slowly. You cannot simply conclude something took a long time because its evidence is preserved in very thick rock sediments. Time and sedimentation are not in lockstep.

To date rocks correctly, one often needs to find evidence based on other techniques, such as radiometrics. Radioactive isotopes are especially well represented in a category of rocks called igneous rocks, rocks generated by molten, deep Earth material called magma. Granites are one kind of igneous rock; volcanic rocks, including lava, basalts, and ashes, are another. The problem here is that igneous rocks are not often found in association with sedimentary rock containing fossils. But there are exceptions. A thick sequence of sedimentary rocks in the Meishan basin of China contains within it lower rocks representing the Permian and upper rocks representing the Triassic. Thus the Meishan sequence straddles the Permian-Triassic boundary and captures the great extinction event within it. Conveniently, it also contains a series of thin volcanic ash beds studded with small crystals, called zircons. Because zircons contain radioactive uranium (U), which decays over time to an isotope of lead (Pb), they can be dated, and thus each ash bed becomes a dateline. Using this evidence, Bowring and his coauthors estimate that the Permian extinction event that wiped out nearly 90 percent of all marine species occurred over a span of less than a hundred thousand years. To us, this may seem like a long time, but from a paleontological perspective it is the blink of an eye. The event could actually have been briefer than this, but the geologic record simply doesn't allow for such a resolution. What is important about this estimate is that it eliminates some explanations for Permian extinction that depend on even longer-term events, such as gradual climate change and a drop in sea level relating to continental drift. But the case is not closed. Scientists who have subsequently studied the Meishan sequence and those from other localities argue that the extinction event could have taken longer than a hundred thousand years.

After the event how long did it take ecosystems to bounce back? The answer here is complex. Some groups rebounded within a comparatively short time, but many did not. Erwin, who has been especially devoted to recording the post-Permian recovery phase, concluded that diverse ecosystems did not reappear until after 5 million years. By contrast, it is thought that recovery of marine ecosystems after the Cretaceous extinction event— commonly but inappropriately called the dinosaur extinction event—took at least 3 million years, though some key recovery events may have taken longer.

Finally, what was responsible for all this biological devastation? Here we are confronted with a plethora of proposals, none of them quite right on the money. There is circumstantial evidence of an inordinate amount of volcanic activity at this time, possibly emanating from the Siberian region, which may have caused a cooling phase, followed by global warming. But massive eruptions are recorded at other times in Earth's history without any indication of mass extinction. Some have proposed that a very high influx of toxic levels of CO_2 in the oceans may have transformed the waters of life into a deadly bicarbonate of soda, and the flooding of coastal areas by shallow anoxic (oxygen-depleted) water may have contributed to the destruction. But anoxia is largely confined to deep ocean water today, and its function in causing large-scale extinction events is poorly understood. There is also evidence of an asteroid impact at about this time. A seventy-five-mile wide crater discovered under western Australia might have been created by an asteroid three miles across. Another possible impact crater lies under Antarctica. But there are as yet no widespread signs of such an aster-oid impact. For instance, there are no geological fingerprints, such as a layer of rare elements that would have been part of the great cloud of ejecta raining back down on Earth immediately after the impact. Such a layer, found in rocks of the same age in many places all over Earth, is critical evi-dence for the Cretaceous asteroid impact, but such evidence is thus far ab-sent in Permian rocks. Perhaps toxic oceans, volcanism, asteroid impact, and some other event combined to deal the deathblow. Unfortunately, we have simply too many reasons for the source of the Permian extinction event, yet no indication in the fossil and geological record of a clear overrid-ing cause.

While not yet solving the Permian extinction mystery, paleontologists have identified some rather stressful conditions during an extended phase from the Late Permian to the Early Triassic, which may have rendered or-ganisms more vulnerable to a global catastrophe when it finally occurred. These conditions may have also delayed the later recovery of the biota. In a paper published in 2005, Raymond B. Huey and Peter D. Ward argue that the extremely low oxygen content in the atmosphere of the Late Per-mian–Early Triassic may have put many species, particularly large land ver-tebrates, in an oxygen-starved state called hypoxia. As we have seen, the Carboniferous was a time of high oxygen levels—more than 30 percent compared with 21 percent today—peaking about 280 million years ago in the Early Permian and declining thereafter; by the Late Permian they were only at about 15 to 18 percent and dipped to about 13 percent in the Early Triassic, an all-time low for the last 500 million years. Thereafter oxygen lev-els began to rise, with a few smaller peaks and dips, up to about 50 million years ago, and then they decreased slightly to the present-day level. Another gas can of course give us additional information about past conditions. CO_2 atmospheric levels usually, but not always, show an opposite trend from those of oxygen. When oxygen levels are low, CO_2 levels tend to be high. That was exactly the situation in the Late Permian–Early Triassic. Because CO_2 is an important greenhouse gas, the climate during this time was un-usually warm.

A 15 percent atmospheric level of oxygen is a serious problem for many organisms, especially big animals, like humans, because such a low concen-tration means that they inhale less oxygen with each breath. The "death zone" above twenty-five thousand feet in the Himalayas is not just a matter of mountaineering lore. Atmospheric pressure is low at this altitude, and oxygen concentration therefore falls to only about two-thirds of what it is at sea level. People do die at that altitude with some frequency because their cells, especially their brain cells, become oxygen-starved. Those places are not places for big organisms. People who, like me, have found it hard to jog or ski at ten thousand feet in the Colorado Rockies also may have enjoyed, when returning to their coastal hometowns, the superhuman pleasure of running on the beach; given the density of the atmosphere at sea level, oxygen concentration there is an invigorating 21 percent. In the Late Permian–Early Triassic, big organisms faced unusual stress, especially because the climate was hot and humid, increasing the need for oxygen, since the heat stress requires faster respiration.

Huey and Ward reconstructed a Late Permian–Early Triassic climate that was truly awful for large animals. Oxygen levels at 15 percent at sea level were equivalent to the kind of hypoxia-promoting conditions one would find at 17,388 feet in today's atmosphere. Given the heat and oppressive humidity, this would have forced big animals to concentrate at lower elevations, where they were actually marginalized, hugging the shores. This altitudinal compression was accompanied by rising sea levels at the end of the Permian, a combination that would have further reduced habitat and needed food resources. Such conditions were doubtless also unfavorable to insects, including the giant dragonflies, which suffered massive losses during the Late Permian. Finally, the intensification of these conditions into the Early Triassic may well have greatly inhibited the recovery of the land ecosystem. In this light, it is interesting that the Triassic marks the appearance of a diversity of advanced synapsids, the therapsids, with skull features that suggest possible adaptations to more efficient respiration and control of body temperature. By the Late Triassic, when dinosaurs and mammals made their entrance, oxygen levels were on the rise again. Giant, oxygen-hungry dinosaurs thrived at a time when oxygen levels were reasonably high—about 18 percent—though not as high as today.

The world's greatest catastrophe thus sets the stage for a regrouping of the surviving species and Earth's eventual transformation to the world of dinosaurs, flowering plants, and the modern ecosystem. However, we must remember that extinction in the Permian seas was worse than that on land, the casualty count possibly exceeding 90 percent of all marine species. This wave of Permian marine extinctions is being mirrored to some extent by the devastation of marine ecosystems today. We have irrefutable evidence of the cause of the modern crime, our own relentless drive to fish out the sea. Nonetheless, synergetic effects like global warming are also playing a part. The Permian extinction event shows just how vulnerable, and how resistant to recovery, marine ecosystems can be in the face of such catastrophe.

Among the major groups of marine species wiped out at the end of the Permian were the multisegmented trilobites, those curious creatures that look like pillbugs with fringe; we see them laid out like ancient coins in wooden trays in fossil and rock shops. Other victims included many kinds of fishes, sand dollars, and small hard-shelled protists known as foraminifera. Perhaps the most dramatic devastation hit the reef systems, which in the Permian were not dominated by corals like those of today. Instead, a variety of other creatures built up the calcium carbonate cement that formed the reef infrastructure. Among them were, notably, bryozoans, the colonial moss animals, algae, and crinoids, animals with feathery, flowerlike appendages propped on top of a narrow stalk covered with scales (thus evoking the name sea lily). Corals too were part of these reefs, and two major coral groups, the horn corals and the tabulate corals, disappeared along with everything else.

Paleontological evidence suggests that such Permian reefs were especially vulnerable to extinction in warm tropical waters. The marked increase in water surface temperature that came with the shift to an extremely hot climate at the end of the Permian may have affected the intricate collaboration of various colonizing organisms. We see the dramatic effects of such temperature changes on today's coral-dominated reefs. Many coral reefs today are severely degraded. Warmer ocean temperatures have killed off tiny dinoflagellates that live in the tissues of corals and lend them their brilliant pigments, leading to a condition known as coral bleaching. Populations of resident corals have been devastated, and in some areas, like the lagoons off the coast of Belize, entire species have been extinguished.

Permian reefs, subjected to even higher water temperatures, were comprehensively destroyed. Some elements of these ecosystems eventually, slowly began to reemerge after 5 million years. But the overwhelming majority of Permian species were not among them. They did not come back. They will never come back. The extinction of a species, whether in a 250-million-year-old reef off Pangea or a present-day reef off Belize, is forever.

8

THE DINOSAURS
OF MIDDLE EARTH

The postapocalyptic phase after the great Permian extinction event is bracketed by the Triassic Period, the first interval of the Mesozoic Era, or the age of "middle life." But as we have seen, the name Mesozoic for this time span is grossly inapt. By the time the Mesozoic dawned, 3.35 billion of life's total 3.5 billion-year-history had already elapsed. Nonetheless, the Mesozoic in at least one sense seems a fitting designation, since it is the interval sandwiched between the two other well-documented phases of Earth's history. It is the bridge between the Paleozoic Era, during which life in the sea dramatically evolved, diversified, and came on land, and the Cenozoic Era, which followed the great dinosaur extinction event. The boundaries of the Mesozoic Era are thus sharply defined, as Phillips recognized, by the two greatest calamities thus far recorded in the history of life.

The first of these, the Permian mass extinction event, was followed in the Triassic by a phase of recovery and transformation. The resilient gymnosperms diversified into species with elaborations of the modes of reproduction seen in ancient ginkgos and cycads. These were the early conifer relatives of pines, auraucarians, larches, firs, and sequoias. Perhaps this changeover had something to do with needs for new reproductive modes under new environmental conditions. Cycads and ginkgos carry their seeds and pollen cones on separate plants; even if the pollen lands close enough to the female ovules, the sperm has to "swim" in water surrounding the other plant. As noted, conifers, by contrast, build pollen and seeds into male and female cones. The pollen with its durable outer casing is spread by wind, not water. Moreover, the pollen grains grow a tube that brings the male sperm straight to the ovule. Conifers therefore thrived in many different habitats. Cycads in contrast were restricted in time to areas where water was abundant and seasonally dependable, such as tropical rainforests.

Patterns of extinction and renewal for insects through the Permian event are less clear to us because the fossil record is sketchy. Nonetheless, enough is known about this record for us to see a major upheaval in insect evolution, much of which accompanied the changeover in plants with diversification of seed-plant lineages such as conifers, seed ferns, and cycads. Gone were many of the archaic lineages that had dominated the Carboniferous coal forests, including many ancient relatives of cockroaches and grasshoppers as well as major groups of alien forms that left no living succes-

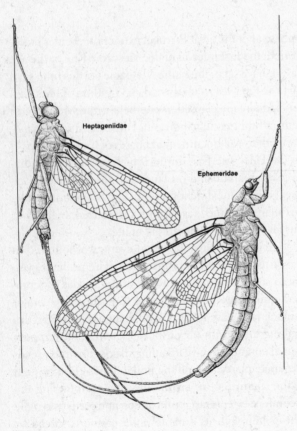

Mayflies, the most primitive living order of flying insects

sors. After the Permian we see an enrichment of hemipteroid insects (known commonly as the true bugs) and holometabolous ones (whose young or larval stage is separated from its anatomically and ecologically distinct adult stage by a nonfeeding stage called the pupa). Freshwater habitats were home to new kinds of insects, including the flies (Diptera), mayflies (Ephemeroptera), and the caddis flies (Trichoptera). With this diversification of seed-plant lineages came a flourish of new plant feeding, including certain beetles and leaf miners. Seed-plant dominance was also coincident with the appearance of new experiments for pollination, such as the mutual dependence between cycads and weevils, which persists today.

This resculpting of the Triassic flora and fauna embraced the rise of an animal group of such awesome size, form, and charisma that it is often difficult to see past them to a more balanced picture of what the ecosystem on land 200 million years ago was like. Dinosaurs were certainly not the only members of that ancient ecosystem, but they are the most entrancing, as well they should be. They're big; they're ugly or beautiful; they conjure and at the same time validate our wildest dreams and nightmares. And they are represented by spectacular preserved fossil skeletons in the world's great museums. Many dinosaur specimens are fragmentary and unimpressive: pieces of jaw, a partial tooth, or a toe bone. Some of these are isolated elements, such as a vertebra, a piece of backbone. But even an isolated element or piece of a stupendous dinosaur is worthy of awe. And then there are the complete specimens.

The first dinosaurs were less dramatic creatures than their successors. They were small theropod dinosaurs, forms that were bipedal—that is, animals propped on their hind limbs with their forelimbs dangling free. Theropods like *Herrerasaurus* as well as the ornithischian dinosaur *Pisanosaurus* have been found in rocks about 228 million years old, from the later phase of the Triassic. A few million years later came the ceratosaurs, primitive forms that include the long-necked dinosaur *Coelophysis*, whose abundant skeletons have been collected from the Ghost Ranch site in northern New Mexico. Other Triassic dinosaurs include the long-necked prosauropods like *Plateosaurus*. Evidence of diverse and abundant Triassic dinosaurs come from numerous remarkably preserved track sites.

Conifers, seed ferns, flies, weevils, dinosaurs: these all were either more populous or in some cases wholly new denizens of the postapocalyptic Triassic world. That world was itself in flux, a stimulus for another tremor in the terrestrial ecosystem. As the Triassic landscapes of lavish seed-fern gardens gave way to conifers, the end of this age was marked by another mass extinction event that affected many species. It left groups such as conifers and dinosaurs traumatized but still standing. The succeeding Jurassic Period could be well named the age of big and tall. Although endless mats of ferns probably accounted for most of their understory, Jurassic forests also had big trees—pine, spruce, and redwood that stretched to the sun—and big dinosaurs to match.

The largest of all dinosaurs were the earthshaking long-necked sauropods such as *Apatosaurus*, *Barosaurus*, *Seismosaurus*, and *Brachiosaurus*. These were the ultimate mass for load-bearing landlubbers. Adults may have ranged between 20 and 70 tons. (Some widely publicized estimates of mass between 90 and 150 tons now seem highly implausible.) From their small, blunt teeth, their massive bodies, and even in a few instances their preserved stomach contents, we can safely conclude that sauropods ate plants. But what kinds of plants? It is popular today to depict these dinosaurs standing erect on dry land, stretching their skyscraper necks to crop the tops of giant pine trees. The image is riveting, familiar, and repeated in just about every dinosaur book around, not to mention movies like *Jurassic Park*. It may therefore seem surprising that this reconstruction is actually a revised way of looking at sauropod behavior that departs from the original traditional view. In books and paintings of a century ago these giants were depicted as lethargic, semiamphibious creatures, tramping through the mud with their heavy tails dragging behind them. In many scenarios they hardly ever left the water, docking their bloated bodies in shallow bays like submarines with only the bridge tower of a neck and head emergent.

By the 1970s paleontologists had taken a swipe at this scenario, basing their new hypothesis on a combination of speculation, analogy, and, albeit rarely, evidence. When it came to such evidence, perhaps most important was the observation that numerous dinosaur track sites rarely showed signs of tail markings. Oddly overlooked for decades, this absence had only one explanation, that the dinosaur tails were held high above the ground in whiplike fashion. Moreover, it was noted that sauropod skeletons were supported by pillarlike legs that could be brought directly under the body. This erect stance was reminiscent of large, active mammals like elephants and rhinos, rather than the usual dinosaur analogy to crocodiles or big lizards, with legs bent out from the flanks in a sprawling posture and gait. Bringing the legs directly under the body makes for more efficiency in walking and a better potential for faster, more agile movement. I've seen two rhinos crashing into each other in the high grasses of Chitwan National Park in Nepal, and their speed as they converged was indeed impressive.

One extrapolation based on the behavior of living animals led to another. If these huge dinosaurs had elevated, whiplike tails and a more stately stance, they were probably more active than we once thought. Their whole metabolism may have been converted into one that sustained more activity. Perhaps they were able to generate and retain their body heat in a closed system that did not rapidly heat up or cool off with changing air temperature. This kind of insulated system for maintaining stable body heat is known as endothermy, commonly and somewhat misleadingly called warm-bloodedness. Mammals and birds are endotherms, but lizards and amphibians are ectotherms, equally misleadingly known as cold-blooded, whose heat source is external and whose body temperatures vary radically with the outside temperature. Thus a lizard will remain as still as a gargoyle on a granite boulder in the cold shade of a Sonora Desert morning until the sun hits it, broils it, and induces it to scamper away. If big dinosaurs were endotherms, we could infer that they would have been very active, cracking their tails, galloping along, raising their heads to the sun, even lifting their front limbs off the ground to balance their frames on a tripod of their hind limbs and the proximal tail. This reconstruction then bears on the question of dinosaur feeding habits. The higher metabolic rate required for keeping the body at an even temperature burns many calories. An active, warm-blooded, and agile behemoth would need a great deal of food.

At the main entrance of the American Museum of Natural History in the great Roosevelt Rotunda, we tip our hat to the active, upright dinosaur scenario: a skeleton of *Barosaurus* is shown rearing up, stretching its neck fifty feet above museum visitors, and waving its forelimbs threateningly at a probing carnivorous *Allosaurus*, in order to defend its young. There is plenty of drama in such a reconstruction. Who can deny the sheer power of a scene involving the largest creatures that ever ambled on land, towering over the forest canopy? Unfortunately, there is an imperfection in this picture. Giant dinosaurs, like many other creatures of the fossil record, have been dead for millions of years. We can't be exactly sure they were capable of any activity like this, just as we can't measure their changing body temperatures or take their pulses.

Yet the fossil record is not completely devoid of clues to the metabolism, posture, locomotion, or other functions and lifestyles of extinct dinosaurs.

And there is evidence, some scientists now argue, that certain dinosaurs were endotherms. The upright postures and agile frames of some predators like *Velociraptor* or even the bulkier *Tyrannosaurus* are part of the evidence. Another clue comes from the examination of the fine structure, or histology, of dinosaur bone. As in a tree trunk cut crosswise, bone often shows a pattern of rings, and just as in trees, these rings represent the variances in rate of growth, depending on season and temperature. At a time when the animal is less active, hibernating or shutting down its metabolism during a prolonged cold snap or winter season, bone growth slows and less bone is laid down. This is reflected in a very thin, slow-growth ring. When its metabolism picks up, driving more blood and nutrients to the bone, a thicker, fast-growth layer of basic minerals and organic material making up the bone is rapidly laid down. In endotherms these rings are not always clearly defined because the animal maintains a rather steady metabolism regardless of external, environmental changes. Interestingly, a similar blurring of growth rings is seen in the limb bones of many dinosaurs. Unfortunately, the match between bone-growth patterns and endothermy or ectothermy is not perfect, so growth patterns and their analysis are complex matters that engender controversy.

Finally, an important insight into the evolution and phylogeny of dinosaurs reveals that at least some of them may have been endotherms. Living birds, undisputed endotherms, belong to a branch of dinosaurs, the theropods, that also includes ancient carnivorous forms like *Oviraptor*, *Velociraptor*, and even *Tyrannosaurus*. So the only living animals in which we can actually observe and record dinosaur metabolism are endotherms. But does this necessarily mean that extinct dinosaurs were endotherms? After all, endothermy could have evolved later in dinosaur history, in the lineage directly ancestral to living birds. Another piece of unexpected evidence directly from the fossil record now suggests that endothermy extended to dinosaurs other than birds. Some newly discovered small, gracile dinosaurs from the Cretaceous fossil lake beds of Liaoning Province, in northern China, are actually preserved with impressions of feathers. These Liaoning dinosaurs represent evolutionary branches outside birds; they are closely related to well-known forms like *Velociraptor* and *Oviraptor*. They are not birds; they are branches off the main limb leading to birds. Moreover, the feathers in these animals are not flight feathers developed on elongate, winglike limbs but instead cover the whole body as small, downy structures or, where the feathers are long or complex, as at the end of the limbs or the tail, suggest some kind of display rather than flying function. So these dinosaurs were not fliers, and most of their feathers functioned to insulate the animal, as down does in living birds. Hatchlings of many bird species are born nearly naked, but these ugly ducklings soon after birth are appointed with fuzzy feathers that give their bodies the miraculous warmth we enjoy in sleeping bags and duvets. Feathers, like hair in mammals, are critical to insulating the body and thus to maintaining the stable internal temperatures we associate with warm-bloodedness. The fact (not supposition!) that we have some fossils of theropod dinosaurs with feathers of the kind that are important in the insulation of living birds leads logically to the conclusion that these forms were probably endotherms.

This conclusion doesn't settle the case for the forest giants of the Jurassic, the sauropods. Not all members of a group have to share every trait. We know, for example, that some dinosaurs, like the duck-billed hadrosaurs as well as sauropod embryos, had a pebbly skin, which is reminiscent more of ectothermic crocodiles and lizards than of feathered birds or hairy mammals.

The conflict in—or worse, the sparseness of—evidence for sauropod posture, locomotion, feeding, and metabolism has not discouraged mounds of published interpretations. Much of the recent literature, of varying quality, turns on the seemingly simple question: Could a sauropod extend its neck vertically while lifting its head? The question has ecological implications of interest to our history of changing ecosystems. If these giants could stretch their necks vertically, they were capable of pruning a forest canopy, and their effect on the forest's floral structure and diversity would be drastically different from one in which they hung close to the ground and grazed on the underbrush.

Some researchers have returned directly to observations of the bones to answer this question, and in the process they have rekindled the issue of sauropod lifestyles. Stevens and Parrish made 3-D computer reconstructions of the skeletons of two sauropods, *Apatosaurus* and *Diplodocus*, which reproduced faithfully the intricate interlocking facets on the neck vertebrae for various positions of the head. Comparing the neck flexion in modern birds, the researchers determined that the facet of one vertebra could slip past another until their overlap was reduced by 50 percent. If the models were read literally, a vertical neck posture in these dinosaurs could be seen

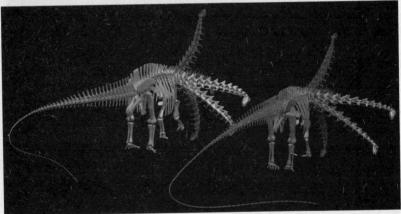

A computer-animated model generated by Kent Stevens (from Stevens, K., and Parrish, M. J., 1999) of neck movements in the sauropods *Apatosaurus louisae* (upper right and lower left) and the more gracile and longer-necked *Diplodocus carnegii* (upper left and lower right). The extremes of flexion of the neck are shown superimposed. *Apatosaurus* could reach higher, despite its shorter neck, due to greater dorsal flexibility at the base of the neck. Both sauropods were clearly specialized for medium to low browsing. The longer neck of *Diplodocus* adds primarily to forward reach, not vertical elevation.

to be impossible because the degree of bending between the lower neck vertebrae at their articulations would be unnaturally extreme. The shorter-necked *Apatosaurus* could raise its head about 6.9 meters (23 feet) above the ground (these measurements were provided by Kent Stevens and are slightly different from those in the original publication). While this is short of a vertical neck extension, which would put the head 10.78 meters (35.35 feet) above ground level, 23 feet are still high enough for cropping the lower foliage branches of large trees and the crowns of small ones. The *Diplodocus* was more ground-based; it could raise its head on its extremely elongated 6.2-meter (20.3-foot) neck only about 5.6 meters (18 feet). Thus, according to Stevens and Parrish, *Diplodocus* carried out its feeding like a horizontal boom crane, not a cherry picker. They concluded that these animals were adapted to ground feeding or low browsing, rather than high browsing.

This clever use of computer technology does not eliminate all sauropods as high-story browsers. The neck vertebrae of the seventy-ton *Brachiosaurus*, one of the biggest of all sauropods, show that its complex articulations conferred great flexibility, facilitating a high vertical head position some thirteen meters (forty-two feet) above ground.

Such a possibility brings us back to the issue of dinosaur body temperature and metabolism and to some recent mathematical modeling based on the known physiology in living animals like giraffes, the anatomy of sauropods, and the law of gravity. Seymour and Lillywhite have argued that a near-vertical head position for *Brachiosaurus* or any other massive sauropod would be physiologically impossible if these animals were endotherms. The central issue here is blood pressure. Seymour and Lillywhite calculated that if they were endotherms, such sauropods would need systolic blood pressure at the heart measuring 700 millimeters (mm Hg) to pump blood to their stratospheric heads. For a number of reasons, they argued this would have made survival uncertain. Giraffes, which have very thick, muscled hearts, have blood pressure levels far below this level, which at 100 mm Hg is only around twice the norm for mammal species. The lethal effects of high blood pressure are well-known in a human population, and in developed countries like the United States millions of people are afflicted with this problem. The whopping 700 mm figure calculated for sauropods would exert so much stress on the dinosaur's heart that it would require a massive pumping organ weighing more than 2 tons, well above 5 percent of the animal's total body weight. This is off the scale. Heart mass is on average about 0.26 percent of body mass in ectothermic amphibians and reptiles and about 0.96 percent of body mass in endothermic birds and mammals. (For comparisons, a fin whale weighing 40 tons, a body weight proposed for *Barosaurus*, has a heart of 190 kilograms [418 pounds], but this represents only 0.5 percent of its body mass.) Thus calculations show that high metabolic rate and upright neck posture would have been mutually exclusive in sauropods, in which case the high browsers of the sauropod group might not aptly fit the picture of active, agile, and warm-blooded beasts. We also remain uncertain about whether these forms were endotherms or ectotherms. (Sauropods might possibly have been able to lift their heads high if they had metabolic rates much lower than expected for endotherms.)

Fossil evidence might give us additional direct insight into the lifestyles, behavior, and food preferences of the largest plant eaters that ever lived on land. A dinosaur like *Brachiosaurus*, with fossil plant material typical of tall trees preserved in its rib cage, would certainly be a clue to a canopy-feeding lifestyle. Following the physiological extrapolations above, such ingested plant remains would also suggest that these animals were, unlike their bird relatives, ectotherms rather than endotherms. Unfortunately, the fossil record is not very much help here. There are few well-known examples of evidence of what dinosaurs ate. Some dinosaur fossils do preserve mats of digested material in a location in the skeleton that once was their abdominal region. Some of the newly discovered fossils from Liaoning show the remains of small mammals in their macerated stomach contents. Some skeletons of the Triassic dinosaur *Coelophysis* from northern New Mexico have smaller skeletons in their abdominal regions. Earlier reports identified these stomach contents as the bones of juvenile individuals of *Coelophysis*, but the alleged cannibalistic behavior of this dinosaur has recently been refuted. One of the digested skeletons appears to have been that of a crocodile, and though other skeletons are indeed juvenile *Coelophysis*, they do not appear to have been digested; instead, the adult skeleton was preserved lying on top of the juvenile skeletons. (A recent description of a dinosaur from Madagascar provides another case, but an ambiguous one, of a dinosaur cannibal.) The fossil of a duck-billed hadrosaur, *Edmontosaurus*, has a mash of fruit seeds, conifers, and flowers in its rib cage. The abdominal region of a sauropod fossil from the Jurassic of Wyoming has macerated plant material—stems, bits of leaves, and other plants—and another sauropod from the same rock unit, the Morrison Formation, has abundant plant fragments, including a plant stem about a half inch long. There is also evidence of smooth stones in the gut region of some dinosaur skeletons, suggesting these animals had grinding mills to help pulverize food, a feature similar to the gizzard stones of living birds. These instances of preservation of food material and stomach stones notwithstanding, the evidence for the feeding habits of most dinosaurs, especially sauropods, is far from definitive.

Indirect evidence of feeding habits comes from the study of teeth. The peglike teeth of *Diplodocus* and *Apatosaurus*, which are not suited for cropping resistant vegetation like cycads or conifers, suggest that these dinosaurs probably grazed on softer plants, such as ferns. Ferns might also have better sustained big herds of large-bodied animals, since ferns can grow and regenerate quickly in comparison to slow-growing cycads and conifers. This offers a vision at odds with one of treetop feeders. Instead, we have a picture of a herd of sauropods cruising the ferns, swinging their long necks out horizontally in semicircular arcs, cropping away at huge expanses of lower-story vegetation.

I relate these various views on the feeding and behavior of the giant sauropods of the forest with caution. The research on this problem depends on analogy with living creatures and on extrapolations from physiology, ecology, and other facets of modern biology. But for all the apparent refinement, we really don't know many of the answers to these questions. Analogy with living creatures may be enlightening, but it has a severe limitation: it can tell us only about what animals do today. We cannot say that it was really impossible for a sauropod to raise its neck or forelimbs without having observed it in action. The fossil record can tell us many amazing things, but not everything.

Whether sauropods were warm- or cold-blooded, or fern or pinecone eaters, is not certain. But we can reconstruct rather confidently one aspect of their behavior. With that gargantuan body size they would have to have been prodigious eaters. We can observe directly the impact of feeding by large mammals, such as elephants, on vegetation. Even in expansive forest or savanna, their harvesting can be spectacular. And as we've seen, in denuded, scattered forestland the harvesting activities of these herds can be disastrous. Graphic data about the consumption patterns of large, living herbivores offer an analogy for the titanic eating requirements of dinosaurs ten times bigger than a full-grown bull elephant.

Small herbivores can subsist on small plants, and many of them specialize on selected species to get nutrients from nectar, fruits, nuts, etc. Large herbivores do not have this luxury; they are bulk feeders that must process massive amounts of food from many different plants. These bulk feeders often have extensive adaptations for oral processing, as well as microfermentation in a gut long enough to break down the food.

Reconstructions of sauropod food intake are rather specific. Some researchers have calculated that daily food intake requirements were 41.4 kilograms (91 pounds) for *Diplodocus* and 186 kilograms (409 pounds) for *Amphicoelis*. This amount would increase significantly if these dinosaurs were endothermic. Also we have strong evidence that the dinosaurs clustered in herds. M. J. Coe and his coauthors estimated that the North American Jurassic ecosystem, such as that represented by the Morrison Formation of the Rocky Mountain region, supported an herbivore biomass of 93,000 kilograms per hectare; this is far greater than the 4,848 kilograms per hectare in the extant Amboseli Plain of the African savanna. In a review of

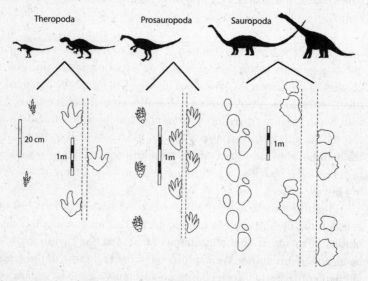

Trackways of three dinosaur groups: from left, the theropods *Grallator* and *Megalosauripus*, the prosauropod *Otozoum* (with a track made by feet close together on left and one made with feet spread apart on right), and the sauropods *Parabrontopodus* and *Brontopodus*. One meter scale bars (20 centimeter bar in the case of *Grallator*) indicate distance between footprints. Note the marked difference in foot shape and track pattern among the groups.

the issue Ralph E. Taggart and Aureal T. Cross argued that such enormous group consumption indicates that these giants relied on rapidly growing ferns for their main food source. Today the ecosystems supporting large mammals tend to be open grasslands with sparse tree cover, as in the African savanna, or the arctic tundra, which lacks trees almost entirely. The productivity of the savanna is about ten times that of the tundra, yet the latter sustains huge herds of caribou and sizable aggregates of other mammals, such as musk oxen. Some 10,000 years ago tundra also supported herds of mammoths and woolly rhinos. Thus productivity is not the only factor that allows for the presence of great numbers of large plant eaters. The constancy of the food resource in these regions, despite the huge demands of large herbivores, is maintained by rapid growth of grasses and the migration of herds with changing seasons and food availability. We know of no Jurassic habitats with these features. Angiosperms, especially grasses, had not evolved at all; grasses did not appear in abundance in the fossil record for another 120 million years. Nonetheless, open, fern-dominated habitats might share much the same general energetics. Interestingly, trackways of the gargantuan sauropods show an organization among large adults and small juveniles that is very like migrating herds of large mammals today. Some of these trackways also show that the dinosaur herds moved in preferential compass direction, clearly suggesting that they may have been migrating, possibly in order to arrive at places with more food.

My favorite time with *Barosaurus* in the rotunda of the museum is early afternoon. Light rays penetrate the south-facing portico windows at the base of the cupola. Apollo becomes an alchemist, turning the lofty head and upper neck of *Barosaurus* to gold. Even without this spectacular pyrotechnic, it is difficult not to be awed by something so big and remote in time. What was it like to witness the forest stomp of a herd of fifty *Barosaurus*? Nothing so gigantic or gluttonous is alive on land today. Yet some of the most amazing living things in those dino-trampled forests were no bigger than your thumb, and those small organisms changed our planet's ecosystems in ways that dinosaurs never could.

9

A FLOWER IN THE FOREST

> Earth laughs in flowers . . .
> —Ralph Waldo Emerson, "Hamatreya"

In the land of dinosaurs, conifers, and fern pastures a quiet revolution was about to take place. Sometime before 130 million years ago—it is not certain exactly when—a small, strange plant was likely making its meek presence known in the shadow of the pines. Unlike the giant trees above, this plant did not expose huge amounts of pollen at the fringes of its cones. Instead, it sprouted a more elaborate and colorful accoutrement, a flower, ready-made for sex of a completely different kind. The little plant did not release its pollen to the wind. Rather, it presented these tiny grains of immortality on thin stalks. The stalks were surrounded by something arresting and beautiful, delicate, brightly colored petals that recall the shape of leaves. Deep within the flower, small, unfertilized eggs were prepared for the arrival of pollen from some other plant or perhaps some other part of the same plant. That auspicious arrival was triggered by a signal: a flash of petal color in sunlight, a flutter of petals in the breeze, a perfumelike scent, or a delirious combination of all these temptations. The signal flashed like the neon sign of a diner: EAT HERE.

The signal was meant to attract not just any wayfarer. Indeed, it might have been intended to catch the eye and the smell receptor of just one species of beetle or fly making its way over the forest clearing. Pollen and perhaps some sweet nectar in the flower were food for this insect traveler. But these items had to be collected. Perhaps as it probed for sweet nectar, the insect inadvertently picked up the sticky pollen on the hairs on its legs or the tips of its antennae. After its brief visit the insect flew on, perhaps in search of more eateries. If the search was successful, the insect made a swap that sparked the critical phase of fertilization for the flower. As the insect picked up new pollen, it kicked off some of the old pollen from the other plant. The alien pollen now invaded its new home, sprouting a tiny tube that penetrated the protective crypt at the base of the flower and then the unfertilized female egg itself.

With this contact, the egg became a seed for the next generation, and one of nature's most explicit and spectacular acts of cooperation was initiated. Tens of millions of years passed before this elaborate system took hold, before flowering plants became a dominant, vital part of the ecosystem. Nonetheless, by 90 million years ago the world was incandescent with fluorescence. The world would become the world as we know it.

This scenario deliberately evades certain details. Withheld are the names of the plant, the insect, the specific color and form of the flower's structure, and the behavior of the insect pollinator. Some studies that reconstruct evolutionary events from the phylogeny of the major lineages of flowering plants and their relatives suggest that flowers first appeared in moist, dimly lit forests, like the one described here. We do not have direct evidence of the first flower on Earth, but we do have pollen grains from Cretaceous rocks about 130 million years old that are remarkably like the pollen of some living flowers, such as magnolias, and distinctly different from the pollen of pines, spruce, ginkgos, cycads, and other gymnosperms. With the presence of this type of pollen, we can safely conclude that plants with flowers that produced the pollen were at least this old. The Cretaceous Period, which lasts from about 140 million years ago to the great extinction event of 65 million years ago, is the third and last interval in Phillips's Mesozoic Era, the last chapter in the age of the nonavian dinosaurs.

Angiosperms, the flowering plants, with their wondrous and effective specializations in reproduction, were the last major group of plants to appear on Earth. Early angiosperm pollen grains are rare but have been found in scattered localities across a broad geographic range: England, Italy, Morocco, Israel, and perhaps Asia. There is, however, no reliable evidence of angiosperm pollen or plants prior to the Cretaceous. An angiosperm reproductive structure from northern China was thought initially to extend the fossil record of angiosperms back to a time just antedating the Jurassic-Cretaceous boundary, at about 140 million years before present. But this fossil flower, *Archaefructus* [ancient fruit] *liaoningensis*, is now believed to be much younger and well within the Early Cretaceous, perhaps about 120 million years before the present. It is of course possible that angiosperms diverged prior to the Cretaceous because fossils tell us only the minimum date of origin for a given group. That is, they indicate only that the flowers are *at least* as old as the Cretaceous. Nonetheless, the fossil record gives us no direct evidence of flowering plants that are any older.

After their appearance, angiosperms gained ground only slowly. The oldest abundant and well-preserved angiosperm fossils of flowers, fruits, seeds, stamens, and other dispersed plant fragments are the floras about 110 million years old from the western Portuguese basin. Many of these fossils are preserved as charcoal from ancient forest fires and retain their three-dimensional shape. By the Middle Cretaceous, between 90 and 100 million years ago, we have evidence of marked increase in diversity, structural complexity, and abundance of angiosperm pollen, leaves, and reproductive parts.

Early Cretaceous flowers lacked the wondrous complexity we see in many modern flowers. To appreciate the difference, we must consider the basic components of the flower itself and understand their significance. A flower is the most elaborate and baroquely decorated sexual organ ever evolved. The word *florid*, obviously inspired by flower structure, is in fact synonymous with words like *baroque, ornate, elaborate, extravagant,* and *fancy*. In most flowers, there are four whorls of sexual or auxiliary structures. The first and outermost are the supporting sepals, which are usually green or sometimes brightly colored, like petals; these make up the outermost "cup" of the flower, the calyx. On the next inner ring are the petals, usually colored brightly to attract pollinating animals to the plant, making up the inner "cup" of the flower, the corolla. Calyx and corolla together constitute the floral envelope, or perianth, which protects the reproductive organs and

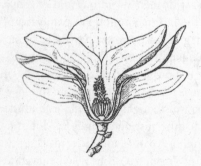 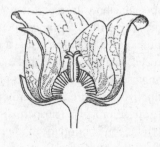

Flowers of *Magnoliaceae* (left) and *Annonaceae*

attracts pollinators. They are both the heraldry and the protective enclosure for the vital reproductive structures within.

Embraced by the perianth are the organs of action. The male components are part of the androecium, which includes the stamens, which vary greatly in number in different species of flowers. Each stamen is divided in turn into two parts: the anther at the top, which produces and releases pollen, and the supporting filament. In the center of the flower are the female parts, the so-called gynoecium, consisting of one or several carpels, each with an ovary at its base containing one or many ovules (the egg plus its protective layers and nutritive tissue). Each carpel is surmounted by a style topped by a stigma, which is specialized to receive the pollen. There are many variations on this plan, including many flowers that are unisexual, containing only the parts of either the androecium or the gynoecium. Moreover, many flowers lack petals or sepals. But the architecture described above is the basic plan of the angiosperm.

How did this marvelous, complex sexual organ evolve? As in many other cases, this essential question was first posed by Charles Darwin, who wondered: What is the relationship between the flower and the reproductive parts of other plants, such as the cones of conifers? He was so preoccupied and perplexed by the sudden appearance and success of flowers that he called it "the abominable mystery." The problem in determining the significance of the similarity between flowers and cones is essentially the same as that concerning the origin of hair in mammals or the relationship between fins in fish and limbs in land vertebrates. For example, how does the spiral-like form of the magnolia flower relate to that of a pinecone? Perhaps most mysterious is the derivation of the angiosperm carpel and the outer layer of the ovule.

Despite obsessive investigation, this "abominable mystery" has not been solved. In the last few years scientists have at least clarified some of the issues by looking at different options for relating angiosperms to other plants. It is a crux issue in the effort to map the tree of life, the origin for the most successful, diverse vegetation on Earth.

What, then, are the nearest relatives of angiosperms? Two competing theories currently command the most attention. The first places angiosperms next to the gnetophytes and the extinct Jurassic bennettitaleans. Many bennettitalean species resemble cycads, but they had elongate trunks and long, leafy plumes. Some species, like *Williamsonia sewardiana*, look like tree people—the ents from Tolkien's trilogy—with their bristly arms and ornate spiky headdresses. The bennettitaleans have fruiting bodies that bear some similarities to magnolia flowers. Fossils of these fruiting bodies show considerable insect damage, suggesting both pollen feeding and pollen dispersal by beetles, which were already common in Jurassic times. If the bennettitaleans are indeed close relatives of angiosperms, their common ancestor was certainly at least as old as the Late Triassic.

Gnetophytes, the other possible close angiosperm relatives, include about seventy species distributed among three taxa. *Gnetum* is either treelike or vinelike, with large leathery leaves. *Ephedra* is highly branched shrubs with small scaly leaves. Many of its thirty-five species live in deserts. *Ephedra* very commonly occurs in the western United States, primarily in the deserts of Utah, Nevada, and surrounding states. It has significant pharmaceutical importance because it is a natural source for the chemical

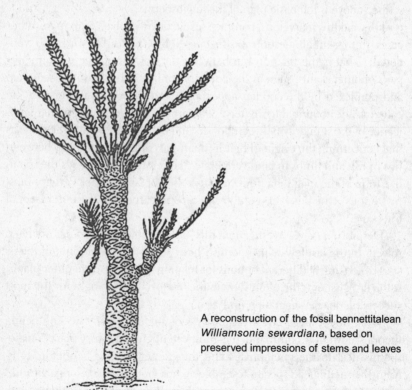

A reconstruction of the fossil bennettitalean *Williamsonia sewardiana*, based on preserved impressions of stems and leaves

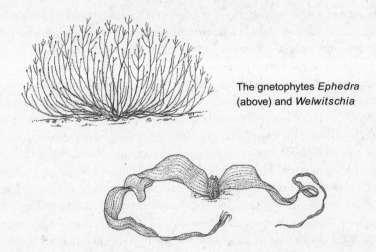

The gnetophytes *Ephedra* (above) and *Welwitschia*

ephedrine. (Though it has been given the name Mormon tea, there are no indications that Mormon settlers ever brewed this scraggly plant for either refreshment or pharmaceutical purpose.) The third gnetophyte is *Welwitschia*, one of the weirdest of all plants. Much of this plant is buried under the sand; what sticks up is a bizarre, woody disk with cone-bearing branches and two long, floppy, extraordinarily dead-looking leaves. The leaves themselves are split from one end to the other, giving the plant the appearance of so much beach trash. *Welwitschia* is limited to the coastal deserts of Namibia, although fossils show that plants of this group may have been widespread on several landmasses during the Mesozoic Era.

Gnetophytes and bennettitaleans are often classified as gymnosperms, along with pines and other conifers as well as ginkgos and cycads. But they do share some features with angiosperms not found in other gymnosperms. One of these, observable only in living gnetophytes, is a peculiar phenomenon called double fertilization. Instead of the simple union of one sperm with one egg as seen in humans and other animals, fertilization in angiosperms and gnetophytes is much more complex. And it involves another critical mark of the angiosperms, a strong emphasis on the sporophyte, or diploid, part of the plant life cycle and an accompanying reduction of the gametophyte, or haploid, phase. The male gametophyte is reduced to just three cells, which form inside the pollen, two sperm cells and one cell that forms the pollen tube. The female gametophyte is reduced to just seven cells, one of which is the egg cell; another cell retains two nuclei, so that the seven cells contain eight nuclei among them. Thus seven gametophyte cells form the crucial component of the ovule, the embryonic sac. The nuclei of all these gametophyte cells are haploid; they all carry just one set of chromosomes. During double fertilization, one of the two sperm nuclei in the pollen fuses with the new egg cell to make a zygote, which develops as an embryo. The other sperm nucleus from the pollen fuses with the gametophyte cell with the two nuclei, thus forming a cell with three sets of chromosomes in it, two from the female gametophyte and one from the entering male gametophyete. This triploid cell multiplies to form the endosperm, which provides the nutritive tissue for the developing embryo. Gnetophytes do show double fertilization, but the process does not unfold in the same way; they do not produce a triploid cell and its eventual product, the endosperm. Why double fertilization evolved in these plants is yet another mystery, but its establishment might relate to the impressive diversity, adaptability, and tenacity of the angiosperms.

Angiosperms and gnetophytes, and perhaps bennettitaleans, share other features that have nothing to do with sex. They have vessels in xylem, a system of water conduction tissues that also serves for food storage, plant support, and the uptake of minerals. Other plants lack these vessels. The gnetophytes and angiosperms also have a distinctive venation in the leaves, in which complex branching and connections are emphasized, often in a reticulate pattern.

These similarities and several others have been used as evidence for a close relationship among the gnetophytes, bennettitaleans, and angiosperms. This hypothesis is important because two major events are then implied concerning the origin of angiosperms. First, it suggests that angiosperms evolved from a branch *within* the gymnosperms—in other words, that they are a subgroup of gymnosperms in much the same way that we have come to recognize birds as a subgroup of dinosaurs. Second, this hypothesis suggests that distinctive features such as xylem vessels, complex leaf venation, reduced gametophytes, and double fertilization are not unique to the common ancestor of all living angiosperms, since they are after all shared with gnetophytes and possibly bennettitaleans (we acknowledge that we cannot determine the presence or absence of some features, such as double fertilization in fossil bennettitaleans). In other words, they must be features of the immediate ancestor leading to *both* angiosperms *and* the gnetophytes and bennettitaleans.

This elegant story also says something about the timing of the origin of the angiosperms. Although no angiosperm fossils are known before those of the Early Cretaceous, about 130 million years ago, bennettitaleans go back much farther, into the Jurassic. There are indeed records of bennettitaleans as old as the Triassic; gnetophytes also have a record, though less complete, extending to the Late Triassic. Now if these groups are close relatives of angiosperms, the immediate common ancestor must be much older than the known angiosperms, at least as old as the Triassic. The branching pattern in the phylogeny brackets the age range of a given group, even when the fossil record is incomplete. And in cases where a phylogeny demonstrates that two fossil groups with drastically different age spans are close relatives, the phylogeny usefully points out how grossly incomplete the fossil record might be.

But how good is this phylogeny for angiosperm relationships? The theory that there is a close link between angiosperms, on the one hand, and gnetophytes and bennettitaleans on the other, known as the anthophyte hypothesis, a hypothesis persuasively based on the visible architecture of the plants, has come under fire. In the last ten years plants of all three categories have been studied for their DNA sequences in a variety of genes. Most of these results are starkly different from those suggested by morphology (although, it should be noted, many of the original arguments based on morphology have been either rejected or modified). They produce a phylogeny that places gnetophytes close to pines, cycads, ginkgos, and other gymnosperms but that excludes angiosperms entirely. In other words, gymnosperms, which include the gnetophtyes and bennettitaleans, are a tight group with a single origin, so the line leading to angiosperms would have to have split off before, not after, the gymnosperm lineages. A study combining both morphology and molecular data also shows this result. Now the oldest remains of known gymnosperms, undoubted cycads from the Permian Period, are much older than the recorded bennettitaleans. Acceptance of the phylogeny based on genes would indicate that our fossil record of angiosperms is either very deficient, or we have failed to recognize the key angiosperm antecedent. (Some molecular results are consistent with a third arrangement, "the gnetophyte hypothesis," wherein gnetophytes and bennettitaleans split off before angiosperms and gymnosperms.)

Which of these stories is correct? Here we have a classic and frustrating case, where studies based on good and plentiful data from two different sources—from genes, on one hand, and from morphology and development, on the other—are in conflict. Most botanists and paleobotanists have

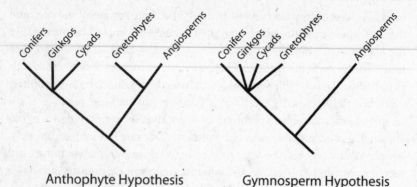

Anthophyte Hypothesis   Gymnosperm Hypothesis

Two opposing hypotheses for angiosperm relationships with other plant groups

not been quick to embrace the gene-based story despite the evidence. They note that most major plant groups are extinct and thus off-limits to gene studies. What if some extinct gymnosperm lineage had a set of genes very like that of angiosperms? Unfortunately, we shall never know. On the other hand, recent gene studies have effectively clarified relationships in groups that have long been difficult to study because their morphological traits are limited or inconsistent or both, and their fossil records are very poor. Moreover, gene-based phylogenies for various subgroups of gymnosperms and angiosperms are in step with the solid results based on classic, exhaustive studies of plant morphology, life cycles, and development. This suggests that in many cases the gene data are reliable. Perhaps in a few years, when we have mapped plants on the tree of life using more genes and more evidence of morphology and development, we will have a satisfactory solution to Darwin's "abominable mystery."

Once we can identify angiosperms as a distinct branch on the tree of life, then describing their basic ancestral features, those shared by hundreds of thousands of angiosperm species, becomes easier. There is far less controversy in this distinction than in the debate over how angiosperms split off from other plants. The angiosperms' several innovations include the refined, reinforced protection of the egg itself. In angiosperms the ovule is completely enclosed within layers of tissue supplied by the parent plant, the carpel, whose margins or flanks overlap and become fused. The closed carpel is a major distinguishing trait of angiosperms; indeed, it inspired the name for the group. Fertilization through this tightly sealed barrier is achieved via a pollen tube that grows out of the pollen. With the sperm nuclei inside it, the pollen tube burrows through the style of the ovule and finally breaches all barriers to fertilize the egg. Angiosperms also show a number of distinguishing details in their flower structure, including the development of stamens each with two pairs of pollen sacs. Another angiosperm feature is the extreme development of double fertilization to the point of forming an additional, critically important structure, the endosperm, which provides the nutritive tissue for the developing embryo.

Other features, if not unique, are particularly prevalent and well expressed in angiosperms. These include deciduousness, or leaf dropping, which may have evolved in the tropics to resist periods of drought but seems particularly advantageous in high-latitude winters. Only ginkgos and a few conifers, such as the larch and the bald cypress, are deciduous. Those angiosperms that complete their life cycle in a year—we call this large group annuals—have a degree of flexibility that opens entirely new horizons for new generations of plants. Weeds and other angiosperms grow extremely rapidly. Angiosperms have the most diverse of habits, growing as lianas, vines, trees, and creepers, and they have good devices for surviving bad times: tubers, rhizomes, bulbs, and so forth. They produce the most delicious fruits and the most robust seeds that can resist drought and winter. This toughness of their seeds, which survive dispersal by wind, water, and animals, combines with precise methods of pollination to ensure that angiosperm reproduction is the most efficient of any plants. Finally, angiosperms are the great pharmacological trove of the biota, producing a huge variety of chemical agents that defend them from disease and plant eaters—elephants to ants.

The amazing adaptability and diversity of angiosperms underscore one fact of life, that sex is a very big deal in evolution. Although bacteria, and some fungi, one-celled protists, flatworms, cnidarians (jellyfish and hydroids), echinoderms, insects, and even some vertebrates can reproduce asexually by simply cloning themselves, reproduction through sex is the norm in life. Sexual reproduction requires the fusion of two gametes—the fertilization of an egg by a sperm cell, for instance. The step toward sex was a big one, perhaps the most important in the history of life itself. From our perspectives as human social creatures, self-aware of our biological urges, our carnal pleasures, and our feeling of fulfillment with procreation, it may seem silly to ask, Why is sex so good? But the question is a serious one, and when we study the process of evolution, it requires an analytical answer. The most straightforward explanation has to do with something farmers and domestic breeders have long called hybrid vigor. Hybrids are formed when individuals of two very different lines or breeds of plant or animal are reproduced through sex. Sometimes these lines are so different that they are thought of as virtually different species; they do not reproduce under the usual conditions. Just the same, their conjugation often leads to robust, vigorous individuals.

Darwin noted the wisdom of horticulturists and stockbreeders and took the principle one step further: if sex is good for the production of hybrids from different lines, it is certainly good for mixing individuals that are not so far apart within a single species. This mixing, he reasoned, produces an enriched variability in organisms upon which natural selection can act, just as a horse breeder mixes individual racers, homing in on the most suitable qualities for a prize steed. But unlike an animal husbandry program, natural selection does not necessarily realize perfection—the fastest stallion, the fattest pig, or the reddest rose. Also, its direction is not readily susceptible to prediction. Climatic conditions, soil composition, or competitors for a food resource may change, and the new conditions exert a new direction in selection; individuals or species that were once on top of the heap may become losers. Dinosaurs are the grandest example of victims of the capricious forces of natural selection. As Darwin himself remarked in *Origin of Species*:

> It is most difficult always to remember that the increase of every creature is constantly being checked by unperceived hostile agencies; and that these same unperceived agencies are amply sufficient to cause rarity and finally extinction. So little is the subject understood, that I have heard surprise repeatedly expressed at such great monsters as the Mastodon and the more ancient Dinosaurians having become extinct; as if mere bodily strength gave victory in the battle of life.

One way to deal with "unperceived hostile agencies" is for a species to maintain a high degree of genetic variation and thus a potential for change and adaptation. Sexual reproduction, which enhances such genetic variation, requires the act of sex. Here nature reveals its creativity in an unbounded, awesome way. Sex generally involves copulation and the fertilization of the egg by the sperm inside the female. This intimate contact, called internal fertilization, is overwhelmingly common in animals. Alternatively, sperm and eggs may be shed and mixed externally. This occurs in some aquatic organisms, including corals, hydras, and many fish, which release great clouds of eggs and sperm, the quantity assuring that at least some will fuse. In addition to these two basic modes, external and internal, of fertilization, a huge array of other structures and behaviors is associated with sex.

At one extreme, the act of sex is in itself an act of mortal sacrifice. In some species of praying mantids, the male may approach the female with caution or stealth, as well he should. The male cannot release the spermatophore, with its sperm, until the female eats off his head. The male head contains a gland whose secretions inhibit copulation. The headless male can continue to copulate and ejaculate because such functions are independently signaled by a nerve ganglion in the abdominal region. The grisly practices in these instances are a bit exaggerated. Most species of praying mantids do not indulge in cannibalistic sex, and when they do, experi-

A praying mantid

ments show that the female consumes the male head and other body parts simply because she is hungry or because she fails to recognize the male as a potential mate rather than a potential predator. In any case, the male seems to make the greatest sacrifice, regardless of what drives the system. This example, by the way, seems wildly incompatible with any notion that the wonders of life, including mantid reproduction, are the act of an intelligent designer.

For the overwhelming majority of animals, the act of sex does not require self-destructive commitment on the part of the male. Nonetheless, sex in many species, including insects, can be ceremonial and prolonged, relying on extremely specialized anatomical structures for the act of copulation. It is no coincidence that one useful way to distinguish two closely related insect species is by studying the lock and key fit of their male and female genitalia. And sex in a wide variety of organisms, including vertebrates, involves elaborate mating practices. It can also be highly selective. Within many species preferences are shown for certain mates of greater strength, size, or beauty. These preferences affect the evolutionary system by skewing influence toward one or a few individuals in the gene pool of the population. Mammals are famous for the frequent instances of polygamy in species, often with rather spectacular extremes. The northern elephant seal, *Mirounga angustirostris*, is perhaps the epitome of the alpha male syndrome. In this species a full-grown bull elephant can weigh as much as twenty-seven hundred kilograms, about three times as heavy as a full-grown, sexually mature female. What elephant seals do in the sea is hard to observe, but the way they spend their time on land is a spectacle of violence. When males land on the beach, ahead of the females, they fight it out for choice spots to build their harems. Fights among the big bulls are injurious and can be deadly. A serious neck wound from the teeth of another bull can ulcerate, rendering the wounded diseased and weakened, highly vulnerable to sharks and killer whales when it returns to the sea after mating season. The winners, the alpha bull and a couple of others slightly lower in rank, begin collecting females landing on the beach. Harems can reach absurd sizes—as many as four hundred females. Sovereignty here is hard won; the bull harem masters are continually beating off male challengers. In all the riotous combat, newborn pups are sometimes inadvertently crushed to death.

Less extreme but nonetheless impressive examples of polygamy are provided by many other mammals, including gazelles, elephants, and lions. Indeed, polygamy in mammals is clearly the normal way of life for more than 90 percent of the living species. Monogamy is rare, occurring mainly in some primates and about 5 percent of other mammals. Polyandry, with the female collecting males, is rare, but does occur. That big hyena I encountered was probably a female; they are much larger than the males that cower before them. Here a role reversal is also extreme. The clitoris in the top-ranked female hyena of a group enlarges at mating time, mimicking a penis in its appearance.

Sexual selection can also occur at a later stage in the process. We are learning of more and more examples of species, including guppies, grasshoppers, fruit flies, and birds, whose females actually select for sperm contributed from two or more individuals after copulation. Here a female mates rather casually with two or more males and then "allows" fertilization only by sperm from one of the males. In many groups—birds, for example—males seem to increase their competitive edge and their chances for being the chosen donor if they have relatively large testes and sperm stores and long spermatozoa. The more successful males also tend to guard their mates zealously, effecting frequent copulations. It is not at all clear what evolutionary benefits are conferred on females that opt for promiscuity. At any rate, sperm selection represents a fascinating variation on approaches to reproduction.

Yet even against this kaleidoscope of modes and methods in sex and reproduction, how much more astonishing is animal pollination of flowering plants! Angiosperms have recruited not only insects but selected vertebrates for this purpose. Some flowers are pollinated by certain species of birds, bats, and an assortment of nonflying mammals (certain primates, marsupials, carnivorans, and rodents). Here we have an act of intimate dependence between different groups of organisms—insects or vertebrates, on one hand; angiosperms, on the other—that belong to separate life kingdoms. This is more than six degrees of separation. It would take a less audacious leap over life's diversity to derive a system in which the sexual reproduction of humans was facilitated by insects. After all, we both are bilateral, skeletonized, encephalized, segmented, and joint-limbed *animals* to boot.

Instead, it seems that evolution has taken the most extreme pathway of all in sustaining the foundation of our modern ecosystem. The cornucopia of life flows from a system in which a flowery species lures other species of a wholly alien group into expending its own energy to help it complete the reproduction process of the first; the flower may provide a reward—namely, food, shelter, or breeding sites—or not. The system seems to require a stupendous number of connections. Although a mere thirty species of mammals facilitate plant pollination, nearly nine hundred bird species (including three hundred hummingbirds!) are pollinators. And the numbers of avian pollinators pale in comparison with the hundreds of thousands, perhaps untold millions, of insect species that get into the act. The exquisite, varied ways that plants allure and insects pollinate offer us some of the most amazing biographies of all.

THE GARDEN OF DELIGHTS

It will be a long, arduous day for a male *Campsoscolia ciliata*. This resolute wasp has been cruising the meadows of a sublime Aegean island for hours. Yet he has not found a mate; sex is indeed hard to come by. Only an hour ago he spied a dazzling female with shimmering blue wings, her robust abdomen with a red hairlike fringe not unlike his own, and her subtle but familiar and alluring perfume. As he began a tremulous, rhythmic contact with her, a larger competitor suddenly materialized and forced him off with dispatch. Now he is catching the scent of another female. Moments later he sees her, dangling from a delicate plant on the slope above a green sea. This

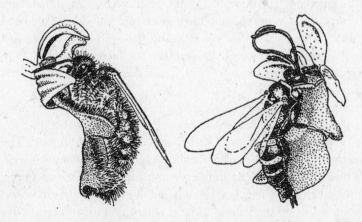

Male wasps visiting orchids: *Campsoscolia* on *Ophrys speculum* (left) and *Argogorytes* on *Ophrys insectifera*

time his visit is not interrupted. He plunges the tip of his abdomen into her long reddish fringe. Within seconds he moves on, but she remains.

All along the Mediterranean region, from the western edge of Portugal to the eastern margin of the Aegean Sea, in the soft warmth of March the sexual trials and triumphs of countless male *Campsoscolia ciliata* are repeated countless times. But such encounters, like any dangerous liaison, have an element of deception. The female wasps in this story—so sought after, so competed for, and eventually so acquiescent—are not what they seem. They are indeed not female wasps or wasps at all. The plant from which this putative female wasp dangles is no mere landing strip; it is an extension of the deceiver itself. The plant in question is *Ophrys speculum*, the mirror orchid. The common name is inspired by the violet mirrorlike sheen of the lip extending from the flower of this orchid, an object that mimics the shimmer of the blue wings of a female *Campsoscolia ciliata* as she alights. Perhaps even more intriguingly, the lip of the orchid is appointed with fringelike red hairs bearing a striking resemblance to those of the wasp visitor and the female of his species. The orchid complements this array of seduction with a scent that triggers a sense of recognition in the male wasp as belonging to the other sex of *Campsoscolia ciliata*.

Why this extraordinary elaboration of disguise? In the act of copulation, the male wasp may or may not have noticed that his antennae picked up two pollinia, two pollen-rich sacs that might easily fall off on his next visit and next copulation. The deception works. Studies have shown that more than 40 percent of *Ophrys speculum* in a given meadow is pollinated and fertilized in this manner. At the same time, the male wasp is ill served when it comes to deriving any direct biological benefits, such as nutrients, from *Ophyrs speculum*, which is so stingy with nectar that the male wasp collects nary a drop during his visit.

Of course the wasp species so energetically pursuant and so outrageously deceived must be perpetuated as well. How does this occur, given the investment of the male wasp in such nonproductive sexual activity? The answer relates to exquisite timing in the life cycles of both *Campsoscolia ciliata* and *Ophrys speculum*. The adult male wasps appear in the early spring, several weeks before the females, which spend most of their lives in underground burrows or on the ground surface. Later in the season productive copulation between male and female wasps ensues; in the meantime, the orchids are ready to exploit the males' ardor. Their success rate is all the more impressive when one considers that unlike many flower species, they are visited and pollinated by only one gender of one insect species. Females of the same wasp species completely ignore the plant. Moreover, the orchid faces little resource drain. Both the male and female wasps visit more generous flowers for nectar.

The mirror orchid does not have a monopoly on elaborate temptation and deception. The misnamed fly orchid, *Ophrys insectivora*, extends a lip topped by two "antennae" formed from two extra petals. This tempting flower then achieves an even more striking verisimilitude of its exclusive visitors, the solitary wasps *Argogorytes mystaceus* and *Argogorytes fagei*, than the mirror orchid does. The early purple orchid, *Orchis mascula*, studied intensively by Darwin, is also miserly with nectar but very free with pollen. Some orchids devoid of nectar merely mimic in shape and color flowers that normally provide nectar. These impostors vary in color and form, making it more difficult for insects to learn to avoid them. Orchids that attract butterflies and moths have, even to our nose, an unusually strong, sweet fragrance. Unlike the orchid deceivers, these species often provide copious amounts of nectar. Not all orchids, however, produce pleasant aromas. The early purple orchid actually has an unpleasant scent that obviously still tempts its insect visitors.

The orchids—with a membership of twenty thousand strong—in many cases epitomize extreme and bizarre approaches to attracting insect pollinators. But this group, though spectacularly diverse, makes up less than 10 percent of the angiosperms. Hundreds of thousands of other angiosperm species depend on insect pollination to sustain their generations. The mirror orchid displays a very special means, but not an apotheosis, to ensuring pollination by insects. Other flower groups offer strategies for attracting particular kinds of pollinators and at the same time for maximizing pollen pickup and transport. Snapdragons and delphiniums hide their nectar in crypts at the bottom of long, tubular petals from whose openings protrude anther-tipped stamens loaded with pollen. This form encourages butterflies, moths, and other insects equipped with long tongues to probe deeply and in the process pick up pollen on their hairy heads. Many species that resemble orchids with the marked enlargement of certain petals also incite "pseudosex": the flower forces the insect visitor to restrict its landing to the enlarged petal, or lip, and align its body to the axis of the petal in a way that offers the best chance for pollen gathering. There are myriad variations of this auspicious, asymmetrical architecture in flowers.

Flowers also lure pollinators with another form of deceit. Some plants offer a false promise of a brooding site, a spot where the visitor might be able to lay its eggs or mate with another of its species. In reality, the flower becomes a cuplike prison that entraps the visitor and holds it long enough to unload the pollen the insect has carried from another plant. In some cases the term of incarceration is short; the insect struggles for minutes, or even hours, but eventually finds its way out of the chamber, picking up the pollen in the process. In the orchid called European lady's slipper, *Cypripedium calceolus*, a small bee plummets down the "slipper" of the bright yellow lip, from which it then struggles frantically to escape. But the ensnared bee cannot leave the way it came; instead, it is guided by a translucent windowpane near the base of the lip to one of two narrow openings on either side of the stamen and stigma. As it makes its escape through this keyhole, the bee does the orchid's bidding. Sticky pollen from the stamen is smeared over the bee's midsection, or thorax. Even more extreme are flower species, such as the Dutchman's-pipe, *Aristolochia sipho*, that retain prisoners for more than a day. With some flowers captivity may last four or more days.

Not all plants that present themselves as brood sites for insects are deceivers. Some actually provide brooding sites. The visiting insects become willing, if temporary, prisoners, and the eggs they lay within the flower represent a critical step in their life cycles. Here the benefits of imprisonment are mutual rather than one-sided, as they allow reproduction to occur in both the plant and the insect visitor. The system in this case may be very simple. The pollinating beetles of the New World tropical palm, *Orbignya phalerata*, lay their eggs in the male flowers of the plant. These flowers drop forty-eight hours after opening, and the beetle eggs within them hatch, become feeding larvae, and emerge as beetles between twelve and fourteen days later. A number of palms as well as cycads provide such brood sites.

At the other end of the spectrum is the highly specialized brood-site mutualism between the edible fig, *Ficus carica*, and the fig wasp, *Blastophaga psenes*. This relationship is incredibly fine-tuned and elaborate, involving flower receptacles for different sexes, different growth stages, and different generations of wasps at different times of the year. Some receptacles, espe-

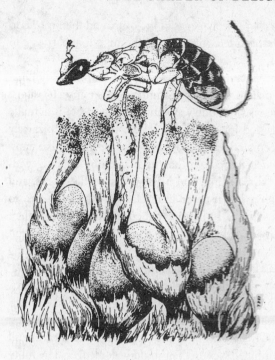

A female fig wasp
(*Ceratosolen*) laying
her eggs on a fig flower
(*Ficus*)

cially in winter, contain sterile female, or neuter, flowers; other receptacles contain male flowers. The early summer is ready for the blossoming of female, seed-producing flowers as well as more neuter flowers. The wasps lay their eggs in the flowers, but only those laid in the neuter flowers develop. The ovary of the neuter flower is incapable of producing a seed but is built to receive a wasp egg. Fig wasps are a type of gall wasp, and the name gall wasp refers to an ingenious structure critical to this complex cycle of reproduction: when the female wasp lays an egg, it injects a drop of a special secretion into the ovary of a neuter flower, and this stimulates a remodeling of the sterile ovule into a spherical gall, which provides food for the developing wasp larvae.

In addition to figs, other plants have established complex mutualistic relationships with insects. Yuccas are a prime example. Here the host plant not only depends on the yucca moth for pollination and fertilization but also gives food to the moth's larvae, which develop within the flower. As usual, the moth collects pollen from one plant and brings it to the flower of another plant. There it first lays its own eggs in the ovary and then moves to the stigma to thrust the transported pollen down the tube that carries it to the ovum of the plant.

There is yet one more variation on the theme of entrapment: pollination in the carnivorous plants, which also imprison their insect visitors. However, in this case the visit is a one-way journey; the insect is actually killed and eventually digested for plant food. These forms have bizarrely modified, often highly decorated leaves that form vase- or pitcherlike traps. The Venus flytrap, *Dionaea muscipula*, has its marvelous opening with, toothlike fringes. It looks threatening, and it is, at least to flies, bees, and other insects small enough to be ensnared in its closing "jaws." It is easy to understand why the Venus flytrap, sometimes called the most famous plant in the world, has crossed over so effectively into pop culture, the inspiration for nightmarish but totally unsubstantiated scenarios involving man-eating plants in such offerings as *The Little Shop of Horrors*. There are indeed a variety of carnivorous plants, and they apprehend and eat in different ways. Sundews (*Drosera*) catch insects on sticky hairs on the tentacle-like leaves. Butterworts (*Pinguicula*) apprehend insects on slimy flypaper-like leaves. Bladderworts (species of *Utricularia*) have showy flowers, but they engulf their tiny aquatic prey in inflated sacs on their underwater leaves. The tubular leaves of pitcher plants (*Sarracenia*) form pitchers with lines, colors, and markings that help lure insects; the potential victim is also enticed by nectar secreted around the mouth of the pitcher. After the fall, downward-directed hairs keep insects from crawling out.

This variety in intricate devices for attracting and eating prey leads to the obvious question: How do the carnivorous plants manage to avoid killing and consuming the insect visitors they require for pollination? In some cases the solution is laborious. Although some pitcher plants devour a variety of mainly small insects, one group of their visitors, the bumblebees, are robust enough to chew their way out of the pitcher wall, carrying away with them pollen they may have picked up earlier from the flower of the plant. In most other cases the solution is a simple matter of timing or plant architecture. The flowers are tall and protruding while the pitcher is short; only insect visitors more curious and exploratory are likely to be devoured, while the more casual visitors provide for pollination. Alternatively, the pitchers are closed during a time or season when the flowers are open and ready for pollination.

Beyond these special devices for luring, entrapping, and eating visiting insects, the color and form of certain groups of flowers make them particularly suited for pollination by particular groups of insects. Bee-pollinated flowers, such as monkshoods, dead nettles, louseworts, and many legumes, are usually brightly colored, favoring yellow, blue, or purple, colors that bees are highly sensitive to, and they are often large, meaning they can accommodate such robust wayfarers as bumblebees. Wasp-pollinated flowers, such as the figwort and the wild cotoneaster, are typically dingy and brown, of short open tubular forms, with easily accessible nectar but meager amounts of pollen (brood site plants are an exceptional case). Flowers pollinated by day-flying butterflies and moths, such as the red valerian, are often brilliant pink or red (less often blue or purple) with a sweet, subtle scent. Flowers pollinated by night-flying moths, such as the honeysuckle, are typically paler and emit a powerful, sweet scent as evening comes. Like their daytime counterparts, they have long tubes, abundant nectar, and assertive pollen-carrying organs.

The flower adaptations I have described thus far all seem aimed at attracting and utilizing particular kinds of insects. It is tempting to claim that strategies such as those in the mirror orchid or the edible fig—intricate and exclusive pairings between insect and plant taken for evolutionary advantage—represent a culmination. But we cannot depict evolution in flowers, as in other organisms, as a one-way march to the best of the best. Many solutions for pollination work very well. Perhaps the flexibility of the system, arrived at through thousands of generations of adaptive experimentation, is a key to the explosive evolutionary success of angiosperms. Similar strategies for attracting pollinators occur in distantly related flower groups. For example, we find enlarged "landing petals" in both orchids and very distantly related members of the buttercup group. In addition, plants that offer false brooding sites have evolved in very different plant groups. Other examples of such evolutionary convergence are numerous: structures and scents that stimulate false copulation, tubular trumpets for long-tongued insects, red flowers, violet flowers, white flowers, and so forth. Similar flower architecture evolved independently in many lineages.

A second argument against equating the best with the most exclusionary is that many flowers have done quite well without picking and choosing insect pollinators at all. In other words, many species are less fastidious about their pollinators and more generous with them. The composites of the family Asteraceae, as well as the dogwoods, hydrangeas, guelder rose, and saxifrage, have showy clusters of simple white flowers that beckon many and varied visitors. Most extreme in this mode is the carrot family, the umbellifers, such as the familiar Queen Anne's Lace, whose flat-topped radiating flower clusters, or umbels, line many a roadside in Europe and North America. The broad surfaces of the umbels offer ample landing space for insects. Moreover, these flowers, unlike many orchids, have loads of nectar. They are gregarious; they bunch together in chromatic galaxies, maximizing the probability that an insect visitor hopping on one flower is likely to visit another a few seconds or minutes later. Meadows enriched with these flowers become shopping malls for many different insect pollinators. More than three hundred insect species visit carrot flowers, a far cry from the mirror orchid's exclusionary seduction of one species of wasp. In this case the availability of flowers of similar dalliance, nectar, and pollen is the key to reproductive success. This strategy calls for a lot of flowers that look the same, rather than fewer flowers that look distinctly different. Small wonder that species of composites are among the most difficult to identify and distinguish, even to a botanist's trained eye.

One cannot imagine more wildly different approaches to procreation than that between certain composites and certain orchids. Yet both strategies seem to work just fine. Composites thrive and dominate in many habitats. In contrast, orchids and other selective flowers often hang out in smaller numbers, sometimes in marginal or restricted habitats, but they also sustain themselves across broad areas, such as the Mediterranean, western Australia, northern Europe, and the tropics of the Old and New World. If diversity is any index of success, it has been clearly demonstrated that there are antithetical means of achieving that success. There are twenty-five thousand species of Asteraceae, or composites, most of which employ the shopping mall strategy for attracting diverse pollinators; this surpasses the twenty thousand species of orchids, with their highly exclusionary, penurious approach to pollinators.

Flowers need insects and have evolved countless fascinating ways to employ them. But insects are of course not simply conscripted passively. They themselves have evolved in ways that have enhanced their role as pollinators and plant feeders. Clearly, these enhancements require special equipment; not all insects are proficient pollinators. There are many groups whose members visit flowers, feed on nectar or pollen, or occasionally transfer pollen, yet are insignificant pollinators. The diverse and intense pollinator behavior that keeps energy flowing, and evolution going, in flowery ecosystems is the mark of only certain insect groups, including members of the stone flies (Plecoptera), bugs (Hemiptera), and scorpion flies (Mecoptera). The groups most committed to pollination are, in very rough order of increasing specialization, the tiny thrips (Thysanoptera), the beetles (Coleoptera), the flies (Diptera), the butterflies and moths (Lepidoptera), and the ants, wasps, and bees (Hymenoptera).

All insect pollinators, whether generalized thrips or baroquely specialized bees, share basic tools necessary to do the job. They have numerous hairs that not only are organs of smell, taste, and touch but collect pollen the way Velcro attracts lint. Their compound eyes produce very fine retinal images, and the multiple facets of these eyes are good for picking up movement and shape. Many insect pollinators have color vision, but most of them, unlike us, are sensitive to the ultraviolet of the color spectrum yet cannot see red; some scarab beetles, most butterflies, and day-flying moths can see red and are accordingly responsible for the flourish of many species of red flowers (though not all, as these are also the targets of hummingbirds and other feathery fliers).

Beyond this rather simplistic description of pollinator skills, much variation abounds. Some insect groups can be at least partially typecast according to their equipment and behavior. The most primitive insects, the collembolans and springtails, are wingless forms that feed on pollen. Although they may have been among the earliest animals to visit flowers, their importance to pollen transfer eons ago was probably insignificant, and it still is today. The primitive winged insects, thrips, or Thysanoptera, have color vision, a keen sense of smell for flowers, and many bristles on the body that offer good attachments for pollen. These tiny insects are usually only around two millimeters long, with narrow bodies and four wings, each with a central shaft and very thin hairs extending fore and aft. The quartet of airfoils has an odd appearance, something like four spindly "feathers" affixed at their junction, on a naked little creature all of whose other "feathers" have fallen off. Thrips are sometimes called thunder flies, as they are annoyances in warm, sultry weather, invading houses and managing to wedge themselves between windowpanes. Their proclivity for pollination is combined with voracious plant feeding en masse, and they are infamous crop pests.

The next group of insect pollinators, the beetles, represent a giant leap forward in refinement. This, the largest group of insects, includes many species adapted for pollination. They are not very important for this job in cold, temperate climates but much more so in arid regions and in the moist tropics. They are instrumental in the pollination of certain primitive flowering plants as well as some gymnosperms, such as cycads. Nonetheless, beetles tend to be less frenetically active as pollinators than flies, wasps, or bees. They are in the main "lazy visitors" to flowers, depending for protection on their hard body coverings and repellent secretions rather than flight. Accordingly, beetles tend to linger, taking their sweet time in foraging for pollen or nectar on a given flower. Beetles are also known as mess and soil pollinators because as they scrounge for pollen and other food in flowers, they pick up pollen on their bodies and transport it to other plants. They rely on smell rather than sight to locate and land on flowers, and they are thus particularly attracted to flowers producing strong odors. Not all these aromas are pleasant to us. Dung-frequenting beetles are drawn by the putrid smell of *Arum nigrum*, a plant relative of the famous lords-and-ladies. Also attractive to beetles is the smell of rotting, fermenting fruit in plants, such as *Calycanthus*.

Another major order of insects, the Diptera, the true flies, if anything show an even greater elaboration of adaptations for pollination than beetles. These are very common visitors to plants that produce carrion scents and force prolonged incarceration in false brooding sites. Many flies, like wasps and bees, have exclusive relationships with certain species of flowers. Others are real generalists. Studies reveal that one species of fly may visit more than twenty species of plants.

The function of flies as pollinators is closely related to their anatomical equipment. The fly is distinguished for the extreme development of its proboscis, a long projection at the tip of the head that is a multitasking organ: sucking instrument, tongue, and, importantly, pollen attachment device. It is not a nose, at least functionally; the organs of scent in butterflies, moths, and for that matter flies are in the antennae. One type of sucking apparatus well-known to us is present in the female mosquito, which can pierce the skin of animals and suck out the nutrients it needs. The other type of sucking machine does not pierce but is limited to mopping up liquids already on a surface. This is the prevalent mode for fly pollinators of flowers; invariably flies that feed on exposed fluids can also eat small particles, including pollen grains.

When it comes to development of apparatus for sucking nectar, the diverse and colorful Lepidoptera, the butterflies and moths, rival and even surpass the flies. Some dipterans, such as the meganosed fly (*Moegistorhynchus longirostris*), have proboscises four inches long, five times the length of their bodies; even greater extremes of Pinocchio-like accoutrement are attained in lepidopterans. In the long-"nosed" lepidopteran, pollination occurs as a side effect. As it probes its sucking proboscis into the tube of the flower for nectar, the visitor makes an involuntary pickup of the much more accessible pollen on its proboscis or bristly head. The tip of the proboscis is often armed as well with fine spines, and it becomes a piercing organ par excellence. It may, for example, pierce the tissue of a nectarless flower and thereby release nutrient sap. Various lepidopterans also use this implement to pierce the skin of fruit, becoming major crop pests.

The proboscis in the lepidopteran is almost always coiled when not in use because of its inconvenient length during other activities, such as flight. It is especially suited for taking in nectar at the bottom of trumpetlike flowers. There seems to be a coordinate relationship between the length of the flower trumpet, or tube, and the length of the proboscis of the feeder, and this doubtless mirrors a trend in evolutionary syncopation. Longer tubes in flowers influenced the emergence of adaptations in lepidopterans for longer proboscises. Then flower tubes evolve to become even longer to safe-

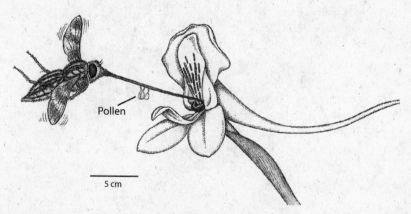

Pollen

5 cm

The meganosed fly, *Moegistorhynchus longirostris*, feeding from an orchid, *Disa draconis*

guard some of their nectar or at least make it more of a challenge to attain. So lepidopterans develop even longer proboscises, and the apparatus escalation continues.

This coevolutionary one-upmanship has been carried to absurd extremes. Darwin observed that some species of Madagascar orchid have a

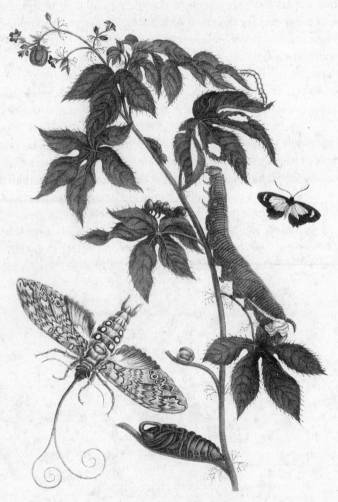

Moths, butterflies, and plants—the exquisite artwork of Maria Sibylla Merian, 1705, *Metamorphosis Insectorum Surinamensium*

flower tube of some thirty-two to forty centimeters (thirteen to seventeen inches), but he knew of no potential pollinator adapted for feeding from this deep portal. However, at the turn of the twentieth century a Madagascar hawk moth (*Xanthopan morgani*) was discovered with a proboscis an impressive twenty-five centimeters (ten inches) in length!

Many of the flowers that attract lepidopterans are blue or deep pink; some, in contrast with those that attract other insects, are brilliant red or scarlet. Their scent is often sweet and heavy: honeysuckle, hyacinth, lilac, wallflower, and carnation. Other familiar flowers that beckon lepidopterans include valerian, ragwort, goldenrod, thistle, wild thyme, and raspberry. Of course, nocturnal moths are attracted to night-blooming flowers. These are mainly pale, distinctly decorated with dark markings and, like their daytime counterparts, highly aromatic.

When it comes to sensory systems in lepidopterans, versatility is the key. These insects have a great spectral range of vision from ultraviolet to red; although the resolving power of the eye is low, the temporal resolution is high. Some butterflies are able to process 150 images per second compared with the measly 40 images per second in humans. Lepidopterans, like flies, also have a very keen sense of smell. Well equipped with sensory organs and a sucking apparatus, the lepidopterans are both superb flower feeders and flower pollinators, not to mention that their presence in fields and forest adds a brilliance of color and beauty that rivals the display of flowers they feed on.

With all the diversity and specializations for flower feeding and pollination exhibited by beetles, flies, butterflies, and moths, one would think that flowering plants would hardly require, or could hardly accommodate, yet another major group of insects. Nonetheless, we have yet to consider the most specialized and spectacular pollinators of all, the hymenopterans. This hugely diverse group includes the primitive sawflies as well as the more advanced ants, wasps, and bees. The lifestyles of different hymenopterans are as varied as their adaptations for pollination.

Except for the sawflies, all members of the Hymenoptera belong to the group Apocrita, which is in turn divided into two groups, those with stings and those without. The latter are mostly parasitic wasps, such as the ichneumon wasps, which are parasitic in the larval stage and are often found on the larvae of other insects. Some Australian orchids tempt ichneumon wasps into pseudocopulation. Other parasitic wasps include the industrious gall wasps, whose life cycles are exquisitely syncopated with those of their host plant species, such as the edible fig.

True wasps, along with ants and bees, are the stinging insects in the group Aculeata. Most true wasps capture insects and spiders, sting and mutilate their prey, store them in prisonlike cells, and feed them to their larvae. Sometimes these predatory wasps as adults feed on the juices of their victims, but for the most part they, like parasitic wasps, sustain themselves on nectar, sap, and honeydew. True wasps are then prominent flower visitors and pollinators. They include the nonsocial wasp *Campsoscolia*, famously duped into pseudosex by the mirror orchids. But wasps show many various habits and preferences, being common visitors to flowers of bramble, thistle, and thornberry. The social wasps, most prominent in the family Vespidae, require flowers that are productive in nectar (wasps do not typically feed on pollen), which they collect and transport to the expectant queen to feed herself and her larvae. Some species are adept at using their mouthparts as a boring tool in order to get at nectar in flowers whose narrow crowns keep the insect from thrusting in its oversize head. Vespids are common pollinators of many orchids and also of cotoneaster, thornberry, gooseberry, buckthorn, figwort, and others.

Another group of social hymenopterans, the ants, is less important as pollinators and more important as great gluttons for nectars. Since ants are wingless, they must crawl from one plant to another, so their role in cross-pollination is limited. One category of flowers highly adapted to pollination by ants is, expectedly, low-growing plants with small, inconspicuous flowers attached closely to the stem. These include a number of small species in temperate meadows of northern Europe, the drier Mediterranean as well as the southwestern deserts of the United States. Otherwise ants are mainly a negative factor in plant sustainability, and many plants have developed defenses against voracious ant feeding, including barriers between the ground and the source of nectar.

But for all the spectacular pollination capabilities exhibited by Hymenoptera, there is one subgroup, the bees, of the family Apidae, that stands out. Bees have an enormous, unchallenged impact as plant pollinators, both in the wild and in agriculture and cultivation. Bees pollinate one or more cultivars of 66 percent of the world's fifteen hundred crop species and are directly or indirectly responsible for 15 to 30 percent of food pro-

Artist's rendering of
the earliest fossil bee,
Cretotrigona

duction. The most important managed pollinator, the honeybee *Apis mellifera*, has an estimated value of five to fourteen billion dollars per year in the United States alone. The decline in the past decade of beekeeping in the United States due to insecticides, disease, land loss, and other human impacts is a serious problem.

The special status of bees as stupendous, economically powerful pollinators relates to a peculiar feature of their life cycle. Unlike other insect larvae, bee larvae depend on copious quantities of flower food brought up to them or stored for them in the nest. As with true wasps, female adults collect the larval food, and they also accomplish a bit of chemistry that is significant to the economies of both natural ecosystems and human beings. In the bee's crop, the nectar's sucrose is converted to a balanced combination of glucose and fructose, better known as honey. To deliver the goods, the industrious female bees of the colony must undertake massive, efficient food gathering. As a result, bees have unusual skills in detecting and discriminating among flowers, foraging, collecting, and navigating.

Some high-tuned equipment makes this all possible. Bees can pick up scents ten to one hundred times fainter than those perceived by humans.

They can also have acute vision, including a sensitivity to ultraviolet even greater than that of many other insects. Thus, where we see two yellow flowers, a bee may see one yellow flower and one flower that is "bee purple," a mixture of light from the opposite ends of the bee-visible spectrum—namely, yellow and ultraviolet. Their collecting machinery is also extreme in refinement. Experiments clearly show that bees can navigate by the sun, allowing for its position and the time of day. This acute set of sensory refinements is accompanied by optimal devices for food gathering. The tongue is even more bristly than that of other hymenopterans. Fragrance, nectar, and pollen can be picked up on the brushes of hairs on the forelegs and then stored in the lower segments, or tibiae, of the hind legs, which are unusually expanded and partitioned into chambers in order to retain liquid. There is a distinctive array of inwardly curving hairs at the base of the abdomen, likewise superb pollen brushes.

Most remarkable of all is that many bees (but not honeybees) collect pollen by vibrating their wings, a process often inaccurately called buzz pollination. This feat is accomplished notably by bumblebees. During flight they use a series of muscles to bend the thorax so as to propel the wings. However, when the bee is alighted and "warming up" or communicating with other bees, the wings are uncoupled from the flight mechanism and hum or buzz at low speed, just like a Cessna taxiing down a runway before it goes full throttle. This induces nearby anthers to resonate in a rhythm that dislodges pollen grains, which stream out of the flower toward the visitor's bristly legs and body.

Added skills are amazing powers of communication, a behavior intensely focused on the goal of efficient food procurement not just for an individual bee but for the entire colony. Some bees will leave a cuing scent on a flower to attract other members of the colony. Social bees in tropical regions, such as the small stingless bees and the sweat bees, are known to transfer information about food—namely, its distance from the colony—by the length of pulses of sound made with the wings. But these various mechanisms of communication pale in comparison with the behavior of the familiar honeybee, which transfers information, a "treasure map" pinpointing productive flowers, to other members of the colony through an intricate dance carried out in the hive. Whereas the nature of the food is transmitted by the scent brought back by the female forager, the direction, distance, and abundance of the food all seem to be indicated by the dance. The first observations of this behavior, made by the famous biologist Karl von Frisch, spawned a whole new area of study in animal communication. Doubts about Frisch's results raised in the 1960s spawned controversy, but many more experiments validated his results. The bee dance is carried out in the darkness on the vertical combs of the hive. The movements of the dancing bee are tracked by a parade of others who bump and jostle her. As she dances, she vibrates her wings, emitting a sound that also seems to cue information, notably the distance of the food source.

In short, bees are virtuosos among virtuosos when it comes to flower feeding and pollination. Keen smell, vision, amazing navigation and time-keeping skills, extremely effective food-gathering devices, avid communication, not to mention social cooperation, add up to a truly remarkable system. Bees are, more than incidentally, the last major group of insect pollinators to appear in the fossil record. This symmetry between extreme specialization and evolutionary youth presents, as we shall see, some fascinating issues.

TOWARD A NEW ECOSYSTEM

Not all the famous sites that preserve important fossils from the Cretaceous, when the modern ecosystem was born, are in classic dinosaur territory of canyon lands and empty deserts. One critically informative site has rather prosaic origins in the middle of what is now urban sprawl. A clay pit in Sayreville, New Jersey, surrounded by housing tracts, railroads, and highways, has produced a remarkable picture of a Cretaceous ecosystem. The pit was mined for gray clay that was baked in a nearby processing plant in one of America's most industrial and polluted epicenters, Perth Amboy, to make road and landscaping material. The source of this fire clay is a layer of sedimentary rock known as the Raritan Formation. The Sayreville pit and a few other outcrops in New Jersey of the Raritan Formation have produced a bounty of ancient life estimated to be (in the cumbersome vernacular) Middle-Late Cretaceous in age. Stems, leaves, twigs, flowers, and other fossilized plant parts reveal a wondrous array of botanica, including bryophytes, pines, and diverse angiosperms — capers, heather, laurels, magnolias, plane trees, roses, sweetgum, witch hazel, and the first recorded fossils of the grass family. Several exquisitely preserved flowers are related to modern families closely associated with bees and other avid pollinators.

Conveniently, Sayreville and other Cretaceous sites in New Jersey also preserve insects superbly. Fossilization in this case resulted from a remarkable process. Ancient insects were engulfed in the resin of ancient conifers. The fossilized form of resin is amber, the dull gold translucent-to-transparent substance of great appeal to artisans, jewelers, and paleontologists. Resin, unlike sap, contains minimal amounts of water, a compound that is vital to living organisms but destructive to dead and buried ones. These amber-encased insects are preserved in astounding detail, down to their appendages, eyes, antennae, genitalia, and bristles. The insect fauna includes ants, ticks, thrips, praying mantids, and, perhaps most important, the earliest-known fossil bee, *Cretotrigona*. This is the only such record of a bee for the Mesozoic. Preservation of *Cretotrigona* is so good that the specimen can be identified as a female worker bee.

There is a certain bitter irony to the urban sprawl engulfing Sayreville and other New Jersey sites. In a feeble gesture to arrest this human progress, the southern part of the Sayreville pit, with its small "lake," has been converted into the John F. Kennedy Park. This is not the most beauteous of America's public lands. It abuts an active mining operation, is stripped of topsoil, and reveals itself as a barren set of eroded gullies sprinkled with scrub oaks and pitch pines. The deposits of amber-bearing lignite were discovered in Kennedy Park just about the time *Jurassic Park* hit the movie theaters. This gave rise to a chaos of prospector interest, a sort of "amber rush." Eager stakeholders dug room-size holes or, in the case of commercial operations, even larger cavities. Intense competition and even a little violence ensued, and there was no authority to manage the whole mess. Frustrated local residents on occasion conscripted bulldozers to cover the holes, or the rains simply flooded them. Eventually new houses buried the most produc-

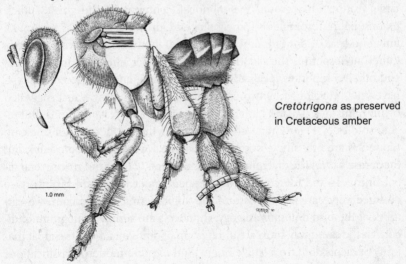

Cretotrigona as preserved in Cretaceous amber

1.0 mm

tive excavations. What a shame! A little forethought might have saved these precious localities for science and the world.

To understand the history behind the rise of flowers and of bees like *Cretotrigona*, we must consider fossils from rocks older than the clays of Sayreville, New Jersey. By the Early Cretaceous, 120 to 110 million years ago, several magnolia-like taxa had been established. In addition, flowers and dispersed pollen appear abundantly in the Early Cretaceous, and they have remained essentially unchanged for more than 100 million years. These resemble the living genus *Hedyosmum*, a group of small trees or shrubs with pointed, serrated leaves and simple flowers known from many species in both the Old and New Worlds.

By the mid-Cretaceous, angiosperm fossils for the first time show a well-differentiated floral envelope of sepals and petals forming a calyx and corolla, respectively. The first records of flowers with an ovary contained within fused carpels and with nectar chambers are of similar antiquity. The evolution of the petals and the fusion of the carpels were major steps in the evolution of angiosperms and may have promoted further enrichment and radiation in the Late Cretaceous, between 80 and 65 million years ago. The great diversity of angiosperms traditionally has been divided into two very large groups. One group, the monocots, comprising the grasses, lilies, onions, palms, bromeliads (pineapples and kin), and irises, has seeds with only one leaf, or cotyledon, which provides food for the growing embryo. Other angiosperms, the dicots—magnolias, water lilies, oak trees, daisies and other composites, peas, roses, birches, cacti, and diverse company—have seeds enveloped by two cotyledons. This division was first suggested by John Ray in 1703 and has endured in many plant classifications. But recent studies by Peter Crane and others show the cut here is not so clean: monocots are of a single origin (the technical word is *monophyletic*), but they arose *within* the diverging lineages of dicots. This is because several dicot lineages—the magnolias, vine peppers, water lilies, and birthworts—diverged very early at the base of the angiosperms. Once again, we can think of the bird-dinosaur analogy: monocots are simply a subgroup of dicots, in the same way birds are a subgroup of dinosaurs. Still, the great majority of dicots do have a single origin, and in more modern classifications these are called eudicots.

To a large extent, the Late Cretaceous flourish of angiosperms also may have been linked to the evolution of specializations that encourage the pickup and dispersal of pollen. For example, many of these ancient flowers had well-developed nectaries, or nectar chambers. There is also clear evidence that fruits and seeds increased in size, a trend that was even more noticeable in flowering plants after the dinosaur age. By the end of the period many of the genera and families of angiosperms that are living today had already appeared. Thus the roots of the modern ecosystems of today were well established in the late age of the dinosaurs.

Of course this system was fueled and cultivated by the marvelous motive' of sexual reproduction depending on pollination. So was it pollination that was the key feature of the angiosperm ancestor, leading to the great success of this group? The answer is well, yes, maybe. There are complicating factors. Pollination via insects or other animals is found in most but not all angiosperms. Some members of the group, particularly the grasses, are wind-pollinated forms. In fact, we don't yet know which came first for them: wind-pollinated or animal-pollinated flowers. The grasses are highly specialized and appeared much later in geologic time, so it is thought that the angiosperm dependence on lofting air is a secondary trait. Yet some wind-pollinated species seem to be very early, like "primitive" branches, such as the Piperaceae, the "true peppers" (not to be confused with the sweet peppers or chilies, which are the fruits of the Solanaceae). Moreover, wind pollination is a primitive, widespread mode for plant reproduction outside of angiosperms. We have seen and noted that in conifers and other gymnosperms. On the other hand, there are also primitive, insect-pollinated flowers with very showy but rather simple forms, such as water lilies, buttercups, and magnolias. Most botanists favor the idea that insect pollination, not wind pollination, was the original enabler of reproduction in angiosperms and that it marked the divergence of this splendiferous group from other plants.

What about the insect side of this equation? How did insects evolve into proficient pollinators? Remarkably, the fossil record itself is good enough to track the trend to an ever more finely tuned relationship between pollinator and pollen provider. Plant pollinators were first myriapods and primitive insects feeding on spores (sporivory), soon after the terrestrial flora took root. Next came a transition to pollen feeding (pollinivory) and, finally, to insect pollination. These events were contemporary with the origin of seeds and involved other transformations, about which we must ask several questions. What would attract a potential pollinator to both ovule-bearing and pollen-bearing organs? Pollen is insect food (or at least larval food) and an obvious reward, but what would attract the insect to the ovule? The possibilities include pollen droplets, sticky secretions that are preserved in the seeds of some plants; secretions of associated glands; and nutritious ovular tissue, as in certain cycads today, which feeds beetle larvae.

The very first insect pollinators of appreciable impact were indeed probably beetles. Evolutionarily, the insect pollination so characteristic of angiosperms was anticipated by millions of years by the cycads and the extinct bennettitaleans. Beetles appeared and diversified early in the fossil record, but the most specialized pollinators, wasps and bees, came later; the first records of them, along with angiosperms, are well into the Cretaceous. Thus the fossil record shows that pollination by insects long preceded the radiation of diverse flowering plants as well as the advent of the real virtuosos at pollination, the wasps and bees.

This asynchrony in timing has stimulated one of the most energetic investigations, and heated debates, in paleobotany. Were the refinements of insect pollination in lockstep with the glorious rise and flourish of the angiosperms? Or did animal pollination, even specialized modes, long precede it? The latter hypothesis, firmly preferred by some, suggests that the extraordinary diversification of flowering plants had little to do with the evolution of pollinating and plant-feeding adaptations in insects. Others disagree. First, they claim that the taxonomic comparisons said to support the asynchrony are too coarse, dealing with the relevant organisms at broad taxonomic levels—families and above—rather than their component genera and species. Yet we know that even family groups of insects and plants show an extreme diversity of modes. Second, the theory underemphasizes the later fossil origin of highly specialized pollinators, such as the bees, which could have been a major stimulus for the Mid- to Late Cretaceous radiation of angiosperms coincident with the bee record. Finally, claims that certain pre-Cretaceous fossils were flower-loving (anthophilous) insects have been called into question.

The weight of current evidence seems to favor the intimate evolutionary syncopation in both pollinators and pollinated. Phase 1, which probably began at or soon after the Jurassic-Cretaceous boundary, saw the arrival of simple flower groups such as magnolias and laurels, whose pollination was promoted by insect generalists—beetles, flies, short-tongued moths, and less specialized wasps. These insect groups may have been important pollinators of cycads and bennettitaleans earlier in the Mesozoic. Phase II in the later Cretaceous, about 90 million years ago, witnessed a wondrous array of flowering, with more heathers, roses, laurels, and liquid amber, as well as a number of extinct groups such as Ericales, that were clearly adapted for specialized insect pollination. This is also the phase that shows the first evidence of a radiation of specialized pollinating insects, including the earliest bee fossils.

Through the remaining 25 million years of the Cretaceous, both flowering plants and specialized insect pollinators dramatically diversified. Angiosperm dazzlers—mimics, seducers, trappers, and devourers—were attracting all manner of six-legged visitors, from rudimentary thrips and slow-moving beetles to long-tongued butterflies and dancing bees. What began as a quiet revolution among a few pioneering flowers in a pine forest was now in full swing.

What accounts for this marvelous makeover? When it comes to playing the evolutionary game, the advantages of insect (and for that matter vertebrate) pollination are clear. In conifers, which rely on wind for pollination, a great deal of energy and resource goes into producing cones, which are shed often without any guarantee of fertilization. And major obstacles confront the heroic windborne pollen: drought, flooding, or simply an ill wind that blows in the wrong direction, away from trees awaiting fertilization. Animal pollination is a solution to some of these problems. There is, for one thing, the matter of precision. An insect-pollinated plant can afford to produce a quite small amount of pollen to be targeted for another appropriate plant, since its mates will be found even in cases where populations are small and scattered. This strategy enhances the survival of species in fragile, isolated habitats and also their diversification in many different ones. Another advantage comes with efficiency; less energy is required to produce the smaller amounts of pollen needed for procreation. Even forms like the composites and the carrots, which, as noted, are pollen- and nectar-rich targets for diverse insect pollinators, do not have to produce pollen in the large amounts known among pines and other plants that actually cast their fate to the wind. In many other flower groups too, pollen and nectar are offered in miserly amounts, though even tiny drops of nectar can release an alluring aroma, the source of many coveted perfumes, a tiny aliquot of sweet essence that is enough to draw insects in abundance. Finally, there are those orchids that seduce the alighting visitor to false copulation, transfer its sticky pollen to its alien "mate," and provide nary a drop of nectar.

Before we fully proclaim the advantages of insect pollination, we should once again return to a repeated adage here: life is not simple. Some environments favor wind pollination, a mode found in ancient conifers and other gymnosperms. Wind may transfer pollen more efficiently than insects in temperate forests or across steppes or savannas, where insect populations are small, scattered, and less available to be conscripted as pollinators. Likewise, species-poor habitats such as salt marshes, where insects are sparse, have many wind-pollinated plants. On the other hand, deserts harbor plants that mainly rely on insect pollination; they burst into flower only after heavy rains, which also stimulate intense insect emergence and activity. Water and warmth are two critical factors. In tropical rainforests, the great sinks of terrestrial biodiversity, warm rains abide, climates are not extreme, insects are diverse and abundant, and plant biomass is fed by great nutrient resource. Here insect pollination has undeniable pluses and is the overwhelmingly dominant mechanism for plant reproduction.

Notwithstanding the advantages of wind pollination to certain plants in certain habitats, we can appreciate that the intimate interconnection between flowers and insects is the primary factor in the great matrix of the modern land ecosystem. Now we come full circle back to contemplating the origin and evolution of this extraordinary act of cooperation between two different kingdoms of life. And we also return to a recurrent theme. Changes occurring during the hundreds of millions of years of plant and insect evolution involve refinements in fertilization and reproduction — namely, attempts at better sex.

You will recall that the first evidence of terrestrial life comes in the form of tiny spores protected in a hard casing for transfer on dry land. A further refinement came later: pollen carried in a protective coating that was transferred through water to make seeds, as in the case of cycads. The next step was the protection of pollen for windborne transport from male cones on a plant to fertilize the seeds in protected ovules in the female cones of the same or another plant. When angiosperms arrived, they neither threw their pollen to the wind nor dropped it to the ground or in the water. Instead, they relied on animals, mostly insects, to pick up, transport, and drop it off. Inducing this laborious assistance from animals required temptations and, usually, rewards, and an infinitely varied palette of color, a complex bouquet of aroma, sweet nectar, and endless ingenuity in flower architecture, in modes of attraction and deception. As for the insect visitors, the system selected for equally spectacular innovations, including elongated probes, acute color vision, hairy appendages for pollen collection, and, in the case of the remarkable bees, the complex language of the dance. This system brings together all the ingenuity that can be mustered by two of the most dominant groups of organisms on Earth. It is difficult to imagine a more radical evolutionary development, a more extraordinary retooling of reproductive strategy, and a more wholesale resculpturing of life on land.

12

DINOSAUR CAMELOT

In 2004, the fifteenth year of the American Museum's paleontological expeditions to the Gobi Desert of Mongolia, we took the new but only partially completed paved road south of the capital city, Ulan Baatar. Our track paralleled the historic Trans-Siberian Railroad, a vital link between sleepy, sparsely populated Mongolia and booming Beijing. After two long days and about three hundred miles we reached the dusty factory town of Saynshand and made a hard right turn into the moonscape west of the tracks. We then found ourselves in Cretaceous rocks. Some hours into this drive we encountered something unusual: people. No, not the rare nomadic family with camels, snarling dogs, a tentlike gur, and a rusty, undersize Korean motorbike. From a distance we could see workmen clustered around heavy trucks like small battalions of ants around sugar crystals. As we came closer, we noticed that one of the nearest groups of men was furiously digging away in the merciless heat in a deep trough that looked like a giant grave. I asked my Mongolian colleague Dashzeveg what was up, and he replied with his customary terseness, "Chinese and Mongolians . . . digging out . . . trees, fossil trees."

"Why?"

"To sell, in China."

This new summer ritual, doubtless at odds with Mongolian law (unlike the United States, Mongolia has strict rules against the exportation of fossils, as do many other countries), is apparently big business. We stopped at a deserted ditch where one gigantic log was still embraced by the rock slope, lying in stately repose. It reminded me of the huge, horizontal, and partially chiseled granite obelisks I once had admired in the ancient Egyptian quarries of Aswan along the Nile. Like many fossil trees, its bark had been replaced by a suite of minerals. These had been big trees, and although evidence of other plant life was scarce to nonexistent, we knew that these were the remaining temple columns of a lost world, an enriched Cretaceous cornucopia filled with trees, dinosaurs, secretive mammals, brilliant flowers, and pollinating bees. Thirty miles to the west we would soon be encamped in another layer of Cretaceous rock, one more generous with fossil dinosaurs and mammals than fossil plants. We would go about our seasonal ritual, finding, digging, hauling out, and eventually understanding more about this remote Cretaceous ecosystem.

Mongolia preserves some of the most complete and diverse Cretaceous dinosaur and mammal sites in the world. This was first disclosed by expeditions in the 1920s sponsored by the American Museum of Natural History. A team led by the colorful, gun-toting Roy Chapman Andrews came upon a rich badland of red sandstones. After one brief afternoon and a few bones in 1922, the team returned to these beds in 1923 to reap the riches of the Gobi. Included among their finds was the first nest of dinosaur eggs known to science, a discovery that rocked the popular as well as scientific world. Andrews named this sandy rampart, which glows brilliantly red in the afternoon sun, the Flaming Cliffs. The place instantly became the stuff of legend and perhaps the most famous (and, at that time, possibly the most remote) dinosaur site in the world.

In the more than eighty years since the Andrews expeditions it is still possible to find new Flaming Cliffs. The Gobi is a vast emptiness, about half a million square miles' worth, and there are isolated Cretaceous outcrops still awaiting discovery by some lucky paleontologist. In the early 1990s our team stumbled into a small valley of reddish brown sandstones, subsequently named Ukhaa Tolgod (Mongolian for "brown hills"), that has yielded scores of dinosaur skeletons and nearly two thousand superb specimens of ancient mammals and lizards. All these smaller fossils are skulls, which are extremely rare in Cretaceous rocks elsewhere in the world. Even more providentially, many of these skulls are attached to whole or partial skeletons. The dinosaurs too are remarkably preserved; they include several *Velociraptor* skeletons, scores of the tanklike, armored ankylosaurs, and the

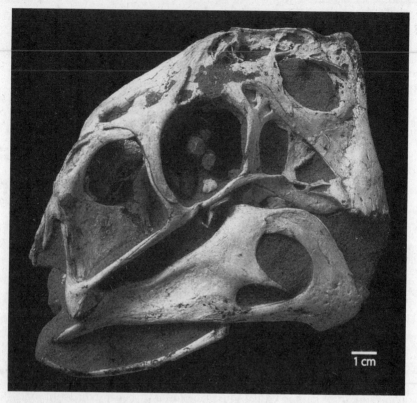

1 cm

The exquisitely preserved skull of the oviraptorid *Citipati*

first embryos in the eggs of meat-eating dinosaurs, members of a bizarre group of long-necked, long-clawed, beak-headed dinosaurs called oviraptorids. In the midst of our delirious days of discovery, we found something even more amazing than the dinosaur embryos. These were adult oviraptorids perfectly preserved, still sitting on their nests of twenty or more unhatched eggs. They gave us for the first time direct evidence of dinosaur-brooding behavior. More than a decade of our team's work at Ukhaa Tolgod has demonstrated that this spot is ne plus ultra for the Gobi Cretaceous, barely matched by any other Cretaceous site in the world.

Despite all this paleontological wealth, we don't know exactly what Ukhaa Tolgod was like 80 million years ago, when the dinosaurs, mammals, and other ancient animals represented by these fossils lived there. The bones at this rich locality are not preserved along with plants. There are no big petrified logs like those being rifled out of Cretaceous beds in the eastern Gobi near the railroad. There are not even subtle clues to ancient plant life — no seeds, no twigs, no leaves, not even microscopic pollen grains. An Australian paleobotanist working in similar red beds near our site lamented in extremely profane terms the lack of such plant fossils in the dinosaur red beds of the Gobi.

Yet a careful look at these sandy sediments gives us some picture of life at Ukhaa Tolgod and other well-known Cretaceous red beds of the Gobi, such as the famous Flaming Cliffs. These places represent oases in a burning desert. Small deposits of shales and clays indicate the presence of small ponds or lakes of standing water at the time the rocks of Ukhaa Tolgod were laid down. The amount of caliche, or hardpan, in certain places also indicates that these lakes were probably ephemeral; they rose in times of rain and shriveled up in times of drought. Extensive red sand layers with very distinctive striations called cross-beds are clearly the remnants of ancient shifting sand dunes. But the fossils are not found in these cross-bedded sands; instead, they occur in sand units with little or no striations, which seem to have been dunes that stabilized and were covered with vegetation. The lack of cross-beds is due to a churning up of the sand dunes by burrowing insects, worms, mammals, and lizards and spreading plant roots.

The animals of Ukhaa Tolgod seem to have congregated near pools and hunkered down in gullies among the stabilized dunes. They were seeking standing water for sustenance and, apparently, for breeding sites. Perhaps the animals migrated here from drier, less sustaining places. Nonetheless, at certain times the communities around these isolated oases were destroyed. At first we, like many workers who had studied other Gobi sites over the decades, believed that the ancient animals had been engulfed in sandstorms. But some of our geologists concluded that what buried and possibly killed the oviraptorids and other inhabitants were sudden, intense rainstorms, which brought huge flows of water-soaked mud from the dunes looming above these fragile colonies. The harsh conditions of an ancient desert climate punctuated by major rainstorms, flash floods, and mud slides may have accounted for the breathless demise of these creatures. This is a unique situation for a Cretaceous fossil site. Outside Mongolia, most dinosaur sites of this age represent wetter environments, not so prone to violent weather cycles or flash floods.

Despite our incomplete picture of the ancient environments of Gobi red beds, we can appreciate one reality: the catastrophe that beset the denizens of this terrain has meant our own great fortune; it yielded superb fossils, a great boon for bone hunting. As I have written before, we're paleontologists; we thrive on carnage.

Only a few hundred miles from Ukhaa Tolgod lies another Cretaceous locality of much recent fame and generosity. This area of widespread ancient lake beds, in the province of Liaoning in northeastern China, offers perhaps one of the world's most brilliant and detailed dioramas of Cretaceous life. Unlike the Gobi red bed sites, these localities have insect and plant fossils as well as vertebrate ones, and the preservation of these fossils is more astonishingly complete than even in places like Ukhaa Tolgod. The Liaoning fossils come from a thick sequence of rock comprising alternating layers of sandstones, shales, and volcanic deposits that represent either lava flows (basalts) or settled clouds of volcanic rocks dust and ash (tuffs). The rock sequence is named the Jehol group, and the fossils preserved in this sequence are therefore called the Jehol biota. Because the sequence contains layers of volcanic rocks, it allows for a bonus: quite precise dating from measurements of the amounts of radioactive argon, uranium, and lead. The Jehol rock sequence and the biota it contains range from about 128 million to about 110 million years in age. In the somewhat complicated vernacular of geochronologists this span is equivalent to the Early to Late Cretaceous. (Original claims that the Jehol biota was about 137 million years old, or Late Jurassic in age, have been convincingly rejected.)

What Jehol brings to us is not just an assemblage of spectacular dinosaurs but a whole ecosystem. The deposit contains a paleobotanical treasure house: delicately preserved bryophytes, lycopsids, ferns, bennettitaleans, ginkgoalians, gnetaleans, and the early angiosperm flower *Archaefructus*. The invertebrate fossils from the Jehol sequence, just as in a living, breathing ecosystem, are by far the most abundant elements in the biota. These include dragonflies, true bugs, flies, and several other insect orders (but no bees!). There are also abundant spiders, crustaceans, bivalves (clams), and gastropods (snails). The discovery of flies with long tubular mouthparts suggests the as yet undiscovered presence of tubular nectar-producing angiosperms and the early establishment of an insect-plant pollinating system, though this claim is disputed by some scientists. Vertebrates too come in a diverse, exquisitely preserved assortment: bony fishes, frogs, salamanders, winged pterosaurs (flying reptiles), and birds and other dinosaurs, including theropods, sauropods, ornithopods, ceratopsians, and ankylosaurs. There are also a half dozen or so mammals, ranging in size from *Jeholodens*, an animal with a head and body about the size of an almond, to *Repenomamus*, a badger-size predator that represents the largest of all Cretaceous mammals. These mammals were early branches on the way to modern marsupials and placentals.

The astounding preservation at Jehol brings a new dimension to our understanding of Cretaceous life. The fossils have not only the usual hard parts—shells, bones, teeth, and other durable elements—but also soft parts like skin, internal organs, feathers, and hair. With respect to those soft parts, fossils of theropod dinosaurs, including birds, have attracted the most attention. In 1996 the paleontological community was set on edge by the disclosure that one of the Jehol theropods, a nonflying form called *Sinosauropteryx* (*Sino*, China; *saur*, lizard; *pteryx*, wing) has a fine coating of filamentous structures clearly reminiscent of the simple downy feathers of birds. Some scientists protested; they claimed that these fuzzy skin features were merely artifacts of misleading preservation. But most experts quickly agreed that these fossils had rudimentary feathers, which indicated a major evolutionary step toward birds.

This argument was bolstered by discoveries of other spectacular Jehol theropods. *Beipiaosaurus* has the simple featherlike structures of *Sinosauropteryx*, while *Caudipteryx* and *Protoarchaeopteryx* have feathers with a central shaft, or rachis, supporting the branching patterns, very much like those in modern birds. Similar kinds of feathers are found in dromaeosaurs related to *Velociraptor*, such as *Sinornithosaurus*. The last few years have brought to light even more amazing theropods. *Microraptor*, a small, gracile dinosaur, is now known to have winglike feathers on both its fore- and hind limbs, suggesting it was an accomplished glider. Even a feathered tyrannosaurid has made its entrance. Who could have imagined the "Terrible Lizard" would eventually qualify in toy stores as another cute, downy-feathered stuffed animal?

The feathered dinosaurs of Jehol have a special status as examples of those highly desirable fossils that bridge an evolutionary gap between groups. These dinosaurs had feathers, but they didn't have wings and therefore didn't fly. Although *Microraptor*, with its strange feathered limbs, was a glider, it likely did not have the true flapping flight of birds. As noted in our

Artist's reconstruction
of a feathered
dinosaur from Jehol

earlier discussion of dinosaur behavior, this leads us to one of those important insights so sought after in paleontology and evolutionary biology: feathers, which in modern birds are so critical to flight, probably served originally as insulating structures, much as hair does in mammals; we know that insulation is the main function of downy feathers in flightless chicks.

Another early purpose is apparent. Some of the feather structures in the Jehol dinosaurs, though elaborate, are not attached to winglike structures. Instead, they form flashy accoutrements at the tip of the limbs or tails. These were probably ornamental devices, used in allowing animals to recognize their kin or in the display among individuals driven by sex and competition for mates. The modification and use of feathers for flight in avian dinosaurs, the birds, came later. One of the great realities we have come to accept in the study of evolution is that a structure with a particular purpose can be co-opted for another purpose. All that is required is time and selective pressure for new adaptations.

Jehol is also enriched with fossils of fully fledged, once flying birds. This is not surprising, since we have long known from the exquisite Jurassic fossils of the bird *Archaeopteryx*, which also preserves impressions of feathers, that birds long antedated the Early Cretaceous. This means that the feathered, nonflying theropods as well go back to the Jurassic. We haven't yet found any of them preserved with feathers. The odds against finding such early fossils of Jehol quality are enormous, but we keep looking.

The birds of Jehol are important because they represent branches of the tree of life located between the most primitive known bird, the Jurassic *Archaeopteryx*, and modern birds. One of these fossils, *Confuciusornis*, is represented by hundreds of specimens, allowing studies of population variation and analyses of sexual dimorphism, or the morphological differences between males and females. Another bird, *Sapeornis*, has extremely elongated forelimbs and was probably a soaring form. *Yanornis* had a specialized and expanded breastbone (sternum) and something like a collarbone (coracoid), typical of deftly flying modern birds.

Feathers are not the only notable soft structure preserved in the Jehol fossils. The fossil mammals have impressions of hair, as we might expect. There is also an intact mass of seeds within the stomach contents of a bird, *Jeholornis*, indicating that this form was a fruit eater (a frugivore). Thus the function of birds in dispersing plant seeds appeared very early in avian history.

Many Jehol theropods contain stomach contents with the jaws, teeth, bits, and skeletal bits of small mammals. The most eye-popping of all fossils in this category is one where the hunted became the hunter. My museum colleague Jin Meng, along with several Chinese collaborators, has described a specimen of a large carnivorous mammal named *Repenomamus* from the lower and older part of the rock sequence, where skeletons are preserved in a three-dimensional aspect. This beast was largish for a Mesozoic mammal, about the size of a badger, but formidable and voracious. Meng and his colleagues found a *Repenomamus* skeleton with a dinosaur in its stomach! Preservation of the specimen was so spectacular that it allowed for a more precise autopsy; the digested animal turned out to be a small, juvenile psittacosaur dinosaur. This discovery reached the scientific press and the world in early 2005. We put the specimen on display in the American Museum's rotunda for our millions of visitors. Above the columned entrance of the museum, we hung a banner that said it all: MAMMAL EATS

DINOSAUR. New discoveries of mammals from the Liaoning sequence continue to verge on the incredible. Meng and his colleagues have now also described a largish mammal, *Castorocauda lutrasimilis*, with a flat, beaver-like tail that was clearly adapted for swimming. Even more astounding is a fossil mammal with impressions of a winglike flap of skin that clearly indicates it could glide, anticipating the origin of gliding and flight in the earliest-known fossil bats and flying lemurs by 75 million years. After inspecting this gorgeous fossil with awe, and a tinge of envy, in Meng's lab, I asked him what he expected to find next: a mammal with two heads?

What natural forces produced such amazing fossils? The fossils themselves were entombed in sediments like those you find on lake bottoms. Usually such bottom sediments are very fine-grained and do not indicate water turbulence or strong currents. Such calm conditions are ideal for keeping the remains of organisms from being broken up or damaged. But there is another factor responsible for the extraordinary preservation of the delicate Jehol fossils. The lake bed sequence includes numerous layers of volcanic tuff, the fine-grained geologic product of ash that falls from volcanic eruptions. It is apparent that many of the Jehol animals were buried under these ash sediments. The poisonous, or anoxic, ash not only made for a powerful stench but prevented the bacterial decay of feathers, hair, and other soft tissue. Anoxic conditions also kept worms and other burrowing organisms from overturning the sediments and scavenging on the organic remains buried within. Gently settling lake sediments along with stinking, poisonous clouds of volcanic ash made for a winning combination and a superb example of preservation in the fossil record.

Collecting the Jehol fossils has not usually entailed the kind of romantic expedition work we think of when we consider paleontology. The Jehol rocks are exposed in gullies and human-hewn quarries at the edges of small forests and agricultural fields. As in much of China, the huge human population has had its effect. The fossils are for the most part collected by local farmers and townspeople. Professional paleontologists (and commercial and sometimes illegal collectors, sadly) then negotiate for the most precious and most important ones. Mark Norell, a colleague and paleontological teammate on the Gobi project, describes in a recent book an expedition to the Jehol as an inebriated wait in a crowded inn engulfed in cigarette smoke. The daily and nightly routine involved drinking copious amounts of beer or baijiu, an execrable sweet sorghum wine, and bartering over fossils.

Completion of the deal usually calls for several baijiu toasts. The intrepid explorers, if they survive with livers and lungs intact, return with very choice fossils.

Outside the treasure fields of northern China and Mongolia, the world of Cretaceous rocks still has much to offer. Fabulous Cretaceous dinosaurs that fill many museums were transported from well-trodden basins in the expansive badlands of Wyoming, Montana, Colorado, Utah, and New Mexico. Similarly spectacular are the Cretaceous beds of Patagonian Argentina, where fossil beasts rival in size and flamboyance the creatures known from Asia and North America. Important sites also include quarries in the mountains and forests of eastern Europe, the edges of the outback in Australia, sand-choked badlands of the African Sahara, the north slope of Alaska, and icebound islands off Antarctica. In recent years Madagascar has yielded a cache of Cretaceous dinosaurs and mammals, and India rather stubbornly has revealed Cretaceous dinosaur bones and isolated teeth of small mammals. Other localities, such as the clay pits of New Jersey, though depauperate in dinosaurs and other vertebrates, show exciting and important organisms that make up Cretaceous ecosystems: fossil leaves, twigs, flowers, and insects in amber.

But our picture of the flowery ecosystem in the late age of the dinosaurs is hard won. To say anything meaningful, you first have to find fossils, and this takes time and sweat of the brow. As a general rule, things get better as we move toward the present; there are more fossils in more rocks in more places. The obvious reason for this youth-oriented bias is the dynamic nature of Earth itself. As time goes by, rocks and their fossils exposed on the surface are split by faults or cooked by hot intruding and extruding rocks as mountains build and volcanoes erupt.

Time takes its toll even where the situation is relatively quiescent. In our Gobi expeditions we have repeatedly visited the famous Flaming Cliffs. We often camp in the very spot that Andrews and his team used. Seventy years of exposure to the extreme heat, cold, aridity, and the occasional downpour, not to mention the omnipresent wind, have made a difference. We can see places where large blocks of sandstone, three times the size of a foundational stone for an Egyptian pyramid, have calved away from the main cliff, like icebergs off an Antarctic shoreline. Nearby, an isolated outcrop named the dinosaur by earlier expeditions no longer looks like one; its sandy head and neck have been wind-severed from the rest of the body.

So time itself is no ally to preservation. The more remote the geologic time, the more elusive the rocks and their fossils. Nonetheless, the Cretaceous Period, despite its great antiquity, is well represented by both, though far from perfectly. The shortcomings affect the way various investigators have reconstructed this key phase of Earth's history. A realistic accounting of the Cretaceous record is particularly important when we consider the great transition on land from the archaic to the modern, flowery ecosystem.

Let us look at dinosaurs with respect to that record. One widely cited cycle in dinosaur history involves the decline of the massive, long-necked sauropods after the Jurassic. By the Early Cretaceous these giants had been crowded out by low-browsing ornithischian dinosaurs—bipedal iguanodontids, dome-headed pachycephalosaurs, and armored nodosaurs. By the Late Cretaceous the herbivorous dinosaurs were dominated by new ornithischians: the parrot-beaked psittacosaurs and protoceratopsians; the bulky, shielded-headed, and horned ceratopsians; the armored, club-tailed ankylosaurs; and the duck-billed hadrosaurs. The arrival of these latter groups coincided with a high-water mark in dinosaur diversity that occurred in the Late Cretaceous. But on a worldwide basis sauropods were a minority component of these diverse Late Cretaceous faunas; they never regained the kind of diversity and dominance they had in the Jurassic.

Timing is everything. The timing of all these toggles in diversity and abundance among the big and the tall may remind you of another concurrent transition—namely, the changeover from a conifer-dominated plant community in the Jurassic to an angiosperm-dominated community in the Cretaceous. This coincidence has not escaped the notice of some paleontologists. Several of them in the late 1970s, prominently Robert Bakker, proposed that the parallel in dinosaur and plant turnover was more than coincidence. Indeed, they claimed, dinosaurs, through their shifting feeding types, may have seeded the great radiation of flowering plants.

The logic of their argument is as follows. Gargantuan sauropods were capable of huge amounts of plant intake. They doubtless had a strong effect on plant communities, just as modern large mammals have in Africa, on the tundra, and elsewhere. Bakker and others have assumed that sauropods were high-canopy browsers, up to forty feet above the ground, and therefore exerted extreme browsing pressure on mature trees. This predilection for

treetops left young and vulnerable gymnosperm saplings relatively un-
scathed. Thus the gymnosperm resource for herbivores was continually replenished by the growth of younger plants and trees. Bakker argued that this
situation drastically changed in the Early Cretaceous, when low-browsing
dinosaurs proliferated, feeding on plants rarely more than three meters (9.9
feet) in height. This low-profile feeding may have increased the pressure on
slow-growing conifers and other gymnosperms, but it opened up opportunities for rapidly growing plants with diverse strategies for pollination and reproduction. In other words, it was a Darwinian bonus to the flowering
angiosperms.

In support of this scenario, David Weishampel and David Norman proposed that dinosaur-feeding innovations also influenced the rise of the
angiosperms. They observed that ornithopods, the dinosaur group that includes the duck-billed hadrosaurs, could move the upper jaw from side to
side while feeding. This sideways, or transverse, jaw movement is similar to
that of herbivorous mammals, a mechanism that might have allowed for sophisticated, varied means of dinosaur browsing. As further support, Scott
Wing and Bruce Tiffney calculated that most Early Cretaceous dinosaur
communities were dominated by animals of 1,000 kilograms (2,200 pounds)
or more. Today large herbivores tend to be generalists in diet; they eat a lot
and aren't picky about what they eat. This daily demand on plant resources
would have further disturbed the resident flora, an unstable regime that
would have favored low-stature, highly productive plants—namely, angiosperms.

Sounds good, doesn't it? Here we have a compelling synchrony between
the changes in dinosaur faunas and floras that indicates interdependence.
Yet the hypothesis has a few serious flaws. First of all, we have already seen
that new studies of sauropod neck architecture suggest that such reputedly
high-browsing forms as *Apatosaurus* and *Diplodocus* could raise their heads
only about 6.9 meters (23 feet) and 5.6 meters (18 feet) respectively—high
enough to crop the low foliage of large trees and crowns of small trees, but
not high enough for the treetop feeding of popular reconstructions. So
these animals may have picked off small trees and saplings, the very trees
that Bakker argued were safe from sauropod harvesting. While some lofty
sauropods, such as *Brachiosaurus*, still might have conducted their topiary
on the crowns of pines, the diverse species of sauropods were probably feeding high and low—on adult trees, saplings, ferns, bushes, and angiosperms.

Another objection relates not to what dinosaurs did but to where they lived. Paul Barrett and Katherine Willis and others point out that the change from sauropod- to ornithischian-dominated communities was not the same in all places. Sauropods precipitously decreased in the northern continents (Laurasia) but held on appreciably in South America in the Early Cretaceous. By the Late Cretaceous, North America shows the presence of only one sauropod genus, *Alamosaurus*. In contrast, in other regions sauropods accounted for about 25 to 35 percent of the total diversity of herbivorous dinosaurs. Unfortunately, the spotty record of localities worldwide, especially in Africa, Australia, and southern Asia, obscures the clear global-scale pattern.

On the other hand, this mosaic in dinosaur distribution might actually provide a welcome test to the theory that dinosaurs incited an angiosperm revolution. Such coevolution might be suggested, for example, if we were to determine that South America, where sauropods persisted, also continued to harbor gymnosperms to the detriment of angiosperms; by contrast, North America, where ornithischians radiated, should show a complementary rise in angiosperms and a marked decline in gymnosperms. Detailed studies, however, fail to demonstrate such a contrast clearly; the explosive radiation of flowery forms took place on both continents at roughly the same time.

Yet it would be unfair to claim that dinosaurs had nothing to do with changes in the plant world. As in the case of many living animals, dinosaurs must have helped to disperse seeds of cycads and other plants in their feces and this would have had marked effects on the distribution of vegetation. Moreover, the great radiation of low-browsing ornithischians in the Late Cretaceous may have created open environments that helped stimulate angiosperm growth. But the theory claiming that dinosaurs were critical to catalyzing the great radiation of angiosperms, though long popular and evocative, is irreparably damaged.

What other possible changes in the ecosystem might be tied to the angiosperm revolution? Another factor has lately gained much attention as a possible instigator. Earlier I noted the marked change in atmospheric carbon dioxide over time. For example, CO_2 levels in the Early Carboniferous, when both coal forests and giant insects proliferated, were nearly four times higher than they are today and then declined rapidly in the Late Carboniferous. As we have seen, the subsequent rise of CO_2 in the Late Permian and Early Triassic once again drove climates to higher temperatures and, in combination with low atmospheric O_2 levels, would have created stressful conditions for large land animals. It also has been observed that such increases in CO_2 seem to occur at times when major plant groups emerge. This is certainly apparent with angiosperms; their time of early radiation is marked by a major swing upward in the amount of CO_2 in the atmosphere.

Why should CO_2 concentrations make a difference to plants? One thought is that the global warming associated with a CO_2 infusion would increase warmer, wetter, and more habitable land areas. CO_2 has been cited as enhancing water and nutrient transport from the soil and stimulating the activities of fungi and nitrogen-fixing bacteria responsible for making nutrients available. Increases in CO_2 concentrations over the past forty years have also been associated with faster growth and shorter generation times in plants, a condition that fosters plant productivity. An important aspect of these influences is that they are global, not local. Plants, if affected at all, would be affected worldwide, regardless of habitat or specific environmental conditions. It seems persuasive that such a global-scale shift may have strongly influenced the rise of the angiosperms.

So was the revolution that brought us the modern ecosystem simply a matter of "something in the air" then? Before we leap to this conclusion, we need to consider a few problems. First, the study of precise historical changes in atmospheric gases and their relationship to ancient climate is still new, and conclusions are frequently revised or corrected. For example, early studies described a warm "greenhouse climate" driven by high levels of atmospheric CO_2 for the Middle to Late Cretaceous, but recent work has demonstrated just the opposite. Second, we can't be certain about the stated relationships between higher CO_2 concentrations and changes in plant physiology, habitat, or plant evolution. These are complex and far from easy to predict.

Studies of the changing atmosphere over time represent an exciting and fruitful area of current research, but they have not yet pinpointed the ultimate cause of the angiosperm revolution. We can only say that the rise of flowering plants and specialized pollinating insects in the Cretaceous had a lot to do with each other. Once plants and insects arrived at their novel solutions for improving their survival and reproduction, the opportunities for adaptation, diversification, and world dominance were overwhelming.

And what of the world once the angiosperms were well on their way to dominating Earth's vegetation? The Late Cretaceous saw an all-time high in dinosaur diversity. The span of roughly 70 to 90 million years ago also saw an acme in the radiation of horned neoceratopsians and duck-billed hadrosaurids. Although sauropod dominance had diminished, especially in the northern continents, these lumbering giants were still present. These radiations were complemented by a diversity of small and large carnivorous theropods, an enriched assemblage well recorded in the Late Cretaceous Gobi localities. And it's important to note that these diverse dinosaurs inhabited a land appointed with a rich flora dominated by magnolias, roses, and other flowering plants as well as enduring pines and other gymnosperms, as so spectacularly preserved in the Jehol sequence or the amber pits of industrial New Jersey. The whole system chugged along with the feverish buzz of bees and other pollinating insects. Dinosaurhood was in flower. But this was also a dinosaur Camelot, a kingdom that teetered between even greater enrichment and disaster.

Part Three

DEATH AND
RESURRECTION

All but Death, can be Adjusted—
Dynasties repaired—
Systems—settled in their Sockets—
Citadels—dissolved—

Wastes of Lives—resown with Colors
By Succeeding Springs—
Death—unto itself—Exception—
Is exempt from Change—

—Emily Dickinson

13

A PUZZLING CATASTROPHE

The town of Gubbio in the hills of Umbria is as splendiferous as many other Italian locales. Its palaces, churches, and walls are hewn from an austere limestone that takes on a sheen of slightly tarnished silver in the late-afternoon sun. Gubbio is a man-made Grand Canyon, with layer upon layer of medieval and Renaissance structures built on a mainly obscured layer of a Roman city. Still older layers are revealed in spotty remnants of an ancient pre-Roman center of Umbria. From the Piazza Quaranta Martiri you can see the whole town, terraced like intricately sculptured stone steps up the foothills of Monte Ingino. At close range the place can seem a bit hectic and confusing, especially to the uninitiated visitor trying to locate a hotel on one of the serpentine streets that switchback up the hill. A tired driver then must deal with the typical incongruity of ancient Europe: quaint cobblestone streets further constricted by pint-size parked cars, which are nonetheless ludicrously large for a passageway built several centuries ago. But the tricky navigation is well worth it. Among other splendors Gubbio has the Palazzo dei Consoli, one of Italy's most beautiful public buildings. The palazzo proudly claims stewardship of seven bronze slabs, the Tablets of Gubbio, which are written in ancient Umbrian language and engraved partly in Etruscan and partly in Roman characters. This, Umbria's very own "Rosetta Stone," discovered by chance in 1444, offers a clear description of religious rites and other aspects of life in the region dating back to the third century B.C.

Not far from Gubbio is another grand facade containing yet another sacred tablet, this one dating back 65 million years. On a fair September morning I drove just north of Gubbio to a narrow defile. This is Gola del Botta-cione, a canyon festooned with pale green poplars and black-green cypresses trying to find holdfasts in the steeply tilted layers of limestone that form the canyon walls. *Bottacione* means "big water barrel" in Italian, a playful name for an aqueduct built in the fourteenth century to bring water down from a mountain spring to the city. I pulled my rented Citroën off on a siding where the canyon took a bend that tracks the river below. A very short scramble up a slope brought me to the base of a layer of tilted limestones. This is a se-quence of rock more than 400 meters (1,312 feet) thick known as the Scaglia Rossa limestone. *Scaglia* is Italian for "scale" or "flake" and refers to the ease with which these rocks are broken into convenient chunks for fine building stones. *Rossa* of course refers to the pinkish red cast of the rock.

The Scaglia Rossa sequence is now perched in a canyon, but it was once below the surface of the sea. It represents an ancient sea bottom from a time when, on land, the last nonavian dinosaurs still romped. The beds near the road are tilted at about forty-five degrees, the result of an active pulse of mountain building that jostled this ancient seafloor. The limestone in this cliff face also has fossils, but these are not so easy to detect as dinosaur bones. I chipped off a small piece of rock and scrutinized it. These rocks contain flecks that, with the aid of a magnifying hand lens, resolved into in-tricate coiled and chambered shells. These are the hard tests of foraminifer-ans, the hugely abundant one-celled creatures that are integral to the food chain and energy cycle in marine ecosystems. The limestone is otherwise devoid of visible creatures; there are no shells of clams or snails. Nor are there lenses (thin layers) of sand or silt that would indicate that these beds were deposited near the mouth of an ancient river. The Scaglia Rossa was not laid down in a shallow ocean bottom offshore; it represents the floor of a deep and abiding sea.

Standing on this little slope above the bend in the road, eyeing these limestones, I knew I was in a place of monumental importance to science. This is one of those rare spots where geology and great insights anneal. The very base of this cliff—the lower section of the Scaglia Rossa limestone—records in its layers a great transition, the boundary between the end of the dinosaur age and the beginning of the age of mammals. The boundary in the Gubbio section is clearly marked out by the disappearance of many kinds of those tiny flecks—namely, Cretaceous microfossils, particularly foraminifer-ans. Such a break is evidence of the Cretaceous/Tertiary, or K/T, boundary.

Diverse species of foraminifera as depicted by Ernst Haeckel, 1904, *Kunstformen der Natur*

The K/T boundary in many locales throughout the world has been dated—by measuring radioactive isotopes and correlating these with time-scales based on both marine fossils and paleomagnetism—at about 65 million years before the present. This date is a very big deal. Paleontologists have recognized for many decades 65 million years before the present as the most infamous, if not devastating, mass extinction event in Earth's long history, the bad day that did in the nonavian dinosaurs as well as erased more than 70 percent of all species on land and in the sea. Certain groups in addition to dinosaurs and tiny foraminiferans seemed particularly susceptible. These included the ornately decorated coiled ammonoids, the ancient relatives of living squidlike shelled nautiloids, as well as many species of corals,

clams, sharks, and marsupial mammals, ancient relatives of the living opossum and kangaroo. The Cretaceous catastrophe is the most recent of the
five great extinction events that have hammered at life's diversity over the
course of the last 500 million years. Its cause and effects are of great importance to our view of the past and future of life on this planet.

Paleontologists had long contemplated the causes of the great dying at
the end of the Cretaceous. Some papers reviewing the subject listed more
than fifty published hypotheses for the demise of the dinosaurs. Perhaps the
most popular of these invoked drastic climate change; other causes included disease, food shortage, intense radiation from the sun, volcanic activity, and changes in plant life that wreaked havoc with dinosaur digestion
(constipation was cited as one culprit!). But there was no direct evidence,
circumstantial or otherwise, from the fossils and the rocks, until something
found in the Cretaceous in that "big water barrel" canyon north of Gubbio
offered the first clue. The discoveries that followed offered an explanation
for one of the most dramatic events in the history of life, an explanation that
cannot easily be dismissed.

At the boundary in the Scaglia Rossa limestone, near the spot where I
stood, is a thin layer of brown and black clay about a centimeter thick. The
section above, below, and within the clay layer is riddled with cylindrical
holes. These holes identify the places where core samples have been removed to be taken off and analyzed in geophysical laboratories at Berkeley,
Amsterdam, and elsewhere. The reason for such energetic hole punching
illustrates both the dogged and the fortuitous nature of scientific inquiry.
This is a detective story, as engagingly told in Walter Alvarez's book *T. Rex
and the Crater of Doom*, that is full of plot twists, false leads, unexpected
clues, and insights about a very ancient and terrible event.

In the early 1970s Alvarez and another geophysicist, Bill Lowrie, sampled the Scaglia Rossa for paleomagnetic signals essentially "frozen" in the
rocks. Here they stumbled upon an unexpected signal that the orientation
of the iron-bearing grains in rocks from some stratigraphic levels were exactly the opposite of that from other levels, evidence that Earth's magnetic
field had flip-flopped more than once during the time of deposition of the
Scaglia Rossa. Such a pattern of "reversed" and "normal" phases could be
correlated with dates of rocks based on radioactive elements and fossils to
form a timescale.

Alvarez and Lowrie found themselves in geologic nirvana. They continued sampling many outcrops in the Apennines. It was clear from their survey that the Scaglia Rossa section near Gubbio and elsewhere straddled the important K/T boundary. As they sampled, the scientists focused on a peculiar break in the sequence, the same one I conveniently observed at the roadside. They noted the profusion of foraminiferans (paleontologists call them forams) in the lower section, the thin layers of clay, and then a drastic change in the forams above this layer. Obviously this sequence revealed a major extinction. Was it the event that had done in the dinosaurs?

And how much time did this break in the sequence represent? Alvarez and Lowrie reasoned that if the duration of the deposition of the clay layer had been short, then the extinction event was rapid and catastrophic; if long, the extinction event would have been gradual, a notion that most paleontologists favored. Unfortunately, the reversal patterns and the changes in foram assemblages were not detailed enough to tell them how long the changeover took. For this, they needed to sample radioactive elements within the sequence, elements that decayed at a rate convenient enough to estimate the duration of the boundary layer.

The next phase of the investigation took a most surprising turn. Alvarez sought help from his father, Luis Alvarez, a physicist at UC Berkeley who had won a Nobel Prize for the detection of an array of subatomic particles. Meanwhile, the younger Alvarez landed a job in 1977 at Berkeley, allowing the father-and-son team to work in convenient proximity. As Walter Alvarez pondered the secrets of the Scaglia Rossa limestones, I was working away on my Ph.D. thesis in the fossil collections in the same building, studying the bones and teeth of both the casualties and survivors of the great Cretaceous extinction event. I was only dimly aware of the rumbling two floors above in the Berkeley geology and geophysics department.

After some trial and error the Alvarez team focused on a very rare element, iridium (Ir), for the purpose of dating the Gubbio samples, and they induced Frank Asaro, a nuclear chemist at Lawrence Berkeley Laboratory, to apply critically needed expertise to the problem. Following several months of analysis, Asaro shared his surprising and troubling results: he had found unusually high concentrations of iridium in the sample taken from the K/T boundary. Besides raising a huge complication for the use of the element as a dating tool, this superabundance of iridium was very odd. Iridium, like other platinum group elements, is concentrated in Earth's core, where it is absorbed by iron, but it is extremely rare in Earth's crustal rocks.

A quantitative sense of what rare means here helps one appreciate the precision of the technology Asaro applied. Iridium is found in crustal rocks at concentration on the order of 0.1 parts per billion! Instead of this expected concentration, Asaro arrived at a figure of 3 ppb and, later, 9 ppb, for the iridium in the Gubbio sample—not a hefty concentration by any measure, but nearly a hundred times the normal.

This unexpected disclosure inspired a wild but brilliant explanation. Such weirdly high concentrations of iridium could only be caused by a highly unusual event. When a bolide, an extraterrestrial object of significant size, such as an asteroid or a comet, collides with Earth, its impact blows out a great fountain of incandescent material, appropriately called ejecta, that rises and then rains back down on the surface of our planet. Iridium occurs in unusually high amounts in this ejecta, because of the ancestry of asteroids and other space objects. These are parts of shattered planets whose cores and mantles, like Earth's interior, concentrate iridium and other members of the platinum family. The higher the concentration and the more widespread the iridium layer, the greater and more widespread the ejecta and the bigger the colliding object and its impact.

These amazing series of clues and deductions brought with them the next dilemma. Was the spike of iridium concentration in the Gubbio section, which became known as an iridium anomaly, merely localized in some Umbrian rocks? Or was the signal widespread, in other regions or continents or ocean basins? An international group of collaborators found the iridium spikes in a chalky Cretaceous limestone cliff in Denmark, a rock section in Caravaca, Spain, a core sample from the deep Pacific, and a clay layer in New Mexico. These samples too could be dated at the K/T boundary. Clearly, something big and impulsive happened somewhere on the planet 65 million years ago.

But where? A search for the source of the ejecta and its iridium ensued. The quest got off to a shaky start. Some doubted that there really was a sudden extinction to explain. A number of influential paleontologists then (and now) viewed extinction of dinosaurs and other Cretaceous species as a gradual change, so they argued that the whole matter of the extraterrestrial impact was irrelevant. Nonetheless, the impact supporters convinced many scientists of the overwhelming evidence for a sudden, widespread shakeup of the biota, especially in marine sequences of fossils and rocks.

Partial victory in this debate did not eliminate other obstacles. An impact crater of the right age proved despairingly elusive. A list of crater sites published in 1982 showed that most of them were too small, and the three that were large enough were the wrong age. Worse, the crater hunters were convinced that the impact occurred on the deep ocean floor, a hunch that later proved wildly incorrect. The deceptive evidence for an oceanic impact came from sand grain–size, rounded objects of peculiar composition called spherules. These tiny grains subsequently became famous as the "smoking gun"—the decisive evidence for the occurrence and the nature of the impact—but when they were discovered by Jan Smit in the Caravaca sequence, they were an enigma. Feathery crystals within the spherules contained a mineral called sanidine, a kind of potassium feldspar uncharacteristic of sedimentary rock found on continents. Subsequently, it was found that the sanidine came from neither the impactor nor the target rock; it was a replacement mineral that had grown later. The minerals in the spherules turned out to be olivines, pyroxenes, and calcium-rich feldspars crystallized from molten rock, the characteristic minerals of basalt, the main rock of the oceanic crust. Since the ocean floor is much younger than the many parts of continental crust, there was a plausible explanation for the elusiveness of the impact crater: some Cretaceous pieces of the ocean floor have already been subducted in trenches, and the crater could have been carried with them. The Alvarez team resigned itself to the possibility that it might never find the devoured crater.

Only later was it discovered that the spherule sample was a weird mixture of chemicals in both continental and oceanic rocks that, when "stirred up," produced a chimera resembling oceanic sediments. The blast had apparently melted sedimentary rock rich in calcium and magnesium together with underlying continental crust. At the moment of impact, the chance mixture formed in molten droplets that were blown right through the atmosphere and launched into space. As these droplets cooled on their return to Earth, they crystallized into olivine, pyroxene, and calcium-rich feldspar.

Unfortunately this elucidating analysis of spherules came after years of wasted effort looking for a crater on the ocean floor. The trek was frustrated by other distractions too. The team started uncovering widespread evidence of another strange feature, grains of quartz damaged in a particular way that showed parallel banding. The grains, known as shocked quartz, had for some time been associated with impact craters. They were found at many sites at the K/T boundary, but it was nonetheless perplexing; shocked quartz is a component of continental, not oceanic, rocks. As the investigators were in hot pursuit of an oceanic site, they wondered why they had discovered shocked quartz at all. Meanwhile, several geologists claimed that the shocked quartz was a sign of massive, widespread volcanic eruptions rather than impacts. After more sparring on the symposium floor, at least some scientists agreed with the Alvarez group that the shocked quartz in the K/T sections differed from those produced by volcanic eruptions.

But resistance to the impact theory was growing in many quarters, while the crater remained unknown. Then, in 1988, Alan Hildebrand, a graduate student in geology at the University of Arizona, proposed that a strange-looking Cretaceous deposit in Texas, in Brazos River drainage, pointed to the location of the crater. This sand unit showed evidence of a massive flow of rock that happened suddenly, a deposit called a turbidite. Another scientist, Jan Smit, had earlier noted that the Brazos sand might have been caused by a tsunami generated when the extraterrestrial object hit Earth. Jody Bourgeois, a sedimentologist at the University of Washington, and her collaborators determined that the Brazos bed was the result of a truly huge tsunami that occurred precisely at the K/T boundary and the time of impact. Hildebrand reasoned that the tsunami had to come from somewhere south of Texas because that was the only place in proximity to the region where there was a body of water; moreover, the force of the tsunami suggested by the beds indicated that the impact occurred not far away.

At last Hildebrand found the likely candidate. Gravity anomaly maps—measurements of gravity that indicate high spots and low spots in Earth's crust—showed the presence of a huge buried crater at the north coast of the Yucatán Peninsula of Mexico. Surprisingly, the crater was covered by continental, not oceanic, crust. But there was an explanation: the impactor had plowed into continental crust, rock of the continental shelf that had underlain a shallow sea off the coast of Cretaceous Mexico. At last there was a clarification for the various bits of conflicting evidence. Hildebrand and several coauthors published what Alvarez called a bombshell in 1991, a paper entitled "The Chicxulub crater: a possible Cretaceous Tertiary boundary impact crater on the Yucatán Peninsula, Mexico."

This brief recap of the discovery of the Chicxulub (cheek-shoe-lube) site overlooks the biggest irony of all in the quest for the crater, an extraordinary feature that had actually been discovered ten years before the publication of the bombshell paper! Was this a case of scientific deception then, of plagiarism cloaked as new and original discovery? Hardly. Two scientists working for the Mexico oil giant PEMEX, Antonio Camargo and Glen Penfield, had analyzed the Chicxulub structure in the late 1970s. This huge, circular "gravity feature" had even been drilled for oil in the 1950s, in probes that were unproductive. The results of those drilling projects, as well as the crater studies of Camargo and Penfield, were the property of a major international corporation that zealously guarded its discoveries, matters not meant for public consumption. Results were shelved, buried just like that gravity depression off the Yucatán. Camargo and Penfield did present a paper and an abstract at a technical meeting in 1981, but it was overlooked by the crater hunters. Only when Hildebrand tracked down the PEMEX duo did one insight converge on another. Fittingly, Camargo and Penfield are among the authors of the milestone 1991 paper.

The Chicxulub depression, detected from seismic and gravity signals, itself is overlain by hundreds of meters of sediments. This broad circular region has gravity values that are lower than is typical for the surroundings, with a gravitational high marking the center of the basin and local highs marking three major rings and possibly a fourth. The structure is strikingly like that of meteor-impact craters on the moon. And Chicxulub is no little blemish: the diameter of the inner basin is estimated to be around 170 kilometers (105 miles), with an excavation depth of approximately 19 kilometers (12 miles). There are only limited data for the fourth, outermost ring of the basin, which probably represents the crater's outer-rim crest in its final form. This suggests that the Chicxulub basin had a maximum diameter somewhere between 170 and 300 kilometers (186 miles). The impact site can

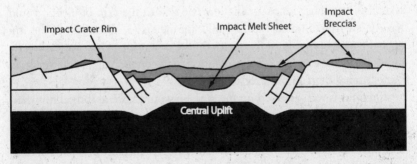

The Chicxulub depression, in cross-section, showing the major features of the buried crater

also be dated, and it conveniently turns out to yield that magic 65-million-year date. The age estimates were derived from many different sources, including stratigraphy, biostratigraphy, radioactive elements, and magnetic measurements that correlate to a date around 64.98 +/- 0.05 million years. The crash site of the killer bolide of the right size, in the right place, and at the right time had been found.

What of the colliding object, the thing itself? We know of course that it came from outer space. It was a bolide, probably an asteroid, a hunk of rock, an inert, bulky combination of iron and other elements. An early supposition that it may have been a renegade comet comprising dust and ice is now deemed unlikely, given what the analysis of the ejecta has revealed. When the object entered Earth's atmosphere 65 million years ago it was probably about 10 kilometers (6 miles) in diameter. With this mass it had tremendous speed, possibly 15 to 20 kilometers a second, or 22,000 miles per hour. When it augured into the crust, its rear end was still sticking out of what we generally call the lower atmosphere, or troposphere.

The shock to that same atmosphere and to Earth below seems to us unimaginable, but scientists have spent a good deal of time trying to reconstruct the devastation. An object 10 kilometers wide entering the atmosphere at such an awesome speed would have had to generate a lot of heat when it hit. The ejecta sprayed out at the moment of impact was projected back into the edge of space. Then, when the tiny particles of the ejecta reentered the atmosphere, they incandesced, burning like trillions of match heads. The atmosphere, which a few seconds before the end of the Cretaceous brought survival and sustenance, became a hell furnace; temperatures in the upper atmosphere might have hovered around 700°C (1,300°F) for several hours. The thermal radiation released in the rain of ejecta would have been concentrated within 6,000 kilometers (3,728 miles) of the impact and at its antipode, at the sector on the other side of the planet directly opposite, but the thermal pulse would have been global, at a power level more intense than an oven set at "broil."

Organisms well exposed on the surface would have absorbed this thermal radiation from the entire visible sky. The scorching and searing of leaf and flesh were prolonged—from one to several hours, sufficient not only to kill well-exposed organisms but also to incinerate virtually every forest, even tropical rainforest, on Earth. The impact released other destructive forces too. The lower densities of water-carrying stoma in the outer layers of fossil leaves can be used to detect higher concentrations of CO_2 in the atmosphere. A recent study applying this method shows a stable amount of CO_2 levels during the last interval of the Cretaceous and a sudden, extraordinarily high surge of CO_2 concentration within an estimated ten thousand years after the Cretaceous extinction event. This upsurge has been attributed either to the great amounts of CO_2 produced by the impact of an asteroid or to the huge amounts of outgassing associated with volcano eruptions.

Other nasty effects included the blast from the impact itself, earthquakes, acid rain caused by the release of sulfur dioxide, nitrous oxide, and other gases, poisoned water and soils, chemicals (pyrotoxins) that cause mutations, loss of plant photosynthetic productivity, and the tsunami. There is even a chance that the intense thermal radiation at the region antipodal to the impact site caused massive volcanic activity. But these destructive agents acted locally or regionally, not globally. The planet-scale destroyer of life was likely that broiler temperature atmosphere.

It would seem, then, that the asteroid impact theory sufficiently accounts for the global destruction at the end of the Cretaceous, and it has found broad acceptance over the past fifteen years. Case closed? Not yet. Skeptics, some asteroid bashers, claim that fossil evidence shows that the extinction of groups like dinosaurs was already marked for some millions of years before this event, caused, most likely, by global cooling and lowering sea levels in the last phase of the Cretaceous. Some, prominently Gerta Keller, a micropaleontologist at Princeton University, argue that the Chicxulub impact occurred three hundred thousand years before the K/T extinction event, too early to be the cause of the mass extinction of the forams in the marine record. Also, she and others argue that the huge amounts of volcanic activity toward the end of the Cretaceous, recorded in piles of lava in places like the Deccan plateau of southwestern India, promoted climate changes, darkness, soot, clouds, and acid rains that more profoundly affected life on land and in the water than any bruise from a renegade asteroid.

Geologists and paleontologists often argue about the origins of rock layers. Their debates can lead to a mighty resolution, when suddenly clarity emerges about the birth of a mountain range, the spread of a seaway, or the drift of a continent. At other times experts can look at the same rock sequence with the same tools and the same sources of information and come up with wildly different conclusions. Like the witnesses to a traffic accident, they just don't see the event the same way. In this case a big Chicxulub-scale accident continues to attract a diversity of perspectives.

Let me add my own view to the mix. Keller's interpretation, though a refinement of the rock sequence that brackets the Chicxulub impact, seems to me weak as a challenge. First, Smit and others have a decent alternative explanation for the sedimentary layer that seems to separate the Chicxulub impact from the die-off of forams at the K/T boundary. This layer could have been caused in a matter of hours and days after the impact, as the sediments washed up against the unstable walls of the newly formed crater. Then how does one explain all those burrows and other signs of colonization in this layer, which seem clearly the result of a prolonged process? Smit says they are not burrows but, instead, traces of decaying modern roots and small weathered fault surfaces. Second, the radiometric dates for the impact, though not perfect, are about as good as we can currently hope for in matching the times of two phenomena: the extinction of Cretaceous life and the impact itself. Moreover, the impact date is 65 million plus or minus about 50,000 years, an error range of less than one-tenth of 1 percent. Now 50,000 years is a long time though less than the 300,000 years that Keller suggests as the crucial stretch of time between the known impact and a second, more lethal one. The remarkable accuracy in pinpointing the coincidence of two different events that occurred 65 million years ago can't be easily dismissed. Also, why is there no evidence for Keller's hypothesis of a possible later impact in the Gulf region or, for that matter, anywhere else? Despite the passionate convictions of Keller and her associates, their argument for multiple causes, including multiple impacts, of Cretaceous extinction fits awkwardly with the available evidence. The Chicxulub impact as the Cretaceous killer fits much better.

A third perspective on the catastrophe and its cause persists. An appreciable number of scientists still argue that climate change in the Late Cretaceous may have promoted extinction in many groups, including dinosaurs, setting in motion an erosion of the biota, merely capped off by the asteroid impact or Deccan eruptions or both. This climate change hypothesis, though old, has recently enjoyed a reprise, which relies on movements of continents and the rise and fall of seaways in the Late Cretaceous. For most of this time, extensive shallow seas fringed the major continents. For example, during the Middle and Late Cretaceous, a seaway essentially bisected North America in the region of what are now the Great Plains, separating the Rocky Mountain region and parts farther west from the rest of the continent. Likewise, large, shallow seaways existed in places that are land today, including North Africa and northwestern South America and a large swath that is now eastern Europe, Russia, and western Asia. These vast incursions of warm, shallow water had major ecological and climatic effects, greatly increasing the surface area for the teeming marine life, ameliorating the effects of seasonality, and reducing the extremes between hot and cold, freezing and thawing—much as the soft air of the Mediterranean allows vineyards to thrive on the rocky hills of the Ligurian coast for much of the year.

By the end of the Cretaceous these warm, shallow seas had been drastically reduced, declining at a rate greater than any in the previous 250 million years. The inland sea of North America retreated south, becoming an embrasure along the continent's southern margin, a mere extension of the present-day Gulf of Mexico. The draining of inland seas also occurred in Europe, western Asia, and the northern edge of Africa. It is therefore not surprising that studies of oxygen isotopes in fossil wood and mollusk shells indicate a slow, steady decline in mean global temperature during the Late Cretaceous, followed by an abrupt drop during the last few million years of the interval.

Could these changes have precipitated the major extinction seen at the K/T boundary? I have described the extinction event as sudden and catastrophic, but let us look once again at various claims that the decimation in the biota took place over a long time, say, a few hundred thousand or even million years, rather than in one incandescent day. Some scientists argue for a stepwise extinction pattern in certain groups, such as planktonic foraminifera. However, a comprehensive analysis by Kiessling and Claeys published in 2002 concluded that most marine organisms did not suffer significant extinction before the K/T boundary.

On land, the situation is much the same. The significant declines in cycads and ferns during the Late Cretaceous were accompanied by a presumably related increase in angiosperm species. Conifers did not show any significant decrease whatsoever before the K/T event. Evidence for a major decline in angiosperm and other plant diversity in the high latitudes of the Arctic is known from the Late Cretaceous. However, angiosperm diversity in these floras was persistently lower than that in tropical and subtropical habitats. Moreover, there is no falloff in floral diversity in Antarctica during the same interval, so the pattern fails to suggests a global-scale decrease. As for insects, the trend through the Cretaceous is one of increase rather than decrease. As noted earlier, bees and a number of other insect groups specialized for pollination appear and diversify during the Middle to Late Cretaceous. One qualification here is that the insect fossil record is currently too sketchy to determine whether or not there was a sharp decline in insect diversity in the last 1 or 2 million years before the K/T event.

Most considerations of Cretaceous extinction patterns on land have focused on vertebrates, especially dinosaurs. Here either the evidence points to no significant decline in diversity for the last phase of the Cretaceous leading up to the extinction event, or the data are insufficient to muster a conclusion. There is unfortunately a glaring gap in the available evidence. Late Cretaceous terrestrial localities with vertebrate fossils are plentiful and widespread on many continents. However, the only localities with enriched terrestrial vertebrate faunas that straddle the K/T boundary are in western North America, notably the Hell Creek rock sequence in Montana. This is a huge handicap for any investigation of terrestrial extinction at the global level. As we shall see, this highly parochial geographic evidence also frustrates our attempts to determine what was exterminated and what survived the K/T extinction event.

At any rate, given our current local knowledge, we find no significant falloff in diversity leading up to the K/T boundary among sharks, bony fishes, amphibians, turtles, lizards, champsosaurs (an extinct group resembling but not closely related to crocodiles), and mammals. The one exception is the true crocodiles, for which there is evidence that a few higher clades may have gone extinct slightly earlier than the end of the Cretaceous. Instead, the Cretaceous saw the radiation, not the decline, of most of the above-noted groups. In fact, this is the very time that saw the appearance of modern clades within sharks, fishes, turtles, lizards, crocodiles, and mammals, clades that survived the K/T extinction event and are with us today. That leaves the most famous casualty, the nonavian dinosaurs. A study of dinosaur abundance and diversity through the last few million years of the Cretaceous as represented by terrestrial sections in the Hell Creek beds showed no statistical difference in diversity and abundance of dinosaurs, though in the main the samples were too small to reach a concrete conclusion.

Given these fossil data for both marine and terrestrial environments, the long-term climate change and extinction scenario seems stretched. It doesn't have the "punch" of huge extraterrestrial objects, especially one that cooked Earth's atmosphere at superbroiler temperature for hours. Nor does it suggest the rapid short-term atmospheric changes that come with volcanic eruptions. No doubt these climatic and geographic factors promoted some changes in the flora and fauna. Moreover, many extinction events in the fossil record do not coincide with asteroid impacts and marked volcanism, and climate may have had a greater effect in these cases. But a scenario for gradual extinction during the Late Cretaceous just doesn't fit the evidence here. Extinction was geologically instantaneous and catastrophic.

What is puzzling about the K/T extinction event is the matter not of its cause but of its inconsistent effect. One would think that giant comets or asteroids, or volcanoes, or a combination would be democratic in their kill targets. The kill effect of the asteroid impact would be sufficient to snuff out just about all life—at least all life exposed on the surface—and not just 70 percent of all species. Yet these events in the fossil record seem discriminating, one might say downright fastidious. To make matters more mysterious, the 70 percent extinction factor is an average across groups but doesn't typify *all* groups. Some lineages, like the ammonites, were roundly whacked, losing nearly all their species at the K/T boundary. Species losses were nearly this devastating—between 75 and 95 percent—for certain kinds of plankton, including planktonic foraminifera, bivalves (clams), corals, and starfish. However, some groups, such as gastropods (snails), crinoids, and certain echinoderms, were hit hard but lost less than 60 percent of their species.

Some groups on land were more vulnerable than others. One caveat here is that global-scale extinction levels are hard to estimate because, as I've noted, the sequences of rocks and fossils necessary for such estimates are virtually confined to western North America. Plants are known from widespread localities, however, and we can gain a broader picture of what happened. A recent energetic study by Peter Wilf and Kirk Johnson considered more than 22,000 plant specimens representing 353 species from 161 localities in a very complete rock sequence from southwestern North Dakota that straddles the K/T boundary. They found that Late Cretaceous floras in the section were dominated in abundance and diversity by angiosperms, with ferns, cycads, ginkgos, and conifers representing less than 10 percent of the flora. The evidence of extinction at the K/T boundary is dramatic, affecting roughly 57 percent of the dominant plant species.

The extinction patterns through this fossil-enriched rock sequence can also be tracked by examining tiny pollen grains. The study of pollen, or palynology, has its own rules because of the peculiar qualities of its subject matter. Different kinds of fossil pollen are difficult to associate with their source plant species; indeed, they could be from either closely related or very distantly related plants. Thus different kinds of fossil pollen, known as palynomorph taxa, might represent different species, genera, or families of the source plant. Extinction levels varied in palynomorph taxa for different major groups of plants. Studies of fossil pollen in the North Dakota section show less devastating losses—about 30 percent—than do the fossil leaf data.

Of all the plant groups, ferns seems to be the most resilient survivor of the K/T extinction event. Above the K/T boundary layer in many western North America sequences, rocks show a sudden disappearance of both leaves and pollen. Only a few centimeters above the layer, however, one finds a great concentration of spores representing fern species. Because of the changes in fossil concentrations through a very thin slice of rock section, this effusion of ferns may have occurred less than a few hundred thousand years after the extinction event. High concentrations of ferns, or fern spikes, are known from lowermost Tertiary rocks in many localities in North America. But this pattern does not hold up everywhere. Some sequences with fossil plants in Japan and New Zealand fail to show fern spikes. Nor do they show the dramatic decimation of plant species recorded in the North America.

This disclosure on post-K/T ferns was immediately provocative. It suggested that at least on land, mass extinction may have not been so marked outside North America. And if there were Early Tertiary plant survivors in these far-flung localities, was there a chance for Early Tertiary dinosaurs? Unfortunately, as we've seen in the case of the Raritan sequence, localities replete with fossil plants often have few or no dinosaurs and other fossil vertebrates. (A dramatic exception is the Cretaceous Jehol Biota in northern China.) Just the same, the plant fossil record opened the possibility that when rock sequences preserving vertebrates are eventually found, they might show that some vertebrates, including dinosaurs, may have hung on a bit longer. Recent studies have somewhat muted this possibility, though. A plant extinction pattern, complete with a post-K/T fern spike, very similar to that in North America has been described now for New Zealand.

Because insects are only sporadically preserved, plotting their extinction across the K/T boundary has been difficult. To judge from the known insect fossils, there are no obvious major losses of species. But an ingenious series of studies of the damage done to fossil leaves by insect feeding offers indirect evidence that insect fauna did change in a major way, at least in western North America. In a paper published in 2006, Peter Wilf, Conrad Labandeira, Kirk Johnson, and Beth Ellis analyzed insect-feeding damage on 14,999 angiosperm leaves from fourteen Latest Cretaceous, Paleocene, and Early Eocene sites in the western interior United States. Latest Cretaceous floras showed both high plant diversity and intense and varied damage from different modes of insect feeding. Most Early Paleocene floras slightly younger than the K/T event showed low richness of plants and little insect damage, suggesting that both these components of the ecosystem were devastated. But one site in southwestern Montana dated at 64.4 million years showed a very low diversity of plants but high intensity of insect damage, especially leaf mining, suggesting the persistence of certain insects in a species-poor flora. In contrast, a locality in the Denver basin dated at 63.8 million years showed a high-diversity flora with little insect damage and virtually no specialized feeding—namely, lots of plants but few insects to feed on them. One striking inference is that the effects of the K/T event on plants and insects were severe and food webs were wildly unbalanced 1 or 2 million years after the K/T extinction. Five to 10 million years later, during the Late Paleocene and Early Eocene, the ecosystem recovered, and both high plant diversity and intense and varied plant damage reflect a lush flora fed on by diverse insects.

Because of those diverting dinosaurs, most interest in K/T extinction on land has focused on vertebrates. Here again we have relied all too heavily on the limited geographic coverage of western North America and more specifically on the classic Hell Creek badland sequence in Montana, where the samples reveal a striking selectivity in extinction. Birds survived the K/T extinction event, but 100 percent of all other (nonavian) dinosaurs were wiped out, and three other groups—sharks and their relatives, marsupials, and lizards—suffered more than 75 percent of species extinction in this area. In contrast, the overlying Tullock Formation in the same northern

Montana section shows that 50 to 100 percent of species of certain groups survived into the earliest Paleocene. These hardier taxa included ray-finned fishes, certain groups of mammals, turtles, champsosaurs, and crocodiles. So whatever took out the dinos and the ammonites effected much less injury to some other prominent organisms.

Why this extreme bias in victimization? It has been popular to suggest that dinosaurs were especially susceptible to terrible things like asteroids and volcanic eruptions because of their large size. This is a very silly argument. Dinosaurs big and small disappeared at the K/T boundary, while conversely, some very big crocodiles (some that doubtless preyed on dinosaurs), champsosaurs, and turtles marched right into the Cenozoic. A more involved explanation offered by D. S. Robinson and several coauthors in 2003 came from their efforts to match the patterns of Cretaceous extinction and survival with the primary kill effect of global-scale thermal radiation. They suggested that certain groups, such as waterbirds and burrowing mammals, may have been protected from the hellish atmospheric heat in the first hours following the impact. Even organisms only ten centimeters (4 inches) below the soil surface would have been well insulated from the blast because soil conducts heat poorly. The insulating capacity of water was also considerable, though not so effective as dense ground. The sea surface might have bubbled like a cauldron, and some of its heat carried deeper by wind-driven currents or turbulence, but just the same, organisms living underwater and underground had a better chance of avoiding catastrophe.

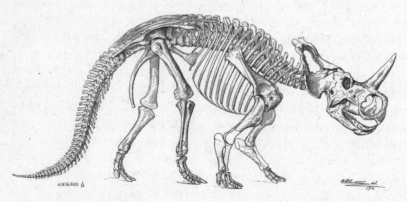

Monoclonius, a K/T casualty

Unfortunately, the complex pattern noted in the fossil record for extinction and survivorship, especially of land animals, does not fit this scenario very well. The authors went to some length to specify which vertebrates would have survived under which conditions, but they had to split hairs. Why did marsupials but no other mammals show high casualties in the K/T boundary in the Hell Creek localities? Were they any less likely than the others to be protected from an incandescent atmosphere? There is no evidence that these small creatures could not burrow any less effectively than the mammalian survivors. What limited clues we can glean from the fossils and the deposits that entomb them do not satisfactorily explain this undemocratic devastation.

I have described the puzzling Cretaceous catastrophe in some detail not only because of the dramatic story of scientific discovery and debate swirling around it but also because this is a pivotal moment in our plot. The epic of the terrestrial and marine ecosystems took a sudden turn 65 million years ago, and things were never quite the same thereafter: no more nonavian dinosaurs, no more ammonites, and no more of certain insects feeding on certain species of flowering plants.

Of course the post-Cretaceous world had its own spectacular events: the rise of advanced mammals and of insects (notably ants and other hymenopterans), angiosperms (notably grasses), snakes, lizards, and birds. About 50 million years ago, tiny photosynthesizing sea algae known as diatoms suddenly increased in abundance and may have triggered an upsurge in atmospheric oxygen that persists today. Then there was that inauspicious appearance, some 7 million years ago, of a certain bipedal relative of the greater apes, whose modern descendant *Homo sapiens* is having its own large global effect on Earth's atmosphere. The Cenozoic was also punctuated by catastrophe, notably the ice age extinctions ranging from about 1 million to 10,000 years ago. Nonetheless, if there is a phenomenon in the last 65 million years to match the trauma, destruction, and transformative power of the K/T extinction event, we haven't found it yet. The only possible competitor, ironically, seems to be the trauma to the biota we are so peculiarly positioned to witness in our own time.

About a mile down Gola del Bottacione from the famous K/T boundary section is one of Umbria's many unassuming restaurants with a delectable daily special. At 2:00 p.m., standard European time, on September 11, 2001, I sat at one of the restaurant's checkered tablecloths, consuming fresh pappardelle smothered with porcinis and a thick sauce that had the color of burnt sienna rather than raw umber. I read the triumphant entries of geologists and geophysicists in the restaurant's guest book, which dated back to the initial discoveries in the 1970s, followed by scores of entries expressing the pleasure of being lucky enough to visit a place of such scientific significance in such a pastoral setting supplied with such good porcinis. There I contemplated, with the help of pasta and an undistinguished but satisfying local table wine, one of Earth's great catastrophes. Later that day I drove to San Gimignano, the "City of Towers." Checking into the hotel, I saw a television flashing with absurdly horrifying images of towers burning in my adult hometown of New York City, and I contemplated catastrophe again.

THE ERA AFTER

The hills around Los Angeles are clad with a scruffy gray-green veneer of thickets called chaparral. This botanical blanket is made up of not a single or a few species but many—creosote, manzanita, sage, juniper, prickly pear cactus, yucca, and other plants both pleasant and punishing. On a late fall day my college friends and I did a "chaparral crawl" from the top of a peak in the Santa Monica Mountains to the mouth of Malibu Canyon near its beach studded with the houses of Hollywood stars. There was no leafy sign of poison oak, but there was plenty of the nasty stuff around in the form of shoots and branches that scratched our arms, legs, and faces. I suffered one of the worst cases of reaction in the annals of Santa Monica Hospital. Chaparral is not the most aesthetic of habitats, but it is the product of favorable, comfortable weather: a dry climate, an easy slide between seasons, and a soft wind from the ocean that brings enough fog and moisture to the hills to sustain the plants in spite of the infrequent rains of southern California. This is what we call in complimentary terms a Mediterranean climate, and chaparral is common to a Mediterranean habitat. Very similar floras are seen in Greece, Italy, coastal Tunisia, southern Chile, and other nice places.

Chaparral has another remarkable characteristic. Many plants in this habitat have seeds that require intense heat to germinate. In a raging fire they explode like popcorn and find holdfasts in the scorched earth, take root, sprout, and grow. Fire, which is decidedly not good for the ranch houses snaking up the hills of Los Angeles, is good for chaparral. This is not an apologia for arsonists. It means only that fire has been part of the life cycle of the chaparral ecosystem long before humans were compelled to contend with it.

I grew up surrounded by hills that were occasionally torched by spectacular chaparral fires. Even from our house in Mar Vista we could watch huge flames leap like solar flares from one chaparral-clad summit to another. Smoke and acrid fumes transformed the already smog-filled sky to an apocalyptic atmosphere. Ashes fell on my turntable and etched out my Jimi Hendrix records. After a fire we drove up the naked roads and took stock of the devastation all around us. The ground was so black it defied any sign of contour; it was merely a shadow that obliterated the texture of hill, gully, and charred stump. And yet within this morgue for an ecosystem were tiny nubbins of green plants struggling upward, preparing the way for insects, birds, ground squirrels, skunks, bobcats, and even mountain lions. I was astonished to see the rebound of the habitat in those once ravaged landscapes within a few years. Still, there were enduring casualties; sage festooned itself around the black, twisted arms of a big manzanita, whose replacement by similar plants must await many years or decades.

After the K/T apocalypse not all organisms were so resilient as chaparral, but the chaparral helps one envision what may have happened in the millennia following that ancient moment of devastation. In the marine environment, as numerous studies have shown, the first colonizers after extinction were groups containing a low diversity of very abundant species, many of them adaptable to a range of conditions—in other words, environmental generalists. Planktonic foraminifera are a striking example. This group was so devastated that only a single species of a form called *Guembelitria* is found in the lowest section of the Paleocene above the K/T boundary. Remarkably, this and one other species gave rise to all the hundreds of younger, including extant, planktonic foraminifera species. During recovery, *Guembelitria* branched off into numerous opportunistic species as environmental conditions improved. A similar pattern is recorded for benthic foraminifera; the shallow bottom of Early Paleocene seas were home for a few most abundant species. One fascinating sequence of changes occurs in strata in Denmark. The first few meters above the K/T boundary preserve only a few species of crinoids, or sea lilies, in addition to forams, but more diverse fossils—bryozoans, or moss animals—gradually appear higher in the section of strata, and some echinoderms start to wane. A similar pattern of delayed succession is seen in lower Paleocene deposits in the Gulf coastal plain of Texas and neighboring states, where the major players are opportunistic mollusks. But once again, there's a problem sorting out global from regional patterns. Studies of other regions show no opportunism in forams whatsoever.

The patterns of recovery following decimation on land are well-known. I have already mentioned the famous fern spikes recorded in the fossil record not long after the K/T event. Ferns with their hardy, plentiful spores apparently played the high-stakes extinction game successfully; some made it through, colonized, and proliferated simply by random chance. It took conifers and angiosperms longer, in North America at least, to establish themselves and regain their pre-K/T glory, but eventually they did. So did insects, invertebrates, microorganisms, mammals, and feathered, flying dinosaurs better known as birds.

How long did all this take? That central question is a current hot topic in paleontology. Scientists have documented the turnover of different species, the amount of diversity and the range of ecological roles in the fossil record. A recent, widely cited study by Kirchner and Weil suggests that it takes at least 10 million years for the recovery of an ecosystem after phases of extinction even less devastating than the Cretaceous or Permian mass extinction events. Others, including Douglas Erwin, question this. The skeptics point out that the succession of crinoids to other echinoderms and bryozoans in the Danish sequence are a matter of "only" about 2 or 3 million years. Reef systems were long thought to recover very slowly from major extinction events because they need a carbonate platform on which to build their coral cities, but recent studies show that many ancient reefs (some of which did not require carbonate platforms) were much more resilient. Studies of carbon isotopes suggest that the productivity of marine environments recovered within a few hundred thousand years of the K/T event, though the flow of organic material to the deep sea and the reestablishment of a new ecosystem of multiple trophic levels may have taken an additional 3 million years.

On land there is a similar mosaic recovery rate. Analysis of organic carbon isotopes in land plants show that recovery of the terrestrial carbon cycle (and thus atmospheric carbon as well) occurred within 130,000 years. We also know that the regrowth of fern gardens may have been a matter of thousands or tens of thousands of years. This is less than 1/65th of the history of life from the K/T event to the present day. A few ferns doth not an ecosystem make, at least not an ecosystem anything like the Cretaceous cornucopia before the disaster. Nonetheless, it seems that conifers and angiosperms were replenished over "only" about 1.5 million years. Analyses of mammalian diversity in North America shows a dip to about fifteen species at the K/T boundary followed by a sharp rise in diversity of more than sixty species within the first few million years of the Paleocene. (These figures are now outdated, because recent papers document even more diversity in Cretaceous mammals.)

Yet by other measures, recovery took a particularly long time. One of the oddest features of the earliest part of the Cenozoic is that there were virtually no big animals around. Body size in North American mammals sharply increased from small, mouse-size forms to medium, squirrel-size forms in the first few million years of the Paleocene, but there was no mammal larger than a badger in body mass for another 15 million years. It was not until 45 million years ago and 20 million years after the K/T event, in the Eocene, the second Tertiary epoch, that truly large mammals evolved. In fact, no mammals stood above two meters (6.7 feet) in height until the Oligocene epoch, about 35 million years ago. Only then do we find evidence of high-browsing animals approaching or exceeding the size of a modern elephant. When we think of the major effects that large mammals have today in cropping and molding vegetation, not to mention that of sauropods and other huge plant-eating dinosaurs in the Mesozoic, we can appreciate the oddity of this 30-million-year lag in trophic opportunism.

When organisms that once flourished go extinct, they do not come back. But extinction also leads to opportunity; it may be one of the major influences of the diversification of life on this planet, and we seem to be living at a time when biological diversity is at a high point. True, the fossil record is imperfect, and it is difficult to measure diversity through time, but despite the many dips in the diversity curve, including precipitous ones caused by the five big extinction events, the overall fossil record shows an upward trend.

Why would this be? Perhaps it is because the organisms that survive extinction flourish to new dimensions, tapping into myriad opportunities opened through the loss of competitors, diversifying into new lineages. A computer simulation analysis showed that under certain conditions 80 percent of the original phylogenetic structure—the larger branches of the tree of life, or clades—can survive a 95 percent loss of their species and become evolutionarily fecund once more. Thus Earth's flowering terrestrial ecosystem was battered but not completely shattered in the thermal blast and global fires of a Cretaceous impact. With the loss of its nonavian dinosaurs and other selected organisms, life on land was not exactly what it was, but it remained on terra firma nonetheless.

We divide the broad phase of Earth's history that followed the K/T extinction event, the Cenozoic Era, into epochs, the Paleocene, Eocene, Oligocene, and so forth. The term *Cenozoic*, as I've noted, means "new life," but the interval is popularly referred to as the age of mammals. After the extinction of the dinosaurs, mammals were the big animals on land and even in the oceans they far surpassed the mass of the largest fish. On the other hand, they don't qualify as "new life," really, since the group is nearly old as the dinosaurs, about 210 million years in age. Other important Cenozoic life also had deep roots into the Mesozoic. Organisms profoundly resculpted the landscape, while new marine and freshwater aquatic creatures changed the composition of the atmosphere, which in turn affected life both on land and in the oceans.

Let's start with the aquatic organisms since their effects were so wide-ranging. You will recall that it took a few million years for the marine ecosystem to reorganize after the K/T extinction event. Among its components was a low-profile but diverse group of tiny one-celled algae called diatoms. These splendid organisms are encased in a covering, or test, made of silica, the basic component of glass. Indeed, diatoms look like tiny glass jewelry boxes filled with precious amber-colored fluids and organelles. Diatoms are autotrophs: they generate their own energy through photosynthesis, not only producing their own fuel but releasing oxygen. This capacity had a mighty importance to the change in the atmosphere 50 million years ago.

Because diatoms depend on sunlight, like many other algae as well as aquatic plants, they inhabit shallow water, down to about two hundred meters (656 feet), where light can penetrate. The pigments of diatom chloroplasts are yellow-brown instead of green, like plants and algae. With their transparent, sparkling silica tests, diatoms are wondrous and varied in architecture: diamond-shaped, ellipsoidal, circular, double-lobed, cup-shaped, pennate, and snowflakelike, with radiating glassy tubes. Some have stacked linear colonies of cells that fit together in a swiveling structure like a carpenter's rule. Others are called art deco diatoms because their colonies take on wonderful radiating glass stems with fan-shaped tops showing striking patterns of yellow, brown, and black, reminiscent of the elaborate ceiling decorations in movie theaters of the twenties. Diatoms live in a secret world that requires a microscope lens to explore, a world deceptively disguised to the naked eye as brown scum. They compose the nondescript slime we see on water plants, on pilings, or around the borders of ponds and lakes.

The unaccountably abundant and hugely diverse diatoms are the passion of only a handful of specialists. At a scientific meeting I ran into a colleague, the diatom expert Edward Theriot, who said that about three-quarters of the entire international community of researchers on diatom systematics, about ten of them, were in attendance. "When we go to meetings, we make sure we take separate planes," he noted. It's too bad that diatoms have not attracted more attention; they tell a very interesting story. They are known from as early as the Jurassic, about 160 million years ago, but at about 50 million years something important happened to them. They diversified and became much more populous in the shallow seas. This diatom bloom clearly boosted the trophic resource for marine creatures and actually fueled marine radiations, and its effect was even more far-ranging. The proliferation of oxygen-producing diatoms stepped up the concentration of that precious element in the atmosphere.

Atmospheric oxygen, just like CO_2 content, has gone up and down over time, as we have seen. The Late Permian, about 280 million years ago, showed an all-time high concentration of oxygen of nearly 40 percent, compared with 21 percent today. But there was not a steady decline, since at the end of the Permian oxygen levels fell as low as 15 percent, a precipitous dip that might have been a major factor in the demise of large oxygen-hungry animals like synapsids and therapsids (mammal-like reptiles) during the Permian extinction event. Oxygen levels fell even lower, to 13 percent, in the Early Triassic and rose only slightly in the Late Triassic; following a few highs and lows in the Jurassic, they remained at a sustained 18 percent through the Cretaceous and across the K/T boundary. The next significant upswing occurred between 50 and 45 million years ago in the Eocene epoch, interestingly at about the same time that shallow near-shore seas were coated with diatoms. Thereafter oxygen remained at roughly 21 percent to the present.

I and several other researchers, an investigative team led by Paul Falkowski at Rutgers University, recently wrote a paper that ties this Eocene oxygen surge to a major change in the ecosystem, one related to the evolution of placental mammals, the diverse group to which we humans belong.

Diatoms, as
depicted by Ernst
Haeckel, 1904,
*Kunstformen der
Natur*

The term *placental mammal* is both misleading and meaningful. The placenta is the complex of multilayered membranes with blood vessels and
nerves that both protect the developing embryo and link it to the maternal
tissue. Placentas are found in many mammals, including the pouched
kangaroo and its opossum relatives among the marsupials, as well as the
monotremes (platypus and echidna), but only in placental mammals are
the membranes around the fetus developed into an extremely elaborate
structure that keeps blood and nutrients flowing to the developing young inside the mother during the prolonged pregnancy, or gestation. Research
shows some inefficiency in the transport of oxygen to the fetus in this system, despite the placenta's complex architecture. A portion of the precious
gas is dissipated through the blood vessel walls before it reaches the fetus.

Under conditions of low atmospheric oxygen levels, this inefficiency could
be dangerous, even lethal to the fetus. The larger the animal and its offspring, the more acute the problem. That is probably why today few placental mammals, and no really large ones, reproduce and live above 4,500
meters, or about 14,700 feet; at this altitude oxygen levels decrease to about
15 percent.

Now let us again consider the very slow increase in body mass in mammals through time. During the Mesozoic mammals evolved for 150 million
years as small creatures, on average being about the size of a house mouse.
There are a few anomalies, like *Repenomamus*, the mammalian predator
from the Early Cretaceous of China that devoured small juvenile dinosaurs, but even this animal was no larger than a badger. In the first few
million years after the K/T event average size increased to squirrel size, but
not until well within the Eocene, about 20 million years after the K/T event,
do we see another upward surge in body mass. This is also when many of
the modern placental mammals, such as horses, tapirs, whales, bats, and
primates, first appear in the fossil record. These "early moderns" include
cloven-hoofed artiodactyls, the ancient relatives of pigs, camels, deer, and
antelope. Artiodactyls are *the* dominant large herbivorous mammals today;

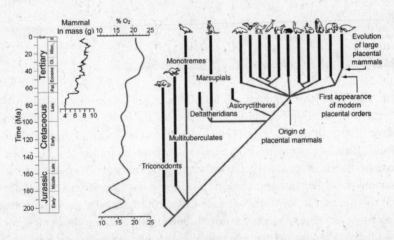

Mammalian body size through time (left), atmospheric oxygen change from the Jurassic to the present (center), and mammalian evolution (right), where thick vertical bars
represent the age range of the mammal groups shown and thin lines represent branching points

think of those herds of tens of thousands of wildebeest and caribou. After the Eocene the average mammalian body size steadily and slowly rose. A sudden, anomalous dip late in the record, about 3 million years ago, may have been related to an extinction event that targeted large herbivorous mammals, and another series of extinction events between fifty thousand and ten thousand years ago wiped out large mammals on most continents but not Africa and southern Asia. We shall consider these developments in due course: they have been tied to the oscillations of an ice age climate or energetic hunting and slaughtering by early humans, or both.

The increase in both atmospheric oxygen and mammalian body size during the Eocene coincided with another dramatic environmental change. The end of the Paleocene and the early phase of the succeeding Eocene marked an all-time high in global temperature for the Cenozoic, the highest in the past 65 million years. During this phase, called the Paleocene-Eocene Thermal Maximum, or PETM, mean atmospheric temperatures have been estimated to have been about 18°C (64°F), compared with the cooler (though rising) average today of 12° to 14°C (54° to 57°F). What biological effects did the heat have? The question has drawn the attention of climatologists as well as paleontologists, who study the PETM in order to reconstruct the ancient global ecosystem and to help predict what might happen to today's ecosystems as our climate changes. During the PETM, warm tropical to subtropical conditions extended nearly to the North Pole. Crocodiles, turtles, lizards, frogs, and broad-leafed trees lived in remote Arctic regions like the Ellesmere Islands, places today encrusted with ice for most of the year and inhabited only by lichen, a few stalwart grasses, tiny nubbins of flowers, polar bears, and wolves.

Another dramatic trend in the Eocene was the rise in insect diversity, as suggested by the increased variety of damage from insect feeding found on fossil leaves. The hothouse Eocene Earth fostered a flourish of new, more diverse, and in some cases, bigger life-forms. But the interval also saw the extinction of many archaic mammals and other lineages, species that were perhaps not adapted for the marked change in climate.

After a few million years these hothouse conditions gave way to a cooling trend that augured yet another fundamental change in the landscape. The cooling was accompanied by increased aridity. In large regions outside the main tropical forests around the equator, rainforests were supplanted by open gallery forests, much like the Serengeti today. In drier areas, gallery forests gave way to wide-open spaces, plains and steppes filled with grasses. If you took a time machine back 25 million years to areas of interior North America, you would be struck by the familiarity of the vegetation and surroundings, rolling terrain broken up by patches of oaks and sycamores in valleys and pines on the hilltops. You might think you were in modern western Nebraska; all that would be missing would be a silo and a barn in the distance. Of course you would at the same time be struck by unfamiliar big mammals—long-necked camels, giant pigs, and elephant-like and rhino-like forms—though they wouldn't be so different from the big, diverse mammals that live today on the African plains.

Indeed, the similarity between that time and ours is enough to prompt some paleontologists to claim that the inauguration of widespread grasslands marks the origin of the true modern ecosystem. The differentiation of other modern habitats, such as tropical rainforest or boreal conifer forest, was also apparently a mid-Cenozoic event. But I prefer to stress the fundamental fabric of these ecosystems, the scaffolding formed by angiosperms and pollinating insects, that gave rise to the modern world more than 100 million years ago.

Grasses are, after all, angiosperms themselves; their history goes back to those 90-million-year old fossils found in the clay pits of New Jersey. But grasses during the first 70 million years of the angiosperm dominion were not flamboyant elements of the vegetation. Diversification of many dominant angiosperm families continued through the period that followed the Cretaceous, through the Tertiary, and up to the present, and among the latecomers were the grasses. Grass-dominated ecosystems were established in many parts of the world in the Miocene and the Pliocene epochs, between 20 and 10 million years ago. It was not until the mid-Cenozoic that they spread and dominated under cooler and especially under drier conditions.

Obviously open spaces and grasses go well together. Grasses are wind-pollinated, a reproductive strategy well suited to the reliably energetic winds of the open plains and savannas. As anyone conscripted to the responsibilities for maintaining a lawn can appreciate, wind pollination works well, and grasses, like invasive wind-pollinated weeds, have a relentless capacity for reproduction and growth. It is also good for grass eaters. Clearly the spread of the grasslands helped sustain the herds of large mammals that depended upon them.

But the spread of grasslands wasn't a boon for all creatures. The Early

Cenozoic was a time of flourish and diversification for many small- to medium-size mammals that spent their lives in or above trees. With the shift to cooler, drier, open habitats in the mid-Cenozoic, the great centers of diversity for these arboreal or aerial groups, as well as for most other animals and plants, were confined to the tropics.

Today, for example, bats are the main reason why the tropics have a much greater diversity of mammals than temperate regions. Bats are another modern group whose origins can be traced back to the Early Eocene. Although there are a number of fruit-eating, primarily Old World, bats, and even nectar-feeding bats that pollinate plants, most of the eleven hundred species of living bats feed on night-flying insects. The insects offer an awesome plenitude of food that is virtually untapped by other vertebrates. Birds, the other dominant vertebrate fliers, are almost all daytime animals heavily dependent on vision. They nab caterpillars off leaves or pull worms from the ground, but for the most part they don't go about picking up insects in midair after dark. The night-flying insects are mostly left to the bats.

And what an opportunistic group bats have turned out to be! Using their astounding sonar system and aerial acrobatics, they zoom in on insects fluttering above water, around lampposts, or in bushes. They also pick off huge swarms of insects that fill the night skies. During every evening of the year Mexican free-tailed bats, *Tadarida braziliensis*, emanate in a horde ten million strong from roosting spots like Bracken Cave in Texas; they gain altitude and then cruise for hours between 40 and 100 kilometers (24 and 60 miles) per hour at more than 3.5 kilometers (11,500 feet), a horde so spectacular in dimensions that it can suddenly light up the Doppler radar of nearby air force control towers. With their large, flabby mouths opened wide, these bats are flying vacuum cleaners, sweeping through great clouds of insects. Their technique is effective; it has been calculated that an adult *Tadarida* can consume a mass of insects equal to its body weight in one night. There is a survival principle working here; flight as a form of animal locomotion requires tremendous energy and thus tremendous amounts of food, much more than one needs for running, digging, or climbing.

In the dense, vegetation-choked tropics, bats fly lower, slower, and even more acrobatically, using their marvelous echolocation signals to close in on their insect prey. The sound emitted by the bat bounces off its flying target and returns to its oversize ears with information on target location, speed, size, even texture. Many night-flying insects have developed erratic flight patterns meant to scramble the bat radar. But these ingenious modes of predator avoidance don't always work, given the extreme refinement of the bat echolocation system. Even if the bat is slightly off target, it may catch its prey—like so much arthropod popcorn—in a wing or the tail flap and simply toss it back to its mouth, all in mid-flight.

Bats are flying virtuosos, but they do need to rest, sleep, and procreate. For these purposes, they congregate in great roosting colonies like Bracken Cave. In the tropics, where there aren't many caves but where trees abound, bats roost within the cavities of trunks or under branches, leaves, or bushes. There they keep company with other arboreal mammals. The cast of hairy characters inhabiting the treetops varies from one continent and region to another. In North America and Europe and northern Asia, we associate tree life with energetic squirrels and chipmunks, but in the tropics arboreal mammals are more diverse and wondrously varied in form and behavior. South and Central American rainforests harbor the bizarre, long-armed, hairy tree sloths, the long-tongued, long-clawed anteaters, and a great variety of opossum-like marsupials. In Africa, in both the jungles and the more open-gallery woodlands, we encounter the tree hyrax; its screeching surely ranks as one of the weirdest and most irritating sounds in nature. The first time I heard a tree hyrax "calling" outside a tent camp in the Serengeti, I was reminded of the line in a movie wherein Woody Allen, as a forlorn Manhattan resident, complains of a noise in the apartment above that sounds like a saw slicing a trombone.

All these forested regions and many more contain a mammal group of special interest to us. In South America the tree-loving primates, the New World monkeys, include the capuchin, with its frothy white and brown head that reminded people of the heads and cloaks of Capuchin monks (who also inspired that indispensable Italian breakfast stimulant cappuccino). The arboreal primate contingent in Africa includes the colobus monkeys, mangabeys, and chimpanzees. Madagascar is home to the svelte and elegant lemurs. In Asia too, primates thrive in the trees, and their number includes the red-shanked douc langur (*Pygathrix nemaeus*), a beautiful colobine monkey found in south-central Vietnam and Laos, now seriously threatened by habitat destruction and hunting for food and traditional medicines. Asian forests also harbor those champion tree swingers the spidery-armed gibbons. A more desultory swinger, the orangutan, inhabits the dark forests of Borneo and Sumatra. *Orangutan* in the Malay language means "man of the forest," a fitting appellation for a creature so alien and secretive and, at the same time, as Queen Victoria noted haughtily, "so disagreeably human."

It is likely that primates evolved in trees for most of their history. The earliest-known mammals thought to be relatives of primates are forms like *Purgatorius*, from the Cretaceous Hell Creek badlands of Montana. These are mouse-size animals known only from their teeth and jaws, so we don't know much about their lifestyle. Primatelike mammals were an important component of the post-K/T rebound and of the diversity of mammals recorded in the earliest Cenozoic. At one time many of these early groups were placed within primates, but their affinities are now hotly debated. Uncontested primates are first known from the Eocene, dating back to about 55 million years ago. Some of these and later species are known from skeletons or skeletal parts that clearly show flexible ankle and wrist joints and elongated fingers and limbs, features that seem especially well adapted to climbing and swinging from branches. The orbits, or eye sockets, in some of the fossil skulls also indicate that the eyes were directed forward, with a strongly overlapping field of vision, a feature that enhances the depth of

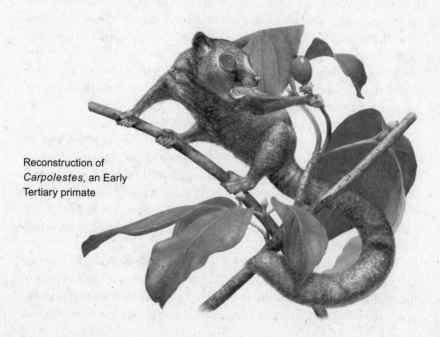

Reconstruction of
Carpolestes, an Early
Tertiary primate

the field and stereoscopic vision, important to life in a complex, three-dimensional world of limbs, branches, and vertical spaces.

Of course not all primates stayed in trees. Baboons are largely landlubbers. While the apes include the arboreal gibbon and orang, other apes are more groundbound. Gorillas and chimpanzees are distinguished by their knuckle walking, although the latter shows great skill moving in trees as well as below them. The locomotory versatility of our nearest relative, the chimpanzee, does not also extend to us. The first members of the human branch, the Hominini, are 5- to 7-million-year-old fossils known primarily from skulls, teeth, and only skeletal fragments; more complete fossil skeletons between 3 and 4 million years old give us evidence and insight on locomotion and lifestyles. The long arms, short legs, grasping feet, and flexible knees in early human species like *Australopithecus afarensis*, represented in part by the famous skeleton Lucy, indicate that this species was probably adept at climbing trees.

Australopithecines do not have a skeletal structure that suggests a full commitment to an arboreal lifestyle, however. Instead, these forms have all the indications of a bipedal (two-legged) upright stance and walk, one feature that separates us from our ape relatives. The hips are short and wide and connect to the backbone in the rear rather than in the front, as in apes. This allows the full weight of the body to pass through a point directly supported by the hip and legs. The sharp angle of the knee allows the inward tilt of the upper legs, bringing the legs together in a supportive stance. In addition, the perpendicular angle between foot and lower leg offers a flat, stable base for the upright legs and body.

Thus the diverse ancient human lineage, as well as modern humans, all included in the group Hominini, has experienced several million years of evolution as earthbound animals. The first fossils of the human lineage are known from Africa, and some of the earliest skeletal evidence suggests that forms like *Australopithecus afarensis* were fairly adept in the trees. Nonetheless, further human evolution is coincident with a shift in the African environment toward increasing aridity, spreading grasslands, and more fragmented forests.

There were some advantages in this land life. One could roam the open spaces between protected areas, tracking the shifting abundance of food or the change in the seasons. But there were also disadvantages to this ambulatory lifestyle. Early humans were, pound for pound, much weaker and

slower than many other animals. They most likely had only rudimentary social organization and virtually no technology (the earliest fossils are not
accompanied by even the most basic of stone implements). The fossil evidence makes it clear that they were easy prey for formidable hunters, such
as the leopard and the hyena. The 7-million-year-old Cenozoic world that
confronted the earliest humanlike species was indeed a terrifying one.

Part Four

TERRA HUMANA

In times of change, learners inherit the Earth, while the learned find themselves beautifully equipped to deal with a world that no longer exists.

—Eric Hoffer

15

WHO THEY WERE

Olduvai Gorge is an impressive canyon incised in the eastern Serengeti of northern Tanzania. From the air this erosional scar has a distinctive Y-shaped outline, formed by the split between the main gorge and its major tributary. From the rim of the gorge one can admire the buff and yellow sandstone outcrops, which alternate with dark volcanic beds of ash and lava that look from a distance like burned coffee grains. The geological features of Olduvai—the sands, the lake bed limestones, and the volcanics—are covered in many places with acacia and other Serengeti scrub, as if there were some attempt by nature to conceal its shocking secrets. In the rocks that are exposed one can dig and find bones of ancient species of hartebeest, waterbuck, pig, and springbok, even as their modern descendants roam the plains above. As a paleontologist I have worked in many canyons deeper, more expansive, more chromatic, and even richer with fossils than Olduvai, but when I visited this site in the late 1990s, it was not for work. Along with several others from the American Museum of Natural History I was there to pay homage to the triumph of science.

Olduvai is what the paleontologist and writer Richard Fortey might call an oxymoronic sacred place for evolution and Earth's history, a place where the antiquity of our human lineage was revealed in decidedly nonbiblical terms. Here over more than half a century of dogged paleontological labor, the famous couple Louis and Mary Leakey and others discovered some of the most important remains of early humanlike and human species, as well as their stone tools, the signature of imaginative new denizens of the African ecosystem.

Olduvai, then, is a site that records our ancestry not in the several hundred years of generations in Genesis, or even in thousands, but in millions of years. Against this expansive timescale what do we mean when we say *human* evolution? What is human and what is not? This question is debated endlessly by paleontologists, physical anthropologists, ethnologists, psychologists, neurobiologists, philosophers, mystics, poets, thinkers, believers. A resolute answer does not seem forthcoming. From a biological perspective the uncertainty seems to reaffirm the closeness of our relationships to the chimpanzee and other living apes. Indeed, only about 1.3 percent of the 3 billion DNA nucleotides in the genomes of chimps and humans differ. This nested affinity is reflected in some scientists' modern classifications of humans and their nearest primate relatives. The group Hominoidea includes both the gibbons, the Hylobatidae, and the greater apes and humans, together the Hominidae; the latter comprises the orangutan, the gorilla, and the group Homininae, which encapsulates chimpanzees, bonobos, and the Hominini; they in turn include several lineages of ancient humanlike forms and the genus *Homo*, the juncture at which the title human is assured.

What about all those other long-lost creatures that parted ways with the line leading to chimpanzees? The human characteristics we emphasize in distinguishing ourselves from the rest of the biological world—higher powers of reasoning, language, technology, complex civilization, and so forth—have nothing to do with the breakaway features we see in the fossil record that segregate the line leading to humans. As we have observed, the most striking and tangible modification that characterizes this evolutionary split is skeletal features that show our antecedents, unlike apes, as having been bipedal, able to walk upright on their hind legs. But what about other *more human* features?

Technology often comes up in this discussion. The attribution "man the toolmaker" is common, and some argue that the appearance of stone tools in the fossil record marks the origin of true humans. However, it is now well understood that chimpanzees among other animal species use grass stems and other objects as tools for various purposes, such as for probing and extracting termites from their nests. The use of tools by humans rather than other primates seems to be a matter of degree—more intensive and eventually more technological usage—rather than pure innovation.

Another oft-cited feature is the development of language. We know that the anatomy of the vocal tract in modern humans—the soft palate, pharynx, epiglottis, and larynx, or voice box—is well suited for producing a great range of complex sounds. The vocal tract in chimpanzees and other apes is not so highly modified. But language and its complexity are doubtless associated with features of the brain too subtle to be detected in the fossilized internal casts of skulls.

One aspect of the brain, however, shows a clear trend in human evolution: its relative size as indexed against body mass. Relativity is important here because it is well-known that larger mammals regardless of their behavioral complexity or simplicity tend to have larger brains. So absolute size of a brain is not as significant for inferring "brainpower" as is the size of the brain in proportion to the size of the entire animal. In early humanlike forms dating back some 7 million years the relative brain size was hardly larger than that of present-day apes. The fossil record between 1.5 and 2 million years before present shows a significant increase in relative brain size. Many scientists consider this a telltale mark of humanity's origin. They argue that some time thereafter a complex language and intricate social organization are at least indirectly indicated by this increased brain capacity. Interestingly, the few differences between the human and chimp genomes are concentrated in genes associated with the brain and cognitive functions. The genus *Homo* includes at least eight extinct species and one surviving one, *Homo sapiens*, that show a larger relative brain size than other hominins, such as *Australopithecus*. Nevertheless, the transition from earliest forms and to *Homo* itself is enough of a blend to use the term *human* rather less formally than the evidence of relative brain size would have it.

This notion of a transition, a blend, between modern humans—namely, us—and our antecedents obscures what really happened. The pattern of human evolution is not simply a single-file procession from some stooping, knuckle-walking ape to Leonardo da Vinci. It is not like those various, sometimes humorous scala naturae imprinted on T-shirts (from ape to modern man to hippie who looks like the ape to knock-kneed tourist with camera and other accoutrements, etc.). Instead, human history was another great evolutionary experiment, a branching bush comprising numerous species that appeared and eventually died out, with only one surviving twig of *Homo sapiens* to represent its endurance.

Olduvai Gorge is important because in it are preserved some key branches on this human evolutionary bush. Sometimes a paleontological site of legend becomes so overnight, as in the case of the Flaming Cliffs of

Mongolia's Gobi Desert. More often, however, the great importance of a place is revealed only through excruciatingly persistent labor over many years. The latter is certainly the case for Olduvai. Discovered when a small collection of bones was found on its slopes in 1911, Olduvai Gorge proved stubbornly resistant to disclosure of its primary treasure, the skeletal remains of some of the oldest human and humanlike species. A large collection of more than seventeen hundred fossil mammal bones collected in 1914 were alleged to include one ancient human skeleton, but this specimen was later revealed to be merely the remains from a recent burial site dug into the older deposits. Louis Leakey began collecting in Olduvai in 1931. Within a few years, the collaboration between him and his wife, the archaeologist Mary Leakey, yielded fossil mammal skeletons and stone tools from several levels in the gorge, and they both became very famous. But it was not until 1959, fully forty-eight years after the discovery of Olduvai, that Mary Leakey found a spectacular fossil of a possible primitive human in Bed 1, the lowest and oldest fossil-producing layer. Nearly three decades of work in the Y-shaped gully by the Leakey team had finally paid off.

This inaugural skull of what is now called *Paranthropus boisei* was later joined by a variety of fossils of an even more advanced humanlike form, *Homo habilis* (handy man), now thought to be the toolmaker, the source of the various stone implements found at Bed I. *Homo habilis* was also recovered at a higher and slightly younger Bed II. Based on potassium-argon dat-

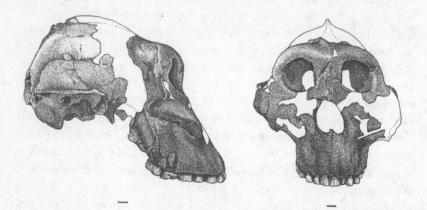

Side and front views of the skull of *Paranthropus (Zinjanthropus) boisei.* Scale bars are 1 centimeter.

ing, the ages of Beds I and lower are estimated to be around 1.8 to 1.6 million years. Higher levels eventually showed more advanced human remains of the species *Homo erectus*. These were associated with more advanced tools, even some resembling hand axes, and widespread evidence of toolmaking, scavenging, hunting, and butchery. Louis Leakey died in 1972, and Mary continued work there until 1984. Subsequently, American teams discovered more important remains of *Homo habilis* from Level I. There is every reason to expect that this great gorge will disgorge spectacular and precious human fossils in the years to come.

Olduvai is not the only crucible for early human evolution. A recent discovery made in Chad by the French paleontologist Michel Brunet revealed the 6- to 7-million-year-old species *Sahelanthropus tchadensis*, which shows a combination of both apelike and humanlike features. Comparable in age is *Orrorin tugenensis*, which also shows a fascinating ape-human mosaic in its anatomy. The 5.8-million-year-old *Ardipithecus kadabba* comes from Ethiopia, as do some of the famous 3-million-year-old fossils, including Lucy, of *Australopithecus afarensis*. The latter may have made the famous 3.5-million-year-old footprints in the ash beds of Laetoli, near Olduvai Gorge.

Localities farther south in Africa have been particularly critical in the study of human origins. Despite Darwin's intuition that Africa was the homeland of humans, no such remains of any antiquity were found in Africa in his time or indeed for many decades thereafter. Instead, fossils of humans from Europe and Java prompted many to surmise that the staging area for human evolution was somewhere outside the African continent. Then, in 1921, a fossilized human cranium was found in a mine near Broken Hill, in what was then Northern Rhodesia. This was followed in 1924 by the discovery of a fossilized skull in a limestone quarry near Taung, in South Africa. The fossil was named *Australopithecus africanus* by Professor Raymond Dart, who emphasized both its ape and humanlike characteristics. Despite its ultimate credibility, his claim was widely doubted for years; many had convinced themselves that humans had their origin in Asia, and many others had been duped by such famous false fossils as the Piltdown Man from England. But soon excavations at Sterkfontein, a network of limestone caves in South Africa, yielded an adult australopithecine skull that closely resembled the juvenile skull from Taung that Dart had described. Soon thereafter Sterkfontein produced

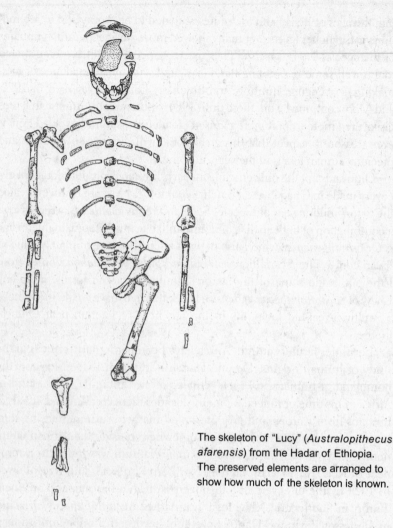

The skeleton of "Lucy" (*Australopithecus afarensis*) from the Hadar of Ethiopia. The preserved elements are arranged to show how much of the skeleton is known.

hundreds of early humanlike fossils, including an australopithecine skeleton dated at 3 million years.

The latest chapter at Sterkfontein concerns a curious early hominin that was just recently scraped out of its hard rock matrix. The talented Ron Clarke was led to this discovery when he stumbled upon some isolated footbones in collections made from Sterkfontein many years earlier. The skeleton, known as Little Foot, is nearly complete, with short arms and short fingers, rather like a modern human. It was bipedal, but the separation of the toes indicates it may have been an adept climber as well. At Swartkrans, another cave site near Sterkfontein, beds between 2 and 1.3 million years old contain fossils of two large hominins, *Paranthropus robustus* and *Paranthropus boisei*, as well as a possible species of *Homo erectus*.

Other early humanlike fossils, including *Australopithecus africanus* and *Paranthropus aethiopus*, demonstrate the striking early diversity and experimentation in our ancestry (the conservative estimate is more than thirteen species predating 2 million years). These fossils have another striking feature in common: they all are from Africa. If our human lineage diverged from the line leading to chimpanzees more than 8 million years ago, the first 6 million years of that history are locked in the rocks of Africa. There is not a scrap of human fossil of this antiquity anywhere else. Indeed, the first evidence of humans outside this continent consists of fossils of *Homo erectus* from the Dmanisi site in the Caucasus region of Georgia, dated at 1.8 million years.

Negative evidence—the absence of evidence to the contrary—is not the most satisfying base on which to build a theory, but in this case we might say the negative evidence is overwhelming. A lot of paleontologists have looked in a lot of places outside Africa (I must confess I am one of them) to find humans that antedate 1.8 million years, but their efforts have thus far come to naught. Meanwhile, the early human record in Africa keeps improving. New research on human fossils from sub-Saharan Africa continues to be consistent with this pattern. Even most studies of genes and DNA in modern humans fit well with the fossil evidence for African origins. The theory holds. Africa is our home, the place where we were born and where we took our first fundamental, and sometimes tentative, evolutionary steps.

Not that Africa was necessarily an accommodating place for human survival, procreation, and evolution. Earlier I remarked that early humans lived in a terrifying world. Here is the evidence for it. In his original studies, Dart noticed that australopithecine remains in the caves were associated with many other animal bones. He reasoned that these early hominins were carnivores who took their kills back to their caves for feasts. Later evidence, though, pointed to the greater likelihood that these human forerunners were prey rather than predators. Their bones were merely among many animal bones that fell in the cave after being killed and eaten by predators such as leopards or hyenas. A fascinating example consistent with this theory is a skull of a young individual of *Australopithecus robustus* from Swartkrans with depressions that match closely in size and spacing the puncture marks made by the canine teeth of a leopard.

Attacks by hungry animals in some areas of Africa today are frequent enough to require routine precaution. When I tried to leave the luxurious, thoroughly civilized Serengeti Sopa Hotel for a jog down the well-graded service road one morning, the staff told me politely but emphatically that such an activity was definitely not allowed, for a beautiful leopard had a favorite hunting spot nearby. Indeed, the leopard is perhaps the most feared of the African predators; its mysterious nocturnal activities include occasional attacks on unwary trekkers and hunters and even invasions of a village. Imagine the horror of the residents helplessly calling out for a child that the stalker has carried back into the night forest. Despite their gracile frame, leopards are spectacularly powerful; they can drag a 150-pound carcass of an antelope or a human twenty feet up into an acacia tree. Lions are of course famous for their ferocity and their historical reputation as man-eaters. Another carnivore, the hyena, is known mainly as a scavenger, but this animal as well is a bona fide predator. Hyenas are often the first to bring down prey; the remains they are seen scavenging often belong to the very animal they originally killed and subsequently relinquished to lions; when the lions have had their fill, the hyenas come back for the table scraps, chomping through bone to the nutrient marrow with their powerful jaws and enormous canines. Hyena attacks on humans also occur in many areas in Africa.

Bones from leopard kills are often found concentrated below their dining trees. One could easily imagine leopards strategically placing themselves in trees near the entrance of caves, nabbing an unwary australopithecine and later dropping the defleshed carcass from the tree to a point on the ground where it easily could either wash or directly fall into the cave entrance. Hyenas, another prospective predator on early humans, are, interestingly, also habitual cave dwellers.

The dispersal of humans from Africa, in the first instance *Homo erectus* or a species close to it, was followed by a number of other milestones occurring over a relatively short period of time: the first firm evidence of domestication of fire (590,000 years ago), the origin of modern *Homo sapiens* (150,000–200,000), the first Australians (50,000), the earliest representational art (35,000), the first Americans (13,000), and the synchrony between these later appearances and the extinction of many large mammals in the same regions. One of the most intriguing chapters in this history is a new discovery. The skeleton of a meter-high human with a brain about the size of a chimpanzee has turned up in Liang Bua Cave on the island of Flores in Indonesia. Despite its rather primitive aspect, this new species, named *Homo floresiensis*, is dated at only 18,000 years, long after *Homo sapiens* was dispersing throughout Asia.

One currently accepted scenario is that, if not some bizarre anomaly, this pygmy "man from Flores" represents a descendant population close to Asian *Homo erectus* or even *H. ergaster* whose isolated evolution on the island led to its diminutive size. Dwarfing of mammals, including elephants, is a frequent phenomenon on islands of various ages and places, including Flores, where fossil elephants that lived alongside *Homo floresiensis* were unusually small. Yet many questions remain concerning this curious pygmy human. How did it get to the island? Were the small, sophisticated stone tools and the evidence of fire the work of these small-brained humans, or was *Homo sapiens* also present in the cave at the same times? Finally, why did *H. floresiensis* go extinct? The story of human evolution often seems more mysteriously tantalizing than definitive.

These and many other milestones in human evolution suggest a complex and uncertain pattern. Clearly, experimentation and diversification occurred early in our ancestry, when many lineages appeared in Africa and eventually died out. Some, like *Homo erectus*, managed to migrate from that continent and spread to others, but this peripatetic species eventually went extinct as well. Then there were the Neanderthals (*Homo neanderthalensis*), robust, large-brained, stout-limbed mammoth hunters in Europe and western Asia that appeared some 200,000 years ago and had gone mysteriously extinct by about 30,000 years ago, some claim because of competitive pressure from the later arriving, more artistic Cro-Magnon *Homo sapiens*. *Homo sapiens* itself seems to have originated in Africa sometime before 150,000 years ago. Studies of DNA variation in modern human populations suggest that *Homo sapiens* numbered in the hundreds of thousands through much of its early evolutionary history but there was also a worldwide population crash—what is technically known as a genetic bottleneck—to about ten thousand individuals at some point. Early populations of *Homo sapiens* were well along on their various migrations out of Africa by about 60,000–70,000 years ago. The diaspora of our own species is thus a very recent blip in the 7-million-year history of our lineage.

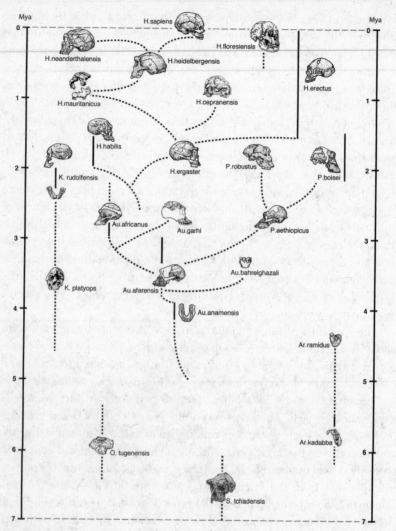

A historical scenario for the major human lineages over several million years (Mya) showing branching events

This network of tangled branches of human evolutionary history can be simplified, at least in one sense, as a general pattern. At least four human-like species were living in Africa 2 million years ago, and there were probably at least three species, one each in Africa, Europe, and East Asia, 100,000 years ago. Today we are alone, an unusual outcome when we think of the prior diversity of our family. Of course this simple fact leads to yet more questions. Between 400,000 and 12,000 years ago the great ice caps spread and retreated several times, bringing with each oscillation traumatic climate and environmental shifts. How did one species out of several survive the gauntlet of changing environments and competition with other species? The fossil record is largely silent on this matter. After all, Neanderthals, a species that seemed well adapted to harsh times of glacial expansion, had brains as large as or larger than those of *Homo sapiens*, and were skilled stone tool makers, but they went extinct after the arrival of *Homo sapiens*.

Nonetheless we see in *Homo sapiens* a refinement in the tools and a new finesse of artistry that were largely unprecedented in the several million years of prior human history. By at least thirteen thousand years ago, when humans likely reached the Americas, the record also shows the various technological and artistic wonders: figurative art, musical instruments, and plaques bearing notation. These were the refinements of a comparatively young and unusually wide-ranging species, earmarks of cooperative hunters and gatherers, with languages, familial and community bonds, a passion and talent for art, and a taste for ritual, possibly myth, and religion.

The unique survival status of our own species does not preclude further speciation of the human lineage in the future, though it seems unlikely. As the history of life and Darwin's explanation for it shows, new species arise most often from isolated populations, clusters of individuals that are prevented by various geographic and ecological barriers from reproducing and swapping genes with other populations of the same original species. These isolated populations increasingly diverge through time; they develop their own evolutionary tendencies and take on new, distinctive characteristics. *Homo sapiens* is, in contrast, a global species; its populations freely and frequently mix genes; genes among individuals around the world are more than 99.9 percent alike. We are currently one evolutionary community. Just the same, we showed early signs in our history of a capacity not only to destroy other species, including perhaps Neanderthals, but also to bash one another's brains out. Fortunately, we also showed an early capacity to cooperate to survive, a quality we doubtless will need to confront our next great evolutionary challenge.

16

THE EXTERMINATORS

Australia was at one time connected with other huge landmasses, including South America and a once decidedly more vernal Antarctica, but you wouldn't know it from much of what you see there now. When we think of the unusual in animal life, we inevitably think of kangaroos, koalas, wombats, and Tasmanian devils. And Australia has many other less familiar but nonetheless unique species. More than 80 percent of its native flowering plants and inshore temperate zone fish are endemic. When humans first arrived on this island continent—we now think some fifty thousand years ago—they encountered a world even stranger still. At that time Australia was a land of giants. Its biggest mammal, *Diprotodon optatum*, was a lumbering one-ton beast that superficially resembled a rhinoceros more than it did its close marsupial relative the kangaroo. There were giant kangaroos too—a massive short-faced animal known as *Procoptodon goliath*—as well as formidable predators like the marsupial "lion" *Thylacoleo carnifex*. Nor did mammals have a monopoly on extra large; other monsters included madtsoiid snakes seven to ten meters (twenty-three to thirty-three feet) long with the "girth of a dinner plate," varanid lizards related to the living six-foot Komodo dragon but twice as large, horned turtles as big as golf carts, and a seven-meter (twenty-three-foot) supercroc of proportions that would have likely intimidated Mr. C. Dundee.

Despite its intimidating ecosystem, as well as its harsh and highly changeable climate, Australia's early human colonizers managed to survive. On the other hand, some of those hulking resident creatures failed to show comparable endurance. Australia still has some fairly large and impressive animals, but nothing to rival its extinct pedigree. It is today devoid of the large browsers and carnivores of its past (no browser weighs more than twenty kilograms). The casualties are estimated to have been at least fifty-five species of vertebrates, including 16 percent of all mammalian fauna, its largest reptiles, and nearly all its large flightless birds. So there was clearly a mass extinction event, or a few such events, on the island continent some thousands of years ago.

Just when did this happen? Unlike the dates for early human sites, the dates for the termination of the vertebrates have not been easy to pin down. But recent research by Tim Flannery and his colleagues suggests a synchrony between the appearance of humans in Australia and the demise of some of its charismatic denizens. The last eggshells of *Genyornis*, a large, flightless bird, are recorded in sites about fifty thousand years old. The extinction of other vertebrates is harder to date, but it seems that fifty thousand years is a compelling figure. Significantly, the best estimate now for the first arrival of humans in Australia is between fifty and sixty thousand years. (The striking coincidence between the appearance of humans and the disappearance of a substantial chunk of the megafauna has problematic exceptions. It has been claimed that at a locality called Cuddie Springs in New South Wales, early humans and a megafauna that included *Diprotodon*, *Genyornis*, and the giant crocodile *Palimnarchus* may have lived side by side for tens of thousands of years. However, the analysis of the Cuddie Springs sequence is preliminary, and the dating may be less than reliable.)

Other places in the world reveal a pattern like Australia's, wherein extinction dates for large animals, particularly mammals, coincide with the first appearance of humans. In North America, not long after the earliest-known human sites—about 13,400 years before present—we see an abrupt termination of woolly (and Columbian) mammoths, mastodons, horses, ground sloths, tapirs, glyptodonts (giant armadillolike creatures), and saber-toothed cats. South America records a similar synchrony, with a great diversity of casualties among sloths, armadillos, and glyptodonts in addition to some peculiar hoofed herbivores known as notoungulates, not known in North America or the Old World. In northern Eurasia the pruning was less intense and more prolonged, but it eventually led to the extermination of the woolly mammoth and woolly rhinoceros, as well as the giant deer (sometimes incorrectly called the Irish elk) *Megaloceras giganteus*; the adult male animal was fully 2.1 meters (7 feet) at the shoulder with a magnificent rack of antlers spanning as much as 3.65 meters (12 feet). The cave bear (*Ursus spelaeus*), once the bane of cave-dwelling Neanderthals and Cro-Magnons, also disappeared. Other Eurasian losses included the straight-tusked elephants (*Palaeoloxodon*), the spotted hyena (*Crocuta crocuta*), the narrow-nosed rhinoceros, and the hippopotamus. (The latter is the subject of a detailed book, *The Hippos*, whose author proclaims convincingly that the demise of hippos several thousand years ago in the British Isles was not due to the asteroid impact!)

Is the coincidence of human appearance and the thinning of the bestiary more than a matter of coincidence? Did indeed humans bash these beasts out of existence? The possibility has attracted one of paleontology's most energetic and sustained debates. In this controversy North America is an especially important testing ground because its fossil record is so rich and has been so carefully analyzed, and it has so many investigators. Although most now favor the death by humans scenario, there are those who strongly reject it.

The skeptics emphasize that the span of time marking the loss of the megafauna in various regions overlaps with a dramatic period of climatic instability. The Pleistocene epoch, extending from 1.8 million years to about 12,000 years ago, was besieged by intervals of enormous glacial expansion from the northern ice cap, which covered much of North America and spread to similarly vast dimensions in Eurasia. These glacial pulses were interspersed by periods of glacial retreat, when conditions warmed and flora and fauna dispersed to higher latitudes. There were as many as twenty glacial advances and retreats in the Northern Hemisphere over the last 2 million years, bringing with them acute oscillations in climate and environment, with major effects on the resident biota.

Scientists have divided this score of freezing and thawing intervals into four major ice ages, named for present-day states covered during maximum expansion of the glaciers: the Nebraskan (2 to 1.7 million years ago), Kansan (1.4 million to 900,000 years ago), Illinoisian (550,000 to 390,000 years ago), and Wisconsinian (180,000 to 12,000 years ago). The last of these produced the enormous Laurentide ice sheet that covered all Canada and Alaska and most of the northern United States. By 10,000 years ago it had retreated dramatically and was largely confined to eastern Canada; by 8,000 years ago it encrusted only the Arctic islands and the area around and over Hudson Bay. We are now living today in this interval of glacial retreat, in the Holocene epoch, extending from 12,000 years ago to the present. This is the time of a big thaw; most glaciers of the world have dwindled dramatically even in recent decades, a trend that has been exacerbated by the global warming resulting from human-induced increases in greenhouse gases. Specialists who study the many environmental and biotic disruptions that occurred between 1.8 million years and the present refer to this interval as the Quaternary Period, which embraces both the Pleistocene and Holocene epochs.

Climate instability was for some time the standard explanation for the Late Pleistocene extinction. Indeed, when Paul Martin first presented in the 1960s a comprehensive theory showing that intense hunting by humans was the cause of these extinctions (often referred to as the overkill theory), his arguments were greeted very skeptically, even derisively. As with many scientific sagas, vindication lagged but nonetheless emerged, and a consensus now identifies humans as the culprits for the decimation of the megafauna in many regions.

Proponents of Martin's overkill theory, like John Alroy, have focused on problems in the traditional climate, or ecological, model. In places like North America, extinctions occurred at the very time when the glacier retreat would have increased, not decreased, habitable areas. Moreover, the very mammals that went extinct were large, mobile species that could have readily moved to more habitable environments if they'd had to. Extinctions occurred at all latitudes, longitudes, and altitudes and under varied ecological conditions and climate, and only a few occurred in other parts of the world that were equally affected by Pleistocene climate instability. Moreover, there is no marked evidence of extinction of large mammals during preceding, equally intense glacial cycles of the earlier Pleistocene. Finally, the fossil record of very high extinction levels for larger species nowhere near approaches the intense selectivity of large mammal extinction at the end of the Pleistocene.

So the overkill theory seems to fit better with the data. Intense hunting by humans is clearly suggested by coincident dates. Extinction of large North American mammals occurred within 1,000 or 2,000 years of the arrival of humans with their weapons, as evidence recovered at sites like the 13,400-year-old Clovis in New Mexico shows. Masses of bones next to stone weapons at several sites of similar age in North America also reveal that Clovis armed hunters preyed on now extinct megafauna, such as mammoths, that were either plausible prey species or carnivores depending on these species for food. Prey species that have survived are the ones inhabiting mountains and tundra that are difficult to hunt (mountain goat, Dall sheep, musk ox, caribou), had huge populations over an enormous range across the Northern Hemisphere, or migrated from Europe much later, suggesting they had adapted to human hunting pressures over a very long time. The predator species that have survived in North America either are small enough to live on a variety of vertebrates (coyote, wolf) or are omnivorous (bear). Finally, the extinctions of large vertebrates in South America, the Caribbean, several Mediterranean islands, Madagascar, Australia, and New Zealand occurred at different times and regardless of local environmental conditions but always when human hunters first appeared. The extinction of birds on Pacific islands and elsewhere coincided with the late arrivals of humans at these isolated habitats.

The skeptics, including climate advocates, have a rejoinder. Extinctions of large mammals in Africa and much of Eurasia were relatively minor, even when human hunters had already appeared there, and Africa and southern Asia were largely spared the exterminations; unlike their northern relatives, elephants, rhinos, and (in Africa) hippos and hyenas managed to survive. Why this erratic pattern of endurance? Darwin himself pondered this question as he stepped off the *Beagle* and scanned the fossil remains of giant ground sloths and glyptodonts that had once inhabited the plains of Patagonia.

There is a reasonable explanation, given by Martin and his supporters. In some places, like Africa, large animals and early humans evolved together over millions of years. The large animals that endured learned how to evade early human hunters often enough to keep their populations intact. In contrast, the large animals of the Northern Hemisphere evolved for millions of years in the absence of strategic human predators and were unprepared for the onslaught of this new, clever, cooperative predatory species. Humans in North America were invaders — the mongooses and zebra clams of ten thousand years ago. Of course we cannot actually demonstrate that large African mammals were resistant to human predation by virtue of their prolonged co-evolution with humans. Yet the theory is plausible and at least consistent with the duration of human presence on different landmasses.

Another oft-raised problem with the overkill theory is that the paleontological evidence shows no further major extinctions of large mammals in North America in the last few thousand years, even though hunting by humans became more intensive. On the other hand, as John Alroy has noted, perhaps the thinning out of large prey species allowed the remaining ones to grow to populations so large that even resolute, widespread hunting could not completely exterminate them. In other words, the lowest species on the preferred hunting list benefited from the loss of their more appealing competitors, and their populations and ranges expanded. Deer are a good example of this pattern.

During the last few centuries prey species have suffered at the hands of a hugely populous and highly technological human species. But in North America alone thirty-two out of forty-four large prey species had already been lost at the end of the Pleistocene. How could a small population of early humans with their rudimentary weapons devastate so many large animals so quickly? One might readily envision such a possibility in smaller areas, such as islands, where animals could be cornered and slaughtered, but not in the vast expanses of the Holarctic region and the Australian continent. This brings us to the argument most frequently leveled against the overkill hypothesis. Some have suggested that the spread of diseases from humans or their domestic species, rather than overexuberant hunting, was the primary reason for the destruction.

Still, analyses by Alroy and others concluded that organized groups of armed hunters could, over sufficient time, do significant damage. As noted, the first evidence for human habitation in North America is about 13,400 years ago; the first extinctions of large mammals occurred about 800 and the last about 1,640 years later. Of course the humans might have been on the continent earlier but with their populations too sparse and widely scattered to leave a record. There are, it should be noted, continual claims for human sites in the Americas that are much older, as much as 50,000 years, but these are rejected by most experts. Although most archaeologists agree that humans might have arrived in the Americas some thousands or tens of thousands of years before the age of the Clovis and other sites contemporary with it, they doubt that their sites have been clearly identified. Assuming a minimal colonization of North America, involving 100 humans and a modest population growth rate of 1.7 percent per year, Alroy calculated that it would take 260 years for the human populations to exceed 1,000 individuals and slightly more than 400 years to exceed 10,000. An archaeological record for the first small clusters of humans would not be likely, so we could safely push the date for initial colonization back to, say, 14,000 years before present.

This recalibration has two important implications. First, it would mean that there was a considerable overlap—as much as 1,200 years—between humans and the prey species they putatively drove to extinction. Second, even if we suppose only a modest population growth rate, humans would have reached a population size in North America of a million people by 13,250 years ago, precisely the time that marked the most intense levels of extinction of large mammals.

There is a problem here, however. Computational simulations of the overkill scenario are strongly affected by the chosen index for hunting ability, a numerical value assigned to an assumption that humans were either good or mediocre at hunting, and this fuzzy parameter can be readily manipulated. Indeed, it seems contrived; it reminds one of attributes assigned to fantasy warriors in strategy and role-playing video games. With a "good hunter" assumption, extinction intensity is achieved at a level consonant with the fossil record; with a "bad hunter" assumption, it is arrived at only later, a few hundred years after the fossil record indicates that most of the victimized large mammals actually disappeared.

Circumstantial evidence suggests that the "good hunter" assumption is the more credible. A number of kill sites in North America and Eurasia recording the mass slaughter of mammoths, bison, and other species effectively document the human capacity for carnage. The stone weapons used for the job were big, destructive ones—the heads of some hand axes were as big as a football—and the hunting approaches were strategic, often involving entrapment and cooperative attack. Archaeological sites also reveal how important these prey were to their human predators: mammoths provided not only meat and clothing but also fuel and building material, as shown by the remarkable huts constructed from piles of mammoth skulls and bones at Mezhirich in Ukraine and other sites in Eurasia.

Evidence from more recent history also demonstrates that small bands of humans with very basic weapons could effectively devastate populations of wild prey. By the time Europeans landed in the New World, Native American tribes along the northeastern coast of the continent were harvesting very large individuals of the cod populations offshore. These extralarge fish are known only from their impressive teeth collected in pre-Columbian archaeological sites. Indeed, it is thought that intensive fishing in coastal waters of northeastern North America extends back into aboriginal times. Only with the development of technology in the last century has such intensive fishing promoted the collapse of cod populations.

Few large animals left in the world will let you walk up to them, and in the rare cases where that is possible you are more likely the prey, or at least the victim, than the predator. Indeed, species like the saola I described in chapter 1 seem to have hung on solely by means of their utter wariness and their secretive behavior. But it is highly likely that woolly mammoths, rhinos, horses, and other casualties of the Northern Hemisphere several thousand years ago lacked these instincts of self-preservation. They were ill adapted and ultimately vulnerable to the human invasion. The actual process of extermination in places like North America, although instantaneous on the scale of geological time, took more than a thousand years. In the experience of early American hunters, the dwindling of their coveted herds of prey was as gradual as the waxing and waning of those glaciers that capped the planet. Doubtless they failed to perceive the impact of their enthusiastic mass slaughter, or what we euphemistically call today their resource consumption. Nonetheless, early humans were capable within a millennium of eliminating most of the large mammals in most of the world. The favored theory for the extinction of large animals at the end of the Pleistocene is not only compelling scientifically but a testament to our destructive power.

There remained only one step for humans to distinguish themselves further from the evolutionary matrix of their primate kin. They next fashioned a way of life in which resources, primarily food, were predictable, thus more dependable in the face of environmental vicissitudes. In some ways, *Homo sapiens* emulated those hugely successful autotrophic species, the bacteria, algae, and plants that act like biofactories producing their own food. In a leap toward evolutionary self-reliance, humans cultivated their own crops and domesticated their own animals. On the basis of archaeological record, these developments arose at least eleven thousand years ago in southwestern Asia and within about five thousand years thereafter in eight other regions of Eurasia, Africa, New Guinea, and the Americas. Food production and domestication soon spread from these early epicenters to become prevalent practice for human populations in all but a few isolated regions where a hunter-gatherer lifestyle persists.

These steps placed human evolution on a new, inexorable course to world domination. Humans were no longer simply parts of environments and ecosystems. They could create their own. In doing so, they changed more than merely the face of a planet.

17

THE CULTIVATORS

Imagine being confronted ten thousand years ago with the daily problem of finding food for yourself, your family, or your community. The once plentiful herds of horses, mammoths, and other sources of meat have become extremely sparse and skittish. Your stone weapons, part of the proud hunting traditions for generations before you, which were once so lethal at close range, are not so effective when the prey are rare and wary. Berries, nuts, seeds, and roots have become important parts of your diet, but even these resources are unpredictable in the face of harsh changes in the seasons, periods of intense drought, and aggressive competition for the same food by an ever-expanding population of human foragers. It is a hard life, a life famously characterized as "nasty, brutish, and short," a life portrayed by some prominent archaeologists, such as Robert Braidwood, in the following terms:

> . . . small groups of people living now in this cave, now in that . . . as they moved after the animals they hunted [with] no time to think of anything but food and protection . . . all in all, a savage's existence, and a very tough one. A man who spends his whole life following animals just to kill them to eat, or moving from one berry patch to another, is really living just like an animal himself.

Yet this portrayal of bare hand-to-mouth existence is, to many other scholars, grossly overstated. Hunters and gatherers might not have been badly off, even when their reliable meat sources were dwindling and wild plants less predictably available. They readily dealt with these challenges.

They made sharper, more lethal weapons. They shifted to smaller prey—fish, fowl, wild pigs, even rodents—that proved as delectable and nutritious as larger targets. And they increasingly depended on new plant sources for food and energy, becoming superbly familiar with the great range of biodiversity that provided potential food.

Foragers tend to find their own secret places in nature, and their traditions persist even in cultures dominated by agriculture and industrialization. Every early fall and spring in our own time mushroom hunters take to the wilds, whether the hills of Piemonte in northern Italy or the coastal rainforests of Oregon, heading for their zealously concealed treasures. Early foragers too had havens for the good and plenty. Hunter-gatherer societies that persist today in places like the deep green core of the Amazon River basin have an extraordinary knowledge of nature in the wild as a source of food and sustenance. They adroitly distinguish among thousands of species of plants and many animals and are remarkably familiar with the multiple uses of these species as food, medicine, or raw materials.

Evidence from both the prehistoric and historic records is clear. Early hunter-gatherers were taller, better nourished, and healthier, with longer life spans than the earliest farmers. For one thing, wild grains have more calories and give more nutrition per effort than farmed crops do. Archaeology confirms that the diet of early North Americans, for example, was extremely diverse, comprising between three thousand and five thousand plant species. Sites in Europe and elsewhere reveal the development of sophisticated technology and culture—superior stones, bone needles and fishhooks, jewelry, and art—anticipating the advent of agriculture by three thousand years. Human culture and its power to help nurture, sustain, and enhance human life were already extraordinarily well developed in these preagricultural societies. It therefore seems highly implausible that hunter-gatherers would voluntarily take on the extra burden of farming in order to improve their lives. We know that in many places in the world people did not take this step. The rich Napa Valley northeast of San Francisco is famous not only for its vineyards but for the fact that it is one of the most bounteous food-producing regions on earth, where food grows all year long. When foreign settlers came upon the original inhabitants of this natural cornucopia, the Wappo Indians, they found hunter-gatherers, not farmers. The best translation for the Wappo word *nappa* is "plenty."

So why, then, did agriculture become the cornerstone of our existence?

Evidence from the past can give us the facts—places of origin, timing, and crops produced—but much of the explanation must be left to inference. One important lesson to derive from the evidence is that agriculture was not a peculiar accident of history. It did not somehow arise in one particular spot on Earth under a unique set of circumstances and then eventually spread around the globe. Yes, from the Fertile Crescent of southwestern Asia agricultural practice and its products spread in many directions. But agriculture and the domestication of animals, another means to cultivate one's own food source, arose independently in at least seven, possibly nine other places. Just as in the independent evolution in disparate organisms of similar anatomical structures, such as wings or eyes, the independent emergence of agriculture in several areas strongly suggests that these diverse early human communities experienced parallel pressures and responses.

The earliest signs of plant and animal domestication in the archaeological record occur in the Fertile Crescent of the Near East about 11,500 years before present. There are indications of agriculture 9,500 and 8,000 years ago in northern and southern China, respectively, and 9,000 years ago in New Guinea. Agriculture came to central Mexico, the Peruvian Andes, and Amazonia at least 5,500 years ago, 5,000 years ago to West Africa, and 4,500 years ago to eastern North America. These early agricultural sites had distinctive qualities relating to their specific environments. Food production was of two basic types. One, seed crops, including cereal grains (wheat, maize, rice), developed in environments of low species diversity, such as grasslands, which were also subject to seasonal or prolonged aridity. The crops were highly productive but unstable, and cultivating them required constant labor: burning, clearing, tilling, weeding, and harvesting. Such practices are well evidenced in the Near East, where the main crops were wheat and barley; in the Far East, with seed crops of rice, soybeans, and millet; and in the New World, with maize (corn in American English), beans, and squashes.

A second type of agriculture, which depended on root crops and tree crops (manioc, yam, taro, avocado, potato), arose in the moist tropical regions of South America, New Guinea, and elsewhere. These crops were less productive but more stable than seed crops and much less labor-intensive. The archaeological record, while good for arid regions with cereal grains and related crops, is poor for the tropics, where much of the sediment and its burial record is erased by rain, flooding, soil erosion, and tumultuous growth, decay, and regrowth of the vegetation itself, and thus direct information on the emergence of the "vegecultures" is not as good as for seed crop cultures.

The origins of animal domestication, as reviewed by Jared Diamond and others, are also distinctively varied. The first domesticates seem to have been dogs, as evidenced by seven-thousand-year-old sites in North America. In the Fertile Crescent the original domesticates were sheep and goats; in China, pigs (but we must also count the silkworm!); in Mesoamerica, turkeys; in the Sahel of Africa, guinea fowl; and in the Andes and Amazonia, llamas and guinea pigs. Within a few thousand years, when the agricultural practices of the Fertile Crescent had spread to Europe, northern Asia, and North Africa and eventually blended with East Asia, Eurasia came to harbor an impressive fourteen species of big herbivorous domesticated animals: sheep, goats, cows (alias oxen, or cattle), pigs, horses, Arabian (one-humped) camels, Bactrian (two-humped) camels, llamas (and alpacas), donkeys, reindeer, water buffalo, yaks, Bali cattle, and mithans (a descendant of the gaur).

Here we must distinguish domesticated animals from merely tamed ones. The former were bred to types that would serve humans as laborers, sources of food, or pets; the latter were taken from the wild and trained but never genetically managed. The elephants that drag logs through the forests of southern Asia today or conveyed Hannibal and his armies across the Alps are tamed, not domesticated, animals.

In many regions, few animals were domesticated, despite an abundance of large mammals. Surprisingly, millennia of human effort led to the domestication of only about 20 percent of existing large mammals. None of Africa's fifty-one large resident mammals was ever domesticated. Is this again a result of the prolonged cohabitation in Africa of large mammals with humans? Attempts to domesticate very temperamental zebras are dangerous and the African buffalo, the most notorious of the continent's wild animals, nearly suicidal. Cheetahs have been desired as pets for centuries (some caliphs kept more than a hundred in their palace bestiaries), but they have continued to elude domestication. Unlike domesticated cats, cheetahs do not breed in captivity; the female seems to require a prolonged chasing by a few males over expansive terrain before she goes into estrus and becomes receptive to mating.

Africa eventually compensated for the failure to domesticate its resident

animals with imports of cattle, sheep, goats, and camels from Eurasia. Elsewhere outside Eurasia animal domestication was similarly constrained. In
the Andes domestication remained limited to the llama, while in eastern
North America only dogs and no herd animals were domesticated. Colonization of many of the Pacific islands between about 3000 B.C. and A.D.
500 brought the ubiquitous pig and the chicken from Taiwan and, ultimately, mainland China. Eurasia's head start and sustained momentum in
domesticating animals have been cited as major factors in what Jared Diamond and others regard as *the* singular event in postagricultural human
history, the spread and dominance of Eurasian cultures elsewhere around
the globe.

Our knowledge of the origin and history of agriculture and domestication
should not lull us into thinking we fully understand the process. Recognizing that critical strides in human evolution were taken independently tells
us of the likelihood of parallel pressures and strategies but does not identify
them. The scholarly literature offers many proposals, scenarios, and hypotheses, some good, some not so good. Here I focus on the four most frequently cited: climate change, human population growth and pressures, the
emergence of concentrated settlements or villages, and the coevolution of
humans and their food sources. The question becomes one of relative influence. Which of these factors was the most important? Which came first?
Which was a prerequisite for the others?

Given the traumatic wobble of climates during the Late Pleistocene and
the Early Holocene, it is understandable that climate change is deemed a
critical element by many scientists. The expansion and contraction of vegetation belts following the last glacial maximum, demonstrated by pollen
sequences, were caused by precipitation as well as by temperature change.
Pluvial (wet) conditions peaked around 11,500 years before present, when
reliable stands of grasses with large grains proliferated. These conditions
may have stimulated a concentration of human populations in areas where
the grains were plentiful and reduced the need and motivation to sustain
the often nomadic ways of hunter-gatherer groups. But a dependency on
wild grain species may have severely limited the options during periods of
high climatic stress, such as the cold, dry spell known as the Younger Dryas,
which occurred between 11,000 and 9,700 years ago. The climatic rigors of

this interval may have promoted the intentional, strategic cultivation of natural stands of cereals in order to have a reliable food source. The record gives some indication that the resulting agricultural practices produced greater yields than those of uncultivated wild cereals.

A second argument proposes that population growth, leading to critically high concentrations of people in a particular area, was what gave rise to agriculture, that population pressures compelled foragers to make the switch because wild plants and animals were depleted and farming became worth the labor. This argument presumes that agriculture has the advantage of providing more food per section of land than foraging. On the debit side of the ledger, however, it entails a higher labor cost per unit of food yield, and as noted, the food produced often has less nutritional value. Just the same, farming protected crops, especially seed crops, from overexploitation, whereas intensified hunting, harvesting, and gathering would have quickly depleted wild resources. In addition, farming was compatible with settlement because it required the stationary cultivation of crops and the harvesting and storage of food. Efficient foraging, by contrast, often (but not always) depends on high mobility—hardly an encouraging lifestyle for development of an organized settlement.

A problem with matching human population increase to agriculture in this way is that the latter came so late. North America harbored a growing population of hunters and foragers for several thousand years before there were any cultivators; even models of modest population growth suggest that thirteen thousand years ago there may have been as many as a million people on the continent and perhaps millions more by about 4,500 B.C., when agriculture is first indicated. (Until recently, frequently published numbers for Native American populations at the time of Columbus were gross underestimates: one or two million instead of a more reasonable figure of twenty million.) In Eurasia and Africa, where Homo sapiens lived for hundreds of thousands of years, one can safely assume that populations were even larger.

Moreover, the growing populations of humans were not randomly distributed. Just as today, humans tended to cluster in areas with abundant water, food, and ameliorating climates. Population pressures could thus have developed rapidly in locales with highly desirable but ultimately limited resources. After all, on many landmasses there were enough humans to promote the extermination of some of these resources—namely, large prey.

Why, then, did it take so long for population pressures to induce humans to shift to agriculture? There must have been other motivations.

Consider the tendency of humans to aggregate in communities. The first human communities were based on a foraging economy, perhaps one that depended on the grains of wild grasses. For such groups, agriculture was simply a solution to the problem of feeding more and more people. As the burgeoning local populations became sedentary, they developed new tools and farming methods and their own art, culture, and means of defense or aggression. Clusters of peoples became settlements with distinctive identities, traditions, and loyalties. Domestication of plants and animals made good sense once the infrastructure of a village was established. Because of the efficiency of farming, more food was produced than was needed just for the farmer and his or her family. Village life offered the opportunity to trade commodities and to build a new economy for survival. Although it does not necessarily take a village to generate agricultural productivity, a sedentary, communicative, and interdependent way of life probably helped to catalyze farming.

And there is a self-reinforcing, mutualistic relationship between cultivator and cultivated. The interplay between improved farming and the evolution of the plants themselves made crops more adaptive and at the same time more beneficial. The word *coevolution* is frequently used for this kind of intertwining, but it is misleading since plant evolution and cultural change are not the same kind of phenomenon. There is no evidence that better farmers have different genes from those who lack the green thumb, but as Darwin recognized, humans themselves have been powerful selective agents, promoting the evolution of new strains of plants and new breeds of animals. At any rate, there is something reminiscent here of the complex connections that gave rise to the modern ecosystem 100 million years before the first cultivated crops took root.

How did the system work? Many wild grasses and legumes have important mechanisms for dispersing their seeds; the pods or stalks containing the seeds actually explode, casting their precious reproductive contents to the wind. But a human forager would find it easier to collect grains from unexploded stalks and pods; if the plants that produced these recalcitrant components did so because of a mutation, then humans were simply selective agents favoring it—that is, favoring the genetic change by cultivating the seeds that were most easily collected. The mutated wild plants that produced these seeds lacked an efficient mode of seed dispersal, and human cultivation became required for their survival. Early farmers likely preferred seeds that sprouted fast and at specific times. Thus the plants most selected were the ones that, by virtue of their syncopated germination, allowed bulk sowing and harvesting.

The selection of plants for the quality of their products—the taste and nutritive value of their seeds, fruits, and oil—was as much a part of early cultivation as it is today. Examples are numerous. Corn is a fleshy, nutritious derivative of teosinte. Olive, flax, and safflower were among many plants selected according to the amount of oil their seeds produced. In some cases the conversion from wild to cultivated was even more radical. The wild forms of some plants, like almonds, are actually poisonous to humans, and farmers chose mutant varieties that lacked the poisons. Mastery in cultivation of fruit and nut plants like figs, olives, and grapes took longer, after it was eventually discovered that cuttings were essential to propagate quality plants successfully. Last to come were cherries, pears, apples, plums, and other fruit trees that required grafting, a refined technique first developed in China and facilitated by the fact that many of these plants carry both sex organs on a single individual; they are, in scientific terms, hermaphrodites capable of self-pollination.

It is most interesting that our appreciation of the critical link in the 100-million-year-old ecosystem—the pollination of plants through the activity of insects—was long delayed. This crucial factor was not understood until millennia after humans first domesticated plants and animals for their own benefit. Cave paintings from 13,000 B.C. demonstrate that beekeeping for the purposes of procuring honey was an ancient practice; the great archaeological record of Egypt shows that beekeeping was well developed along the Nile in ancient times. Aristotle studied bees and their behavior, and Virgil wrote at length about bees and their honey. But people of classical times were content with concluding that the visits bees paid to flowers and their production of honey were not connected and, moreover, had no significance for the plants. Greek and Roman authors did recognize the importance of insects in the pollination of date palms, but notable scholars, including Theophrastus, the greatest botanist of classical times, had only a vague notion of the process of sex and fertilization in plants.

In the seventeenth century A.D., enlightened naturalists discovered the nature of sexual reproduction in plants and in particular learned that stamens, pollen, and other flower features were part of their sexual organs. The idea seems to have come independently and nearly simultaneously in the 1682 work of Nehemiah Grew and observations of his contemporary and colleague Thomas Millington. A few years later the influential naturalist John Ray accepted Grew's fundamental discovery, albeit with reservations. These studies were followed by a spate of papers from scholars in England and on the continent, including Christian Konrad Sprengel. By the time Darwin published his important observations of flower pollination around the same time as his publication of the *Origin of Species*, the subject had already attracted a diverse literature. Nonetheless, Darwin gave one of nature's most complex feats of cooperation a new evolutionary spin, and his

The title page of a monograph on insect pollination by Christian Konrad Sprengel, 1793

work launched a phase of intense interest. As research on pollination progressed through the twentieth century, newly appointed with the insights of genetics and ecology, scientific knowledge could be transferred to enlightened agriculture. Today beekeeping is a practice not primarily intended for honey production but for the managed pollination of such important crops as alfalfa, red clover, tomato, watermelon, and avocado.

Despite the apparent inevitability of agriculture and its independent appearance in several regions, some cultures and regions developed faster and more elaborately than others. Australia and the far north of Eurasia and North America harbored enduring populations of hunter-gatherers who adopted only minimal techniques and practices for cultivation. In dramatic contrast, the Fertile Crescent became a great staging area for the spread of farming and domestication, and it had a more stable and more enriched food resource that became abundant and varied as did the human population centers, tools, and technology.

Why this checkered geographic pattern of haves and have-nots? In his influential book *Guns, Germs, and Steel* Jared Diamond concludes that geography itself, and the environmental factors related to it, were primary. The Fertile Crescent was nicely ensconced in an equable, gentle Mediterranean climate that encouraged both dense human habitation and farming. The environment made for a comfortable incubator for the cultivation of diverse plants. Farming practices developed there spread easily to southern Europe and northern Africa, regions along the namesake sea blessed with a similar clime. By the time farming spread to the harsh seasonal climates of northern Eurasia, many plant strains had become well adapted to human cultivation.

Another geographic factor may have been critical. Eurasia's major axis, its largest dimension, runs east-west. Thus humans and their plants could readily migrate over vast expanses under similar climatic and environmental conditions at the same latitude. This serendipitous shape of a landmass likely spurred the eventual coalescence of the two food production centers in the Fertile Crescent and China. In contrast, landmasses like the Americas and Africa that did not advance so rapidly in food production, writing, or metallurgy were arrayed along a north-south axis, so that changes in environments with latitude, not to mention other barriers—the corrugated terrain of the Andes and the Cordillera of North America and the narrow tropical bridge of Central America—impeded the spread of agriculture.

What about even more static areas like Australia? Diamond argues that this island continent was so isolated and its environments were by and large so harsh that agriculture simply did not have any momentum for nearly fifty thousand years. Australia did not accommodate crop cultivation until Europeans came on the scene in the eighteenth century. Much of Australia, a land where only the indigenous aboriginal hunter-gatherers have mastered a way of life, is still harsh and arid.

The rest, as people say, is history. Eurasian cultures, with their massive human populations, bounteous food production, advanced weapons, and, not trivially, infectious diseases that were likely incubated over millennia of cohabitation of humans with their domesticated animals, became the great invaders, conquerors, and colonizers. As Diamond succinctly puts it, they had the "guns, germs, and steel" to overwhelm the cultures of Australia, Africa, and, especially, in the most one-sided act of domination in history, the New World.

Diamond's emphasis on the fundamental, primary function of geography and climate is a direct attack on the once prevalent traditional Eurocentric belief in the innate superiority of the conquerors over the conquered, and it has many compelling elements. Yet there are uncertainties. Many exceptional patterns of food cultivation require an elaborate explanation. It seems odd, for example, that New Guinea's level of agriculture and technology in early times was strikingly more advanced than neighboring Australia's. The two landmasses were separated by a water barrier no greater than that faced by colonizers of southeastern Asia heading toward Australia some fifty thousand years ago. And the mountains, jungles, and other geographic barriers that putatively blocked the spread of food culture in the Americas were not so formidable that they kept the first human colonizers of America some thirteen thousand years ago from quickly spreading from the Bering land bridge to the southern tip of Patagonia. Why would these barriers later be so inimical to the migration of humans with their food products? My criticisms notwithstanding, Diamond's work represents a powerful summary of a major interpretation of human history from a scientific, biological, and certainly geographic perspective.

Eventually neither geography, climate, cultural isolation, nor other factors halted the global spread of agriculture and animal domestication. This historic change redefined the relationship between humans and their natural environments. Farming required certain kinds of crop plants that could easily outcompete wild strains. Domesticated animals as a food resource required populations of cattle, sheep, and pigs that far outnumbered anything in the wild. Mongolia, a country nearly one-fifth the land area of the contiguous United States and as large as Mexico, has only 2.5 million people but more than 33.6 million cattle, goats, sheep, horses, and camels. Both crops and domesticated animals required huge amounts of land. The conversion of natural terra firma into land for food production and human habitation changed, in a few short centuries, the wild earth so fundamentally that it is virtually impossible to find a crypt of nature untouched by humans.

"Man the hunter" connotes many images, and among them is man the aggressor or the destroyer—not an unrealistic impression, given what we know. On the other hand, "Man the farmer" brings to mind an entirely different, benign image, of one who cherishes the land, respectfully cultivates it, and leaves it flourishing and productive for future generations. How ironic it is, then, that our activity as agriculturists has devastated Earth's natural environments much more than our selective overkill of some large animals. I am not a moralist here. Hunting is not intrinsically bad, nor is farming. (It has been for some time my unrequited desire to cultivate a small vineyard.) I merely report the fact that plant cultivation and animal domestication in combination with unbridled human population growth, consumption, and lack of foresight got us to the dismal place where we are today. Don't shoot the messenger.

18

LAND RUSH

For an interlude of my young life my parents transported our family from the metropolis of Los Angeles to the isolated north woods of Wisconsin. This was an experiment in migration; my parents actually kept their house in Los Angeles, and after three short years we moved back to the big city. But the temporary uprooting left its mark on me, because I was suddenly transported to a world where it seemed to me time was motionless. My father worked as a musician in one of those refugia for jazz in the age of Elvis, a smoky supper club where glasses of scotch and soda on ice clinked against the aural swirl of Sinatra-era ballads. He worked and we lived near Three Lakes, a rural town that could claim barely three hundred residents. It had an old-fashioned soda shop, a country store, and a county fair. A girl in my fifth-grade class sometimes rode her horse to school.

This travel back in time was for me not just an encounter with ancient culture. The north woods seemed a primeval and relatively pristine landscape. I explored stretches of uninhabited dark forests that hid tiny lakes that had names but nearly no other mark of humans. In search of bullfrogs and on the lookout for eagles, I trudged through swamps festooned with pine and spruce that had never been thinned by the Europeans who came to settle and farm. Wolverines, gray foxes, and a legendary albino deer lurked in woods around the abandoned lakeside lodge aptly named White Buck, which served as our home during the summer months.

Tourist pamphlets touted the claim that Three Lakes was a town with twenty-seven lakes interconnected in a chain, the biggest chain of inland lakes in all Oneida County or, for that matter, in the entire world. This assertion bothered me a bit; I kept asking elders how they knew this. No one seemed shaken by my doubts, which were readily dismissed. As far as I know, the reputation has held.

The lake country of Wisconsin and neighboring states, as well as the Great Lakes that embrace it, are a gift of the last great extravagance of ice on Earth. About a hundred thousand years ago, the Laurentide ice sheet began to form. At its peak, eighteen thousand years ago, this ice empire covered all Canada, New England, the northeastern United States down to present-day New York City and northern New Jersey, the Midwest down to present-day Ohio and Chicago, and areas of Montana and other western states. The ice sheet reached a maximum thickness of nearly 4 kilometers (2.58 miles), or only 1,000 feet shy of the top of Mount Whitney, the highest mountain in the contiguous United States. Its glaciers plowed down former mountains and scoured out low, flat areas into deep valleys and basins. When the Laurentide ice shield retreated, starting about twelve thousand years ago with the warming trend that heralded the Holocene epoch, its melting ice and moving rocks left huge gouges in the terrain. Water streaming into five of the largest gouges formed the five Great Lakes.

Nine thousand years ago those Great Lakes were even greater, covering a much larger expanse. By seven thousand years ago the northward retreat of Laurentide glaciers was enough to open passage to the Atlantic through what is now the St. Lawrence Seaway. The outflow of water through this seaway dropped the lakes to their present levels. Niagara Falls, still a natural wonder, is one of Earth's most spectacular drainpipes.

The retreat of the Laurentide ice sheet left behind other notable mementos—the rolling hills of Michigan, the Finger Lakes of upstate New York, and the impressive piles of rock that form the headlands of Cape Cod and the island-strewn coast of Maine—but nowhere is the Laurentide footprint so vivid as in Wisconsin. When glaciers melt at their retreating edges, cobbles, sand, and rocks form massive ridges of geological garbage called moraines. Ice continually flows to these edges and then melts. Sometimes the flow slows or stops, however, and debris buries massive chunks of ice. Pockmarks form when the buried ice melts, a terrain feature called a kettle. Those lakes I explored as a boy were a few of the fourteen thousand Wisconsin glacial lakes formed from water-filled kettles. There were also bogs, marshes, and forests replete with flora and wildlife. The result was a magical green landscape sprinkled with countless lakes, many connected like sapphires in a tortuously twisted necklace. The beauteous region attracted the Oneida, or Onyotaa, aka the People of the Standing Stone, whose ancestors settled here more than eight thousand years before Europeans arrived.

In those harsh Wisconsin winters much of nature fell away as we retreated to town, which itself became dormant during the siege of cold and snow. We inhabited the upper floors of a blocky, unadorned stucco building that on its ground floor housed a radio and TV repair shop. This was the last structure of any heft on the northern edge of town. I used to look out the large picture window to the north and ponder a barren patch of land, about the size of two football fields, colored either the dull brown of frozen ground or the steel sheen of snow under a relentless winter overcast. I had read in my geography book that those two football fields of waste could be measured scientifically: they were about 1 hectare, or about 2.4 acres. Those open fields to the north were indeed the corpses of deserted farmland, an early agricultural aspiration full of hope that had turned to dismal folly. I tried to look in my mind far beyond that unappealing foreground, then past a horizon of fir trees blackened by the dim winter light, to an imagined Arctic, where a narwhal was using its one tusk to break through plates of sea ice as the atmosphere shimmered with the fluorescent green of the aurora borealis. But imagination did not completely transport me from that adjacent hectare of degraded habitat that seemed so oddly out of place in this natural wonderland.

My parents told me that those ravaged fields had probably been tilled and overtilled before the farmers knew about modern fertilizer, the new wonder substance for crops, whose fumes engulfed the countryside in spring, making the air smell like one enormous latrine. We made jokes about fertilizer and its smell. When my brother got in the car after a walk at the strawberry farm, his shoes smelled like fertilizer. The girl who rode a horse to school smelled like fertilizer. But we all recognized that fertilizer helped plant growth and ultimately ensured food on the table.

The fields near our apartment were forever disqualified, it seemed. This was land that was to all appearances totally useless, in summer choked with weeds that turned as tawny and dry as straw in the fall, coated with snow in winter and implacable mud in the spring. This was scarred ground, good for burying things, like my poor dead dog, but not much else. There was one expression of unspoiled nature, a single oak tree that had turned its contorted gray limbs to embrace the blasts of cold north wind. That tree entranced me for reasons I did not then subject to examination. But now I think of it metaphorically. Was that tree a symbol of nature's defiance against the onslaught that had left the Earth otherwise barren and dead? Or was it simply a living fossil—a vestige of an ancient, vastly more enriched, and fleeting nature—that would soon vanish with the rest?

Of course all over the world there are great farms where there are no trees in any direction as far as one can see. These farms grow the corn, wheat, alfalfa, potatoes, beans, and pasture that sustain humanity. From an airplane flying seven miles above the midwestern United States, the patchwork quilt of agriculture and domestication seems to cover the entire Earth below, bending with the curved edge of the planet. Croplands and pastures now occupy 40 percent of Earth's land surface. Punctuate those fields with the anastomoses of cities and suburbs, and one contemplates a world of utter human domination.

It is hard to imagine a more radical reshaping, a more profound reworking of a landscape. During the Carboniferous Period, 350 million years ago, dense fern, cycad, and gymnosperm forests blanketed the megacontinent of Pangea, thriving in a worldwide climate of warmth and moisture. Likewise in the Early Eocene, about 55 million years ago, when oxygen in the sultry atmosphere took a big jump, angiosperm-dominated forests extended from the equator to the poles. Even in the early age of humans, these great crucibles of diversity dominated the landscape. It is estimated that just preceding the advent of human agriculture 11,000 years ago, forests occupied nearly 70 percent of the land's surface. But in subsequent centuries energetic cultivation thinned out those forests. There were dubious milestones; the Epic of Gilgamesh contains history's first tale of deforestation, the stripping away of mountain woodlands in the third millennium B.C. by Mesopotamia's first kings and their minions. The destruction of forestland in China and other parts of Asia may have been earlier still; centuries of land conversion make Asia the continent with the greatest overall loss in forestland.

But these historical losses pale in comparison with those caused by the spread of agriculture in response to rapid world population growth during the past two hundred years. Forests now account for only about 30 percent of the land surface (about 4 billion hectares, or 8 billion football fields' worth). The global annual rate of forest loss is about 7.3 million hectares (13 million hectares a year if you count land converted into tree farms and other "pseudo" forests). The pace of destruction for tropical rainforests, estimated to account for 40 to 50 percent of Earth's total biodiversity, is worse. Between 2000 and 2005 South America and Africa, where tropical rainforests represent the dominant forestland, each lost about 21 million hectares. Even these figures fail to indicate the depth of the wound in some regions. Nigeria is destroying its forests at a whopping 11 percent per year and very soon will lose all its primary forests, those that show minimal or no indications of past or present human activity. Elsewhere destruction is also rampant. Madagascar and Thailand now show an annual rate of loss of roughly 8 percent. As I said in chapter 1, between 2000 and 2005 Vietnam lost a horrifying 51 percent of its primary forests. Cambodia during the same time lost 29 percent, and Sri Lanka and Malawi, 15 percent. At this pace, much of the tropical rainforest ecosystems of the world will vanish within the next few decades. It is highly doubtful that there will be anything left of even the largest tracts of the Amazon or the Congo by the end of this century.

Well, some people say that's progress. Humans need to eat to survive, and there are a lot of humans. To accomplish this gargantuan task, we need to convert forestland to cropland. And we have successfully produced enormous amounts of food in the process; world grain harvests, for example, now weigh in at more than two billion tons a year. This does represent at least in one dimension, a real achievement, a doubling of food production in the past four decades. Part of this upswing is due to a 12 percent increase in world cropland area, but much of it is due to the green revolution technologies that depend on a 700 percent increase in global fertilizer use and a 70 percent increase in irrigated cropland. The modern miracle of food production has been a forty-year spike in productivity built on many millennia of slow improvement. Since the emergence of agriculture eleven thousand years ago, few things have been more important to human communities than soil. Our effort to maximize yield from the land has been in lockstep with the rise of civilizations.

Today the vast, intensive human effort to produce food is enough to feed most of the world's 6 billion people. Yet unfortunately, for a variety of reasons—political instability, poverty, poor infrastructure, war, desertification, and degraded land—about 845 million people, 14 percent of the global human population, are chronically or acutely malnourished. The justifiable goal to reduce world hunger as well as keep up with expanding human population puts further pressure on food production and on the land responsible for it. We are stretched between two opposing needs: more humans, not to mention those who are still starving, require more food; at the same time, growing that food means altering natural habitats and losing natural resources that confer many benefits, including, ironically, sustainable well-managed environments for food production. Our flowery, angiosperm-dominated cropland relies on pollination from wild bees that live in nearby natural habitats. Thick forests on hillsides harbor the watersheds that soak up rain and prevent surges of runoff that might otherwise flood and erode the farms and fields below. Yet these areas are vulnerable, simply because more land must be cultivated to feed the exploding world population.

Experts say that such a simple tension—one between the need for more food production and the need to preserve natural habitat—can be strategically and successfully managed. Sadly, the problem is exacerbated by a history of real achievement in agriculture frustrated by phases of utter mismanagement. The long-ago development of great river settlements along the Tigris and Euphrates of the Fertile Crescent, as well as the growth of river cities in China and Central America, made for expanded populations, an invasion of virgin habitat, and an increased rate of soil erosion. People moved upriver, cutting down trees and tilling the soils in the mountainous watersheds at the sources of those rivers. In this newly stripped landscape, heavy and often destructive seasonal rains aggravated erosion. The loess of northern China is a product of this kind of ancient, enduring human invasion. The winnowed, tawny soils of the Loess Plateau have been transformed to dust and silt borne by strong winds or carried downstream. The loess is what engulfs the city of Beijing in clouds of dust every spring and gives the Yellow River its name. The result is an enormous annual loss of loess, some 1.6 billion tons, an erosional impact greater than any other in the world.

For centuries this expansion of activity had only modestly negative side effects. The human population and demands for food were limited. People could move to greener pastures when soils became unproductive. But then this changed radically between the sixteenth and the nineteenth centuries, when the rapid acceleration of global human population put major pressures on food production and forced many into hunger, malnutrition, and misery. Demand drove farmers to cultivate huge expanses of steppe, prairies, and pampas. New tools made semiarid regions feasible for agriculture, but farmers more accustomed to humid climates and damp soils now worked in dry environments that were much more susceptible to erosion. Soil fertility thence became a rational obsession.

The next stage in human effect on soils both preceded and accompanied the rise in the use of artificial fertilizers in the 1950s. By the early twentieth century, human population's growth had accelerated, partly because modern medicine had slowed infant mortality and increased life spans. Ever-larger populations in need of food slashed and burned tropical rainforests and other pristine habitats. Soil erosion in such places became acute. Not only were bounteous forests obliterated, but their conversion to useful farmland was short-lived because overplowing left little vegetation to hold water. During the Great Depression era the dust bowls of the Great Plains saw eight years of "black blizzards" that blocked the sun, buried farms, and drove families from their lands.

These agricultural, economic, and social disasters stimulated improvements in North America and Europe, such as the U.S. Department of Agriculture's Soil Conservation Service. The clock was slowed down but not turned back. Worldwide surveys in the 1980s showed that of the 11.5 billion hectares of vegetated terrain on earth, 17 percent was degraded by erosion. Of this land, 1 in 6 hectares no longer could sustain crops. Other estimates showed that at least 40 percent of the world's croplands might be experiencing some degree of erosion, reduced fertility, and overgrazing. After some debate over high and low estimates of soil erosion, the experts agree that there is at least 11 billion tons of human-induced erosion yearly, about 1.5 million hectares' worth, representing at least eleven billion U.S. dollars annually in lost crop production.

There is some ambiguity in these numbers. Dire predictions in the 1980s that soil erosion would lead to horrific food shortages and human mortality of Malthusian proportions have not been borne out. Even the pessimists agree that on a global scale soil loss is not likely have such a calamitous impact on food sustainability. But tell that to people in certain regions of the world! In Haiti only 3 percent of the nation's once lush tropical forests remain, and one-third of all formerly productive land, about nine hundred thousand hectares, has been destroyed through loss of topsoil. Desertification is rampant in the Sahel of Africa, Kazakhstan, Uzbekistan, and northern China, where the desert is growing by about thirty-six thousand square kilometers a year, an area about the size of Rhode Island. Effects of soil loss are still serious and destabilizing in countries where extreme environmental conditions, poverty, and rudimentary farming practices persist.

Ironically, many regions struck by these insidious forces are places where per capita consumption is quite low. Demand for food and other resources is very unevenly distributed on a global scale. Seventy percent of the world's fossil fuels and 80 percent of its chemical products are consumed by only 25 percent of its population. We are still a world of haves and have-nots. No one, however, is completely insulated from the environmental problems of our changing landscape. Those of us who enjoy the often illusionary comforts of a wealthy nation may not have any sense of the daily struggles of a poor farmer in the Sahel, but overconsumption, deforestation, and soil misman-agement are problems that ultimately affect us all. Those problems relate to the availability of another substance that is inseparable from soil, that life-giving, rare, and hugely threatened commodity—freshwater.

I have spent many summers thinking a lot about water. That is because our team of paleontologists works in one of Earth's driest places, the Gobi Desert, where spots for drinking water are as few and scattered as rubies on a beach. Our main excavation site at Ukhaa Tolgod is eighty very hard miles from the nearest freshwater, a place called Narun Bulak, which means "sunny spring." Here icy water streams from an old Russian galvanized pipe driven into the gravel and sand at the top of a hill. Its meandering runoff creates a small natural pasture so brilliantly and oddly green that it can be seen from several miles away against the surrounding bleached-white cliffs. Naran Bulak is a sanctuary and a place of sustenance for nomads and their camels and horses, as well as for thirsty paleontologists. It lies not far west of a major trade route between Mongolia and China that was used by cen-turies of camel caravans heading for the Silk Road far to the southwest. Genghis Khan was probably here too. We know he took a route through this region as he rode south with his troops to put down and punish some ram-bunctious Chinese cities. There are not many other places he could have stopped for water.

In the early 1990s, when we first visited the sunny spring, the water flowed in lavish volume from the Russian pipe. Now that flow is alarmingly slowed. In some dry years it becomes a mere trickle. The spring has been heavily used by more and more expeditions, tourists, herdsmen, miners, and farmers. But the major factor in this decline is the seven-year drought, a likely manifestation of the worldwide phenomena of desertification. The water table in the Gobi and all Central Asia is sinking.

Global statistics illustrate why such a decline in freshwater is not just a local phenomenon. Mexico City, with its huge population, has drawn on its aquifer so heavily that the land it rests on has sunk thirty feet over the last hundred years. One of the greatest environmental disasters of all time was the draining starting in 1918 of the Aral Sea, the world's fourth-largest inland body of water, in order to irrigate cotton crops in Central Asia. This conversion left skeletons of old fishing boats out on a basin of evaporated salt and dust. More than half of Europe's cities are using groundwater at unsustainable rates. More than thirteen African countries suffer from water stress and scarcity, and that number may double by 2025. North America's largest aquifer, stretching from South Dakota to Texas, is being depleted at a rate of twelve billion cubic meters a year. Worldwide, agriculture accounts for 70 percent of the total consumption of freshwater, while industry comes in at about 22 percent, and domestic use at only 8 percent. Water and soil go together. You can't have productive soils without water because you can't have organisms without water.

Freshwater is of course only a small fraction of the water on the planet. Ninety-seven percent of Earth's water is in the salty oceans, leaving only 3 percent with salinity levels that can be tolerated for consumption by humans and most nonmarine organisms. Most of this freshwater is locked up, however, in the vast ice repositories of Antarctica and Greenland. Nonetheless, there is enough freshwater currently to sustain all life, including humans. But there is a twofold problem: its uneven distribution and uneven consumption. Most of the available freshwater belongs to a fortunate few large countries laced with rivers: Canada, Brazil, Russia, Indonesia, China, and Colombia. (The United States ranks seventh in freshwater holdings behind these.) Smaller countries that are similarly endowed include Iceland, Finland (which has the highest water quality index), Suriname, Guyana, and Papua New Guinea. Outside the Gobi Desert, Mongolia is well fed by freshwater lakes, streams, and rivers, enough certainly to sustain its small population. Water-deprived countries include rich ones, like Saudi Arabia and other oil producers in the Middle East, and poor ones, like Mali, Sudan, Chad, and other dehydrated countries in Saharan and sub-Saharan northern Africa.

The accelerated use of freshwater in both water-rich and water-poor countries is causing a serious problem. Between 1990 and 1995 the global consumption of water increased sixfold, more than double the rate in world population growth. Americans buy six billion gallons of bottled water every year, 40 percent of which is actually just tap water, although the price of bottled water is more than a hundred times that of tap water, currently accounting for sales of $4 billion a year. As of this writing, the unstable and rising price of a forty-two-gallon barrel of crude oil hovers around $75 U.S. At $1.29 a half liter, bottled water is clearly more precious, or at least more expensive, than oil; it costs about $400 per forty-two-gallon barrel.

Consumption is tied to wealth, and wealthy countries seem both to be thirstier and to have the capacity to acquire resources. But asset distribution is not the only factor. Sometimes demand is skewed in a way one would not expect given the wealth curve. Stark differences in consumption exist even among well-endowed countries. Average water consumption per person in the United States is about 2,300 cubic meters (m^3) per year (about 607,600 gallons), in Canada 1,500 m^3, and in the European Union an average of 1,250 m^3 a year. But we find that on a global scale water use and wealth match up pretty well. In developing countries, the use of water falls precipitously, to about 20 to 40 m^3 per person per year. Similar relationships obtain for the use of forest products and many other commodities.

The big worldwide gulp that puts heavy pressure on freshwater resources has collateral casualties. Freshwater ecosystems support a diversity of birds, fish, plants, frogs, turtles, insects, aquatic invertebrates, protists, and bacteria, many of which are highly sensitive to environmental disruption. But the use of water for irrigation, drinking, waste disposal, damming for reservoirs and hydroelectric power, and the introduction of alien species can mean anything from a serious threat to these habitats to the utter destruction of many species in them.

This focus on freshwater should not lead one to conclude that other ecosystems are healthy in comparison. As we have seen, tropical rainforests are being rapidly destroyed, and as go the forests, so go the species that live there. The simple rate of destruction of these habitats is what gives our current estimates of extinction their heft. Take away these environments— remember our march to total deforestation of tropical regions may well be complete within this century—and one is left with a projection that between 30 and 50 percent of *all* species will be gone by mid-century. The projection is all the more disturbing when we consider that biodiversity is the source of sustenance that makes our lives not only interesting but also livable. Diverse species are the potential source of tens of thousands of different food products, and thousands of species serve as sources for medicines. How much more would we have available if we managed forests in ways that allowed for continued mining of their treasures?

It is fair to recognize the mitigating trends. In southeastern Asia reforestation has had a good, albeit limited, affect. In Europe, inadvertent nitrogen fertilization, peatland drainage, and direct management have increased biomass by 40 percent. Replenished forests not only provide more sustainable resource but mitigate the radical disruption of the atmosphere by absorbing carbon dioxide and acting as carbon sinks. So why not be satisfied with these measures? What's so bad about the restored, productive forests of Europe and Asia? After all, if the world is given to humans, some compromise between meeting our needs and maintaining the forest primeval must be achieved. For example, of the hundreds of thousands of species that are potential sources of food, humans have come to rely on about fifteen hundred plant species, only twelve of which provide three-quarters of the world's food. Why, then, is maintaining the biodiversity of the original forest so important when restored forests give us boosted amounts of some of its core products? The answer lies once again in what we call ecosystem services and the many benefits conferred by having a wide variety of species. Restored forests fail to perform many ecosystem services that original forests can because they usually fail to maintain the original number of species.

To illustrate, let us return to the very fiber of the modern land ecosystem, the intricate, complex, intertwining of angiosperms and pollinating insects that emerged in the dinosaur age. Here we confront the worldwide problem concerning those most proficient of pollinators the bees. We have learned that the first-known fossils of bees are from the mid-Cretaceous, about 100 million years ago, an entry synchronous with the birth date for the full-fledged flowering ecosystem. The pedigree of the ancient bee that gathered pollen for honey, pollinated fields of Cretaceous flowers, and buzzed the head of T. rex is in deep trouble. There is currently a crisis in bee management.

When we considered earlier the power of pollinators, we learned that animals, prominently bees, pollinate one or more of the cultivars of 66 percent of the world's fifteen hundred crop species and are directly or indirectly responsible for an estimated 15 to 30 percent of food production. Hence it is easy to see why the most important managed pollinator, the honeybee *Apis mellifera*, has an estimated annual value between $5 billion and $14 billion in the United States alone. But this service is diminishing; beekeeping has declined by 50 percent in the United States, possibly as the result of insecticide poisoning, disease, and loss of subsidies. The aggressive and unmanageable African strain of A. *mellifera* is now hybridizing with managed colonies and spreading from Brazil to the southeastern United States. In May 2007 came news that the problem is even worse. Managed honeybees worldwide are abandoning their hives, and the direct causes of this global exodus are far from clear.

Perhaps there is a ready insurance policy here. It comes from our surprisingly strong dependence on wild species of bees for agriculture. Only eleven out of the estimated twenty to thirty thousand species of bees are used in managed bee farms and greenhouses. We know little about the nature and magnitude of the pollination services that wild bees perform for crops, but indications are that they are substantial. A 2002 study by Claire Kremen, Neal Williams, and Robin Thorp focused on farming in the Central Valley of the Coast Range in Yolo County, California. Here watermelon crops depend heavily on pollination. They require multiple visits by bees and the deposition of five hundred to one thousand pollen grains on the stigma of the flower for the yield of marketable fruit. To this end, the watermelon fields are visited not only by honeybees but also by thirty-nine species of wild or native bee species. The study by Kremen and coauthors showed that native bees could provide a pollination service equivalent to that of managed honeybees. And there was a bonus: half the native bee species that visit watermelons also visit other crops. Some of these, such as cherry tomatoes, require lower levels of pollination than watermelons, and it turns out that managed honeybees do not service many of these crops. The wild bees are doing the work.

Unfortunately, numerous areas, including farmlands studied by Kremen and colleagues, show a decline in native bee populations because of degraded habitats and the use of pesticides. Varying agricultural practices produced strikingly different results. Organic farms near more or less unmolested native habitat could acquire full pollination service for heavily pollinated crops like watermelon without even recruiting managed honeybees. They profited from a flexibility that comes with the availability of wild species. But nonorganic farms distant from native habitats could not sustain these crops on native bees alone. Their dependency on managed honeybees made them vulnerable to the variation between high and low yields from year to year because of the decline in successful beekeeping. The California study is not unique. A 2004 review by T. H. Ricketts and several coauthors showed that coffee farms within a kilometer of healthy forest benefited substantially from wild pollinators, which increased crop yield by 20 percent and reduced the proportion of shriveled, unusable coffee beans by 27 percent.

Pollination has kept our modern land ecosystem going for 100 million years. Now the loss of natural habitat and their myriad species threatens the very core of what has been built by evolution. We may be able to get by for a time on artificial systems that mimic the original—managed honeybees and expansive industrialized farms—but we are already seeing the limits to this emulation. Opening the door to reviving the system by putting more of natural biodiversity into service could stop the trend, improve crop yield, and save more than a little beautiful land in the process. But how wide is this window of opportunity?

DARK FORCES

The role of humans in reshaping the modern global ecosystem is doubtless most vividly seen in the massive conversion of forest to cropland and in their unquenchable thirst for precious and fleeting freshwater. But other forces are in play. In the beginning of this book I characterized the Four Horsemen as land degradation, overexploitation (whether through hunting or harvesting), invasive species, and pollution. Of course this is an oversimplification, and these broad categories are not independent of one another. For example, the huge output of CO_2 and other greenhouse gases in modern times, only one type of air pollution, carries a special potential for havoc since it can contribute through acid rain to the denudation of forests or through global warming to flooding coastal cities and spreading invasive species adapted to warm climates, such as malaria-carrying mosquitoes. The interplay of these forces and their effects on nature constitute a vast and complex subject, and we must ask, Is the intensity of actions we are experiencing unprecedented? Or are there events in the past that foreshadow the current situation and inform us about things to come? How do these forces link to the current and projected loss of species? Also, how does the current shakeup of the biota compare to the scale of earlier great and tumultuous events in the history of life?

As we have seen, once humans established a toehold on several landmasses, they were devastatingly effective in mowing down mammals, birds, and other wildlife. This began nearly fifty thousand years ago with the human invasion of Australia and continued with the first colonizers of North America twelve thousand years ago. Islands were particularly vulnerable, and many of their mass extinction events even more recent. Human colonization, over-hunting, and introduction of human-borne or domestic fowl–borne disease on Pacific islands starting about three thousand years ago are thought to have promoted the extinction of more than two thousand species of birds. Many large marine creatures—species of whales, sharks, seals, and turtles—had been nearly exterminated by the early 1800s. Today massive overhunting and poaching for food, medicine, and cultural fashion persist in Africa, Asia, and other regions. The Asian demand for tiger parts for traditional medicines has nearly driven this species to extinction, and that is only one example of the problem. Recent efforts to control such activity must confront trends established over a long history of exploitation and devastation.

Nonetheless, in some cases the picture has improved, although results are mixed. We have seen that the increase and expansion of range for elephants now put heavy stress on already denuded, confined habitats. Controlled hunting, along with ivory export, is now allowed in several countries. It is important to stress, though, that the resurgence of the elephant is not uniform across Africa: eastern and southern Africa account for most of the estimated four hundred thousand individuals, while in western Africa there are only about five thousand, living in small, genetically isolated groups encroached upon by expanding human settlement. Recent studies suggest that these western elephants diverged perhaps as long as 2 million years ago from other African populations, and there are proposals to recognize them as a separate species, which could mean much greater scrutiny and conservation regulation.

Other cases of exploitation have notably improved but are not fully corrected. The horn of the rhinoceros has long been an item treasured by certain cultures for its ornamental and medicinal powers. On an expedition to Yemen in the late 1980s I wandered through the maze of streets in Sanaa's market, where scents of coriander, aniseed, cloves, and cumin mixed with the incisive smell of burning keratin, which is much like the unpleasant odor in a dentist's office. The fumes emanated from several workshops where artisans were shaping the handles of the jambia, the curved dagger ceremoniously worn by Yemeni males, from the horns of African black rhinos (*Diceros bicornis*). Before international restrictions, Yemen was importing about 6,000 pounds of black rhino horns a year. An average rhino horn weighs 5.5 pounds, so the annual contraband amounted to more than 1,000 individuals. The government of Yemen had banned the import of rhino horns in 1982, but as I witnessed, this had not entirely shut down the industry. In the mid-1990's as much as 150 pounds of rhino horn per year were being smuggled in. Ornamental prestige has a price both in dollars and in natural devastation. On the black market at the time a single pound of rhino horn could cost twenty-seven thousand dollars. The demand for rhino horn for both daggers and traditional medicines over the last sixty years reduced the African black rhino populations by more than 90 percent to a record low of 2,410 for the whole continent in 1995. Stricter enforcement and conservation measures have somewhat reversed this disastrous decline; steady increases resulted in a continental-level population of 3,100 by 2001. But gone are the rhinos that once foraged over large swaths of territory; most populations are now confined to managed areas in or near wildlife parks. Poaching and illegal export persists, especially with continuing demand and new money for traditional medicines, which affect both black rhinos and, even more, Indian rhinos.

Rhino and tiger products are emblematic of the wasteful exploitation that comes with an adherence to traditions and practices now wildly out of sync with the preservation of nature in the modern world. But some people have genuine survival needs that force them to overhunt, drawing on ever-diminishing natural resources as their populations grow and expand. In Africa, bushmeat—the meat derived from wild pigs, elephants, antelopes, apes and other primates, crocodiles, porcupines, pangolins, lizards, guinea fowl, and many other species that inhabit the bush or forest—attracts a multibillion-dollar trade. Recent studies show the demand for bushmeat markedly increases at times when the availability of fish, the other major protein source, is low. From 1965 to 1998 in Ghana the supply of harvested fish ranged from 230,000 to 480,000 tons annually, with marked fluctuations of as much as 24 percent between consecutive years. Regional bushmeat trade, which averaged about 400,000 tons for the same years, deviated significantly above or below this figure depending on the availability of fish. Years of intense bushmeat trade when fish stocks were low promoted intense hunting, including invasion of natural reserves. Thus declining fish stocks, more than 50 percent cumulatively in biomass since 1977, could have a ripple effect on land, further exacerbating the decline of vertebrates. Between 1970 and 1998 forty-one species of mammals in natural reserves in Ghana declined in biomass by 76 percent, and depending on the particular area, 16 to 45 percent of them have become locally extinct.

Unfortunately, solutions to such complicated problems require real international leadership. The reduction in offshore fish populations is due not only to the growth of West African commercial fleets but also to a dramatic increase in foreign fleets, especially from the European Union. Foreign enterprise has artificially increased the profitability of fishing in spite of declining fish stocks. The result is a perfect storm. The drastic decline in fisheries along the African coast and of wildlife in western Africa are on a collision course with a threefold increase in human populations in the region since 1970. A collapse of the fish and wildlife stocks in West Africa in the very near future is a real possibility.

We must also consider a related problem in the marine realm, for in the seas the problem of overexploitation of natural species is paramount. Terrestrial wildlife can no longer be the primary source of human food, nor can it possibly sustain a hungry global population that long ago shifted to a dependency on cultivated crops and domesticated animals; the major threat to terrestrial wildlife and biodiversity remains the loss of habitat resulting from the human need for more land. But the marine realm is a different story. We humans have invaded virtually every habitat on land, yet our impact could be greatest in the sea, the one Earth environment for which we have no natural adaptation. More than 450 million years ago the vertebrate line leading to humans abandoned fins for limbs and eventually shed even its embryonic gills for a life fully committed to land. Some of our tetrapod lineage—whales and sea snakes, for instance—returned to the sea, and so in effect did we, in machines that are specialized for marine life, from open boats to submarines.

Among these craft are myriad fishing vessels: Chinese junks with sails that look like pleated fans, elegant Indonesian sloops with sloping crescent sails, elongate Aleut canoes covered with sealskin to repel the icy water of the Bering Sea, charming little top-heavy Maine fishing boats stacked with lobster cages, and giant factory ships with five-story hulls that make a cavernous yawn to swallow whole whales, as if to act in retribution on behalf of Jonah himself. Fish is the primary food source for as much as one-fifth of the world's population. In most developed countries, where meat products are also heavily consumed, the demand for high-quality seafood has had an especially bad effect on delectable species, which are now seriously depleted. The marine harvest is monolithically huge and calamitous, and it is hardly mitigated by the nascent rise of aquaculture, which largely entails a net loss. Vast amounts of wild species are required to feed farmed fish and shrimp. Meanwhile, experts claim, overfishing has been far more devastating to marine ecosystems than either habitat destruction or pollution.

Surveys of world fisheries populations, which have been recorded since the 1980s, show a steady decline, followed by a precipitous fall. (For some years the global average of fish populations seemed to be on hold, but this illusion was recently ascribed to underestimates of decline given in sketchy reports by some fishing fleets.) Arresting data include, as I've mentioned, the 95 percent reduction of the North Atlantic's cod and several other species and the 90 percent loss of bluefin tuna.

Recent conservation measures have at least reversed some trends. On the famous and once fabulously enriched Georges Bank, off New England, haddock has slowly climbed back to an estimated annual biomass of a hundred thousand tons from a historical low in the early 1990s of about twenty thousand tons. Stiffer fisheries management has allowed for the rebuilding of stocks of black sea bass, scup, summer flounder, sea scallops, yellowtail flounder, and king mackerel. The United States has been a leader in developing and enforcing laws to reduce overfishing, notably through the Sustainable Fisheries Act of 1996, an update of the Magnuson-Stevens Act of 1976. Here the goal is maximum sustainable yield (MSY), leaving population levels that would permit lucrative fishing indefinitely. The law calls for the end of overfishing and a rebuilding of populations over a ten-year period, unless biologists indicate that more time is required. The implementation of this act by the National Marine Fisheries Service of the National Oceanic and Atmospheric Administration (NOAA) is already responsible for the restoration of populations of several key species, such as haddock. A key strategy here is to reduce fishing automatically when a population declines below a sustainable level. This practice follows a simple rule: more fishing when there are more fish; less fishing when there are fewer fish. Some organizations, such as the North Pacific Fishery Management Council, have adopted this practice to lucrative end.

Lest we enthusiastically return to the sea in ships, it is important to remember that these success stories are the exception, not the rule. Much more rebuilding will have to be done before we can claim to have restored any semblance of what once thrived in the oceans of the world. For example, though haddock populations have recently risen, biomass for this species was once far greater; in the mid-1960s Georges Bank populations were estimated to weigh in yearly at about five hundred thousand tons. The once populous Atlantic cod continues to decline in the same region, hovering at an annual biomass of only a few thousand tons, compared with highs of nearly two hundred thousand tons in the late 1970s. As fisheries scientists point out, the story for haddock is better because fishing pressures on this species were abruptly reduced, whereas reductions in fishing for cod were phased in only slowly.

Scientists, knowing the brutal facts, often recommend actions that are not quickly adopted. A maelstrom of conflicting commercial interests, consumer demand, and shifts in government policy and attitude suddenly engulfs their recommendations. Currently in the United States there is much pressure to emaciate the Sustainable Fisheries Act. Congressional legislation has been introduced that replaces the phrase *end overfishing* with *address overfishing*. In March 2005 a federal court made the paradoxical ruling that overfishing could be allowed so long as the population was rebuilt by the end of the ten-year period! NOAA has recently published complex guidelines relating to spawning and generation time for different species of fish that are difficult to project. In some cases, they would shorten recovery periods, but in others prolong them.

In December 2006 the Senate adopted a measure that reauthorized the Magnuson-Stevens Act through 2013 and called for more science-based management of the nation's fisheries and stiffer penalties for illegal fishing in international waters. This legislation was hailed by many environmentalists as a step in the right direction, but it provided no additional funds to support these measures and could not directly attack the global problem of overfishing. The world's oceans are not owned by any single country after all, and the U.S. Sustainable Fisheries Act and its implementation are unique developments, stark exceptions to the general rule in other countries. Of all nations, Japan continues to be unmatched in its drain upon the global marine ecosystem and remains the world's biggest consumer of fish. According to the Food and Agriculture Organization, the Japanese consume 30 percent of the world's fresh fish. Between 1999 and 2001 the average annual per capita consumption of fish and shellfish in Japan was an impressive 145.7 pounds (by comparison, Americans' average annual per capita consumption for the same period was 47 pounds).

Japan is a major force in the decimation of bluefin tuna and has resisted international pressure to change its practices. A well-publicized, inflamma-

tory example is its slow adaptation to a world without whaling. Still, Japan
has exacting measures and requires reports on its fisheries hauls and environmental impacts, and it has cut its commercial fishing fleet by a quarter.
Unfortunately, its huge appetite for seafood is served by many other nations—South Korea, Indonesia, Chile, Peru, Taiwan, and Russia—that ag-
gressively compete to satisfy its demand, especially for the rare and tenuous
bluefin tuna. Various organizations and scientists accuse these suppliers of
violating accepted procedures for sustainable fisheries management, and illegal fishing has surged in reaction to the stricter rules that limit catches.
Only a few countries have clamped down on this practice. Notwithstanding
regional sea conventions and international collaborations, more than 70
percent of the world's fish species are either fully exploited or depleted. Desertification on land now has its counterpart in the sea.

To humans accustomed to seeing fresh, luminous fish on ice-caked slabs at
the seafood market, this crisis in the deep may have a curious, questionable
reality. The sea still seems such a vast realm surrounding our islands of habitation. Its expanse and volume are incomprehensible; its life is mostly unseen and mysterious, its wealth of food constantly retrieved in great mass for
us by means we know little about. Few of us have witnessed firsthand the
wreckage to the sea bottom evidenced by the tracks of a dredger.

Things are different on land. We can quickly see the scarring of the earth.
I have witnessed with dismay over just a few years what now seems the "natural" succession of eastern Long Island. A woodland filled with cedars,
pines, and stately elms was some time ago converted to farmland, then further homogenized to a sod farm, making way for a new real estate deal and
the sprouting of uniform white stucco condominiums stretching eave to
eave to the horizon. We have a palpable sense of loss when wildlife disappears during such development. Your grandparents might remember when
the last black bear, red fox, or eagle was seen in the place where you grew up,
but these are only stories in your childhood. You yourself might remember
when as a child you swam with impunity in a lake now smelling of sulfur and
choked with slime and weeds. One of the most arresting and rapidly effective forces of destruction does not at first seem to emanate directly from humans at all. It comes with the appearance of an intruder species that
somehow stakes a claim on new land, proliferates in huge numbers, and de-

stroys what it encounters. We can witness the effectiveness of such organisms at a small scale—too many weeds in the lawn or too many snails in the garden—but invasive species also do their damage at impressively large scales.

One of Earth's most potent invaders hardly seems terrifying at all. The zebra mussel (*Dreissena polymorpha*) is a small mollusk in a small shell. Adults of the species grow to about the size of a thumbnail. The shell itself is slightly rounded above and flattened below and ornamented with a series of gently curved, wavy stripes, usually various shades of brown, that look like muted Chinese watercolors. In their native habitat in Eastern Europe, zebra mussels, if not completely welcome, are hardly unmanageable. Factories were designed with pairs of pipe systems: one system could continue to function while the other was temporarily shut down to be demusseled. In Western Europe, where the zebra mussel has lived for two hundred years, these little mollusks are controlled because they serve as food for diving ducks and other birds. Apparently other species, including humans, have learned to live with zebra mussels in their original habitat.

Zebra mussels are a new visitor to North America, but anyone observing a distribution map of the current range of the species would find that hard to believe. As of 2007, these little creatures can be found in huge, unfettered masses along the shores of all the Great Lakes and in major river systems in eastern North America, all the way down to the mouth of the Mississippi River. If you sucked the Great Lakes dry, you could still see the outlines of Lake Michigan, Ontario, and Erie in zebra mussels. Soon the other lakes will be similarly festooned. Likewise, the distribution of these creatures mark with near-perfect verisimilitude the sinewy course of the upper Mississippi, the Ohio, the Illinois, the Hudson, the Cumberland, the Tennessee, and the lower Arkansas rivers. Zebra mussels have been found recently in new places, in Haymarket, Virginia, in 2002, in El Dorado Lake in southeastern Kansas in 2003, and in the middle Missouri River in northeastern Nebraska in 2004. Following the route of Lewis and Clark, the zebra mussel is striking westward. It is known to have been carried on trailers and boat hulls to several rivers in California, Arizona, Utah, Colorado, Washington state, and Montana. The species has not established any significant presence in these locations yet, but a full invasion seems only a matter of brief time.

Less than twenty-five years ago not a single zebra mussel could be found in any of these places. The early larval stage of this animal, the microscopic veliger larva, probably first entered American waters in 1985 or 1986, but the animal was actually first discovered in 1988 in Lake St. Clair, a small body of water between Lake Erie and Lake Huron. The adult mussels, carried on the hulls of boats and aided by a capacity for explosive population growth (females can produce a hundred thousand eggs and grow to reproductive maturity within a year), spread around the shallow bottom of shores of adjoining lakes in a matter of years.

In their new habitat zebra mussels can wreak awesome environmental damage. They scarf up microscopic zooplankton and algae that would otherwise feed larval and juvenile fishes and other animals. Attaching with their strong byssus threads to virtually any hard substance—glass, rock, wood, rubber, fiberglass, metal, gravel, and other mussels—they encrust water intake pipes, irrigation ducts, boat hulls, boat-cooling systems, docks, and screens for drinking water facilities. Zebra mussels can also accumulate dangerous pollutants like PCB and deposit them as waste that small scavengers devour and pass up the food chain to various consumers, including us. They are a recreational hazard at many beaches, where their sharp-edged shells cut the feet of swimmers.

The mechanism for the zebra mussel's devastating arrival in the Great Lakes is a form of pollution, one of the dark forces ravaging the ecosystems of Earth. Those inconspicuous veliger larvae were originally taken aboard freighters in European waters as a tiny component of ballast water, carried by ships to maintain buoyancy. Ballast water is routinely taken on or pumped out depending on the load. Ships sailing westward across the Atlantic also penetrate deep into North America via the St. Lawrence Seaway and to the Great Lakes. Sometime in 1986 or 1985 a few ships, or maybe only one, happened to have the critical number of zebra mussel larvae in the ballast water released into Lake St. Clair. The startling progression of invasion and devastation that followed probably came from a founder population of larvae that wouldn't fill an eyedropper of ballast water.

There is nothing new about the invasion of alien species. For hundreds of millions of years species have migrated to new areas and arrived better adapted and more competitive than the residents. Doubtless, some of the earliest land plants succumbed to new plants when they arrived, as well as did the earliest insects, vertebrates, and other animals when their successors arrived. Three million years ago, when the Isthmus of Panama emerged from a shrinking sea and brought together the two great continents of the

Western Hemisphere, a grand isolated bestiary on South America—giant ground sloths, armadillos, saber-toothed marsupials, long-necked camel-like herbivores, and massive long-snouted rhinolike creatures—was wiped out by invading species from the north. Only a few southern species, like the giant ground sloth (which went extinct by ten thousand years) and the armadillo, successfully migrated in the opposite direction and established themselves in North America.

Why were the North American immigrants—wolves, mountain lions, bears, true saber-toothed cats, true camels, and horses—so much more successful? Paleontologists have surmised that the North American species had become evolutionarily seasoned by the waves of invasion that came over time when North America was connected with Asia and Europe by land

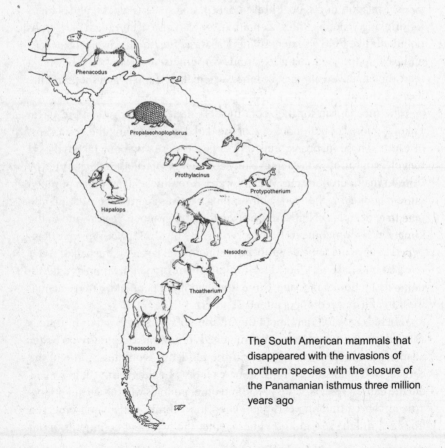

The South American mammals that disappeared with the invasions of northern species with the closure of the Panamanian isthmus three million years ago

bridges. They were species experienced at being at the crossroads. South America, by contrast, was essentially the world's largest island for about 100 million years before the Panamanian bridge. By 3 million years ago, with the coalescence of North and South America, the island became larger.

A later recession of the oceans and the uncovering of the Bering land bridge prompted the next wave of invasion. Among the species marching across was *Homo sapiens*, whose appearance was coincident with the massive demise of large mammals. And humans did not travel alone. The presence of the dog, probably the first domesticated animal, doubtlessly thinned small wild prey and marginalized wolves, coyotes, and other wild predators. Humans arriving on the Pacific islands also carried with them domesticates, including fowl. Diseases that spread from these domesticates, in combination with human overhunting, are thought to have promoted the extinction of nearly two thousand native Pacific island birds.

Today invasive species are facilitated by tramp steamers, horticultural importers, pet dealers, and even misguided land managers seeking to "improve" conditions in the original habitat. There is, like the worst-dressed list, a negative honor roll call of the worst invasive species, and all their successes are promoted, directly or indirectly, by the worst invader of all, humans. A list of the hundred worst invasive species issued by the Global Invasive Species Program (GISP) includes mammals, birds, fishes, snakes, frogs, ants, beetles, mosquitoes, aphids, moths, termites, snails, wasps, flies, flatworms, crabs, clams, mussels, weeds, guava, tamarisk tree, raspberry, mimosa, mesquite, kudzu, ginger, reed, prickly pear cactus, pepper tree, tulip tree, hyacinth, seaweed, root rot, elm disease, chestnut blight, frog chytrid fungus, crayfish plague, and microorganisms like the ravaging *Plasmodium relictum*, responsible for avian malaria.

The devastation wrought by invaders like the zebra mussel is tied to some basic biological characteristics that are not unique to this animal. Based on the spectacular examples in GISP reports, virtually all invasive species share these qualities:

They tend to be highly opportunistic. Invasive species gain footholds in places under sometimes highly unusual or unlikely conditions. The notorious brown tree snake (*Boiga irregularis*), which eradicated Guam's native forest birds, got to this isolated Pacific island from its original habitat in Australia, New Guinea, and the Solomon Islands probably by hitchhiking on military aircraft in the 1940s and early 1950s. Avian malaria, which has wiped out many of Hawaii's native birds, was carried to these islands in 1826 by the southern house mosquito (*Culex quiquefasciatus*) living inside water barrels on a sailing ship.

They tend to reproduce at very high levels. The aesthetically attractive native South American water hyacinth (*Eichhornia crassipes*), with its beautiful indigo and violet flowers, was originally much desired as an ornamental plant. But water hyacinth can reproduce wildly, doubling its population in as little as twelve days. It is now choking lakes and waterways in more than fifty countries on five continents. Another would-be ornamental from South America, the miconia tree (*Miconia calvescens*), was introduced in 1937 as a specimen to a botanical garden in Tahiti. On that island it became a desirable garden plant, and its fruits were spread into the wild by birds. Within decades it made prodigious inroads and is now the dominant macrovegetation on more than half of this large island.

They tend to be extraordinarily destructive very quickly. After its arrival on Christmas Island in the Indian Ocean, the yellow crazy ant (*Anoplolepis gracilipes*) was able to kill three million red land crabs (*Gecarcoidea natalis*) in eighteen months. Since its introduction to Lake Victoria in Africa in 1954, the Nile perch (*Lates niloticus*) has promoted the extinction of more than two hundred native fish species that were unique, or endemic, to this magnificent but now radically degraded body of water.

Some of these introductions, such as the brown tree snake on Guam and avian malaria in Hawaii, were accidental. But sadly and ironically, humans deliberately introduced many of these harmful species with some misguided sense of their efficacious effect. Mongooses (*Herpestes javanicus*) were brought to Mauritius, Fiji, the West Indies, and Hawaii in order to control rats. They did, and they kept eating, causing the extinction of several other native mammals as well as birds, reptiles, and amphibians. The notorious Nile perch was brought to Lake Victoria in order to supplement native stocks that were being overfished. Instead, this robust alien species itself contributed to the overfishing and extermination, wiping out many of the lake's residents. Its oily skin, which requires a more concerted drying of the catch, in turn caused massive deforestation around the lake for the necessary firewood. Erosion, runoff, and excess nutrient loads then opened the gate for algae and water hyacinths. Commercial exploitation of the Nile perch altered and marginalized the livelihood of local families dependent on more traditional and less destructive fishing and commerce.

The Lake Victoria ecosystem was once a sapphire set within an emerald, a brilliant blue body of clear water with an unbroken surface set in an unblemished forest. The lake harbored two hundred more species of cichlid fish than it does today. The first cichlids' original colonization of this lake probably occurred only a few thousand years ago, and the explosive diversification of these beautiful and varied creatures over the very short period of millennia since then is an icon for the amazing potentials of adaptation and evolution. But with the introduction of Nile perch and the invasion of water hyacinth, this wondrous ecosystem unraveled almost immediately—within one or two decades. Lake Victoria is now iconic for something else, a big human-promoted ecological mess and a disaster for the people whose livelihoods depend upon it.

From a purely scientific perspective, the amazing powers of alien invaders like the Nile perch and the water hyacinth are impressive and intriguing. What make these and other alien species so inordinately successful in their new habitat? Opportunism, explosive population growth, mobility, and voraciousness only describe their power; they do not explain it. Classically, biologists have ascribed to these invaders some quality of robustness, of toughness tempered by long stretches of competition with their enemies. They are not necessarily dominant forms in their original habitat, merely additional cogs in the ecological wheel. When they find new territory, they escape the competitors and predators that held them in check and can tap into their full evolutionary potential. Much the same argument has been used to explain the dramatic invasions of whole faunas and floras at certain times over the past millions of years as, for example, when the marauding mammals from North America, tempered by the competition from migrating mammals across Holarctica, were suddenly unleashed on the southern denizens after the rise of the Isthmus of Panama.

Some ecologists have mustered evidence that evolutionary robustness alone may not explain fully the intricacies of invasion or the ingenious ways in which the intruders exploit their new habitat. Two scientists at the University of Montana, Ragan Callaway and Erik Aschehoug, studied a noxious North American weed, the diffuse knapweed (*Centaurea diffusa*), and noted that it was much less destructive to grass species in its native habitat in Eurasia. What accounted for the difference? Through a series of experiments Callaway and Aschehoug determined that in its Eurasian habitat *Centaurea* had evolved with a number of grass species over a long period in such a way that the chemicals produced by its roots did not harm the other plants. For example, the uptake of nutrients like phosphorus (32 P) by other grasses was not hampered by the presence of *Centaurea*. Indeed, several species including *Centaurea* were highly successful collaborators in Eurasia, together boosting the overall efficiency of nutrient uptake and plant growth. In North America, by contrast, phosphorus uptake was severely limited in grass species being overrun by *Centaurea*. Here evolution is apparently not working in the same way. The chemicals released by *Centaurea* detrimentally affected some of the resident grasses. The explanation is straightforward. In a given habitat plant species may interact as a coevolving functional unit, depending to an extent on one another to maximize efficiency in acquiring nutrients and ensuring plant growth. Take one of these species out of the system and move it elsewhere, and the results cannot be predictably the same. The invader comes with functions beneficial in its original habitat but harmful in its new one, bearing what were originally symbiotic gifts now transformed to weapons, and the resident species are ill adapted to defend themselves.

We must do much more to refine and expand on these coevolutionary studies. In the meantime, we can rely on the general indications that the sudden introduction of new species to new areas is bad news, unless proved otherwise. Checkpoints and control for introduced species, whether through inadvertent hitchhiking, trade, or deliberate action, have increased. Unfortunately, many efforts to mitigate the effects of invasive species come after the fact. The momentum of invasion is tremendously difficult to halt. Around Chesapeake Bay are huge expanses of feathery reeds known to botanists as *Phragmites australis*. This plant, originally introduced from Europe about a century ago, invades marshes, crowding out native cattails and other desirable denizens. Since 2000 National Park Service strike teams have been on a *Phragmites* search and destroy mission. Armed with Global Positioning System receivers, all-terrain vehicles, and a nasty herbicide (dyed fluorescent blue to indicate the plants sprayed), the teams have successfully treated some 270,000 hectares. But victory may be short-lived. Studies show that most invasive plants return in decades or even years. Some ecologists accordingly question the rationale for the *Phragmites* obliteration program, noting that the herbicide itself, released in amounts yearly of thousands of gallons, has its own insidious and hazardous effects.

To be sure, we need to exercise retrospective control of diverse invaders, whether water hyacinths or brown tree snakes. Yearly the United States spends more than a billion dollars spraying *Phragmites*, holding back the leafy spurge (*Euphorbia esula*), taking chain saws to the salt cedar (various species of *Tamarix*), and using sundry methods to control other invaders. But the forces in play make yet more invasions imminent. The growth of international travel and shipping has been very good for the global spread of seeds, cuttings, and species. But frenzied commercial exchanges mean more unwanted marine organisms in ballast water and more insects in shipping crates.

Because these unwanted newcomers seem to find "greener pastures" in new places, invasion is often a two-way street. To check erosion along its rivers, China in 1979 introduced a native marsh grass of eastern North America (*Spartina alterniflora*). At home it is manageable, even vulnerable to invaders like *Phragmites*, but in China it has been spectacularly, if predictably, successful, choking estuaries, crowding out native grasses, and reducing food and habitat for migratory birds and fishes. In return, China has contributed, albeit unintentionally, the Asian long-horned beetle (*Anoplophora glabripennis*), an unwelcome hitchhiker in shipping crates, responsible for devastating maples, elms, and willows in eastern North American woodlands. Both old and new invaders are an enormous liability, an estimated drain on the U.S. economy of about $137 billion a year.

As in dealing with other dark forces, we hear of some success stories in control and mitigation. China and the United States, which represent huge crossing conduits of trade, have moved toward building a cooperative strategy. Some of these ironically involve the more scientifically informed introduction of alien species, such as an international swap of insects that devour invasive plant species. A leaf beetle (*Diorhabda elongata*) that is native to western China seems to be effective in attacking those destructive salt cedars, or tamarisk trees, that are the scourge of the arid western United States, where they crowd the edges of scattered and fleeting waterways and suck up precious groundwater. Moreover, the leaf beetles do the job without promoting serious ecological side effects. This remediation experiment is now about eight years old, and still doing well. Conversely, beetles from South America have been successfully introduced into China to control the water hyacinth.

Indeed, controlling invasive species is now an intensive, global, and hugely expensive effort. In the United States the one-billion-dollar program in cargo and baggage inspections is thought to have intercepted more than fifty-three thousand arthropods, plants, and pathogens each year. Unfortunately, this represents only a tiny fraction of what's out there, moving to and fro in a global biotic ecosystem now tightly interconnected with human activity and commerce. And people are sometimes reluctant to clamp down on some of this promotional trade for aesthetic reasons. Horticulturalists, for example, are slow to abandon the exchange of exotic plants, which has been a tradition for centuries. Many new invaders will proliferate, but we must remember why invasive species are now thought to be only second to land use in driving myriad natural species to extinction.

THE WASTE OF A WORLD

Now, all the plagues that in the pendulous air . . .
　　　　　　　　　　　　　　　　—*King Lear*, Act III, Scene 4

Our family house in Mar Vista, on the edge of Los Angeles, was one of the first topographic high points east of the Pacific shore four miles away. Theoretically, as the name of our town suggested, we could see the sea. But on many days this was not possible. Not just smog or rare stormy weather blocked the view. I remember days in the early 1960s when the smog in the air over L.A. mixed with a fog from the coast to create a cool inversion layer of toxic soup. The kids in the neighborhood used to play a morbid game. We would try to breathe in this filth as deeply as possible, filling every available air sac in our lungs, without coughing. The feat was virtually impossible, a lesson that failed to register on us with the appropriate alarm as we biked up the hilly streets of Mar Vista. Conditions in Los Angeles have since improved with the implementation of the Clean Air Act of 1963 and the adoption of unleaded fuel and catalytic converters, but the people and their cars keep proliferating. California has the toughest emissions standards in the United States, but it also still has one of its worst air pollution problems. Los Angeles, Bakersfield, and Fresno are at the top of the worst rankings. Air pollution is also the scourge of many smaller cities. Logan, Utah, Eugene, Oregon, and Visalia, California, each with less than one-fortieth the population of New York, rank far worse than this megalopolis.

We tend to think of pollution as largely a problem that emerged with the great Industrial Revolution of the nineteenth and early twentieth centuries. Our current situation is doubtless part of this phase, but pollution has been a human contribution to the environment for millennia, only worsening as more and more people get involved. As early as sixty thousand years ago humans were burning out large tracts of forestland and generating smoke, ash, and greenhouse gases in the process. Toolmaking in the Iron Age (twelfth to seventh century B.C.) involved metal grinding, which created piles of materials that were small and easily dispersed in the environment, but pollution was a genuine problem even in ancient times. By 3000 B.C., with the development of smelting techniques, copper production appeared. The Greek physician Galen in 200 B.C. observed the health hazards of the acid mists generated in copper mines. Ancient Rome was not only the largest but also the most polluted city in the world. Horace disparaged "the smoke, the wealth, and the noise" there. Romans called their polluted air *gravioris caeli* (heavy heaven) or *infamis aer* (infamous air), and they set new standards for public health. Streets were cleaned, and public hospitals built. The Cloaca Maxima, an ancient sewer that had been built in Rome by the Etruscans in 500 B.C., became the main channel for a complex network of tributary sewers.

Still, these measures were not enough. Mines were small in scale, but metal production required the smelting of large amounts of ores in open fires, and this produced high concentrations of toxic emissions. Occupational hazards persisted in lead and mercury mines and smelters. Pliny the Elder (A.D. 23–79) wrote that workers at zinc smelters used bladders as respirators. These problems led to the suspension of some mining operations near certain cities, a very early expression of environmental management. Just the same, industrial activity persisted because lead, zinc, and copper all were necessary to sustain the empire's large affluent society and high standard of living. Wildly unsafe practices infiltrated refinements of the good life. It is thought that ingestion of lead acetate, used as a sweetener in wine, contributed to health problems among Roman aristocrats and assisted in the decline of Rome itself.

During medieval times the effects of pollution in Europe, mostly smoke from burning wood and wastes dumped into water used for drinking and cleaning, were scattered and usually local. The reopening of ancient mining districts in central Europe in the eleventh century increased pollution levels, and pollution certainly helped spread the bubonic plague, or Black Death, in the fourteenth century. As cities industrialized, the rising levels of pollution elicited some measures of self-control. In 1361, King Edward I of England banned coal burning in smoky, sooty London. The development of large furnaces with tall stacks in the sixteenth century, which insidiously spread air pollutants over much greater areas, came in tandem with water contamination and other forms of pollution. Massive dumping of sewage and other impurities putrefied the Thames River in 1858, an event known as the Great Stink, stimulating the belated construction of London's modern sewage system.

We know about metal pollution in ancient times not just through the written historical record. A study by Sungmin Hong and several coauthors in 1996 of copper impurities preserved in Greenland ice layers reveals a peak in copper-borne air pollution around 500 B.C., when copper production in the Roman Empire was greatly stimulated by new demands for military equipment, tools, and especially coinage. Copper-smelting centers of the empire included districts in Spain, Cyprus, and central Europe. The decline of Rome is similarly reflected in the Greenland ice layers with a drop in copper concentrations at roughly A.D. 500. Another peak in concentration corresponds to high copper pollution levels about A.D. 1100, when smelters were proliferating in China thanks to the activities of the superb artisans, toolmakers, and craftsmen of the Sung dynasty. High levels did not occur again until the early Industrial Revolution, though copper emissions from ancient Roman and Chinese periods were higher than those from the mid-eighteenth and nineteenth centuries and equal to nearly 10 percent of such emission levels today.

Similar patterns are seen for profiles of lead (Pb) concentrations as recorded in Arctic ice layers. Between 500 B.C. and A.D. 300 high lead concentrations correlated with the activity in those centuries of Roman mines and smelters. Levels declined with the fall of the Roman Empire and then began to rise again with the reemergence of mining. The ice record shows that by the 1700s lead pollution in the air over northern Europe was very high. It reached a peak in the 1990s, and has sharply decreased, by sevenfold, since drivers began to use unleaded gasoline.

To insist on unleaded fuel is one of several steps taken to reverse the rise of air pollution caused by automobiles. Since the late 1950s clean air acts in the United States and elsewhere were inspired by a series of atmospheric disasters. The Great Smog of 1952 in London was responsible for four thousand deaths. In 1948 an air pollution inversion such as I knew from my childhood in Los Angeles—with pollution in the lower layers of cool air trapped by the warmer air above it—enshrouded the town of Donora, Pennsylvania, killed twenty people, and caused illness in 40 percent of the town's fourteen thousand people. Smog events responsible for deaths and severe illnesses were recorded in Los Angeles throughout the 1940s and 1950s.

Modern pollution disasters come in all forms: from sewage, oil spills, toxic runoff, smoke, smog, greenhouse gases, and nuclear meltdowns. We associate the names of Love Canal, Three Mile Island, Bhopal, and Chernobyl with human casualty and misery, but the devastating pollution crises in each of those places also had disastrous impact on the entire ecosystem. When pollutants degrade the quality of life or even kill organisms, including humans, they function like an indicator, a litmus test, to show us just how much we and other species depend on clean water, soil, and air. That we must secure an untainted environment seems self-evident, but for some reason this premise seems to place impossibly rigorous expectations on us. We are reluctant to deprive ourselves of other things we find equally or more valuable — that is, until we foul our surroundings irreparably and intolerably.

In April 2006 newspapers reported that Baiyangdian Lake, northern China's largest inland body of freshwater, was "choking itself to death." Problematically, the lake, about 360 kilometers square, is only 130 kilometers (78 miles) south of Beijing and even closer to the Baoding City region, a basin teeming with ten million people. Once known as the pearl of the north, Baiyangdian Lake has been a natural reserve for birds and other wild species for millennia and a site for productive fish farming for decades. Fish farming brings excess fish waste and discarded fish food, but this polluting human activity is now a casualty of an even greater environmental disaster. Sewage loaded with excess phosphorus and nitrogen from Baoding City has flowed into the lake like an unrelenting black tide, sucking the oxygen right out of the water and making it incapable of harboring sustainable levels of fish populations. Mass fish kills have occurred for decades; drought and falling water levels have only worsened them. In the winter of 2005–2006 the oxygen-starved lake went through its usual freeze. When the lake thawed in the spring, the dead fish simply floated up to the surface. Only those farmers who had harvested fish in October 2005 had a crop; those who had waited to harvest larger fish that survived the winter barely found any survivors.

Lack of fish is not the only blow to the quality of life around Baiyangdian Lake. One farmer in his early forties noted, "When we were kids, we used to drink the water straight from the lake. Now we can't even cook with it." The Chinese government has declared an all-out effort to restore the lake to its original condition. Acknowledging that there are too many fish farmers, the government has already registered many of them for relocation. But such official declarations and plans have been issued over the years, and the fish farmers are skeptical. "It's a lot of talk," one commented. "As long as they keep talking, we keep farming our fish. We can't do anything else."

Baiyangdian Lake is a perfect example of a water pollution crisis in a human-dominated ecosystem. China is in our sights because its economic growth, huge population, and enthusiastic embrace of many of the technologies that have created environmental problems in the West make an especially worrisome augury for the global future. China now releases forty to sixty billion tons of wastewater and sewage into its rivers and lakes each year. About 30 percent of China's rivers are too dirty for agricultural or industrial use, and at least three hundred million people, roughly the size of the entire U.S. population, have no access to clean drinking water. The polluted waters are creating international tensions because they flow into Russia and other countries in Asia. But it is unfair to single out China; the situation at Baiyangdian Lake is replicated by many other disasters elsewhere in the world.

Another great danger to both marine and freshwater ecosystems is the deliberate or accidental spillage of industrial products, especially oil. On March 29, 1989, when the U.S. tanker *Exxon Valdez* hit an undersea reef and spilled more than ten million gallons of oil into Alaska's deep blue Prince William Sound, the resultant kill-off of marine mammals, plants, seabirds, and other organisms was enormous. As many as 250,000 seabirds, 2,800 sea otters, 300 harbor seals, 250 bald eagles, up to 22 orcas, and billions of salmon and herring eggs were lost. This shocking event catalyzed the American environmental movement, and a thorough cleanup ensued, but while the sound has slowly recovered, the terrible effects remain.

There have been worse oil spills, though not always in such pristine coastal habitats. Two ships collided in 1979 near the island of Tobago, spilling about 46 million gallons of oil, and when one of the two, the *Atlantic Empress*, was towed away, it spilled an additional 41 million gallons off Barbados. Seven major oil spills from refineries and tank farms around New Orleans caused by Hurricane Katrina in 2005 totaled 6.7 million gallons. In January 1991 the Iraqi military deliberately released 460 million gallons of oil into the Persian Gulf. The collaboration between war and environmental destruction has unmatched power.

Runoffs of the kind suffered at Baiyangdian Lake affect not only waters but solid ground. China and much of the rest of Asia are hotspots for ground pollution. Substances that contaminate the substrate include an impressive variety of chemicals: pesticides and herbicides, like DDT; fertilizers, with their toxic side effects, loading the soil with excess nitrogen, phosphorus, and other chemicals; petroleum derivatives like benzene and toluene; solvents and cleaning agents (acetone, trichloroethylene, formaldehyde); contaminants from solid wastes (mercury, cadmium, lead, bacteria); organic pollutants (dioxins, such as PCBs); depositing of chemicals in dust from smelting operations and coal-burning power plants (cadmium, lead, zinc); and leakage from transformers (PCBs). The health hazards of dioxins and other contaminants have been well documented. Many are carcinogens; others affect the central nervous system or promote kidney damage (although recent reports call for a better scientific assessment of the actual risks, especially for those who do not work in industries using such chemicals). Damage to ecosystems rooted in the soil substrate is likewise catastrophic. As we have seen, recent DNA studies have shown much lower levels of microbial diversity in soils contaminated with pollutants.

Cleaning up this dirty mess is necessarily expensive and prolonged. U.S. agencies have identified more than six hundred thousand known or suspected sites in need of soil cleanup, some of which will cost millions of dollars. Among them are the 1,612 places identified by the EPA's National Priority List as Superfund sites because they seriously threaten human health. More than sixty-five million Americans, one in every four, live within four miles of a Superfund site. A study in California showed that women living within a quarter mile of a Superfund site were at significantly high risk for birth defects, heart disease, and neurological problems. In the effort to remediate, Americans were quite progressive for a while. For years after the Superfund program was initiated in 1980, polluting industries were taxed to pay for the cleanup. Then, in 1995, the Superfund tax was suspended, and polluters enjoyed a reprieve worth billions of dollars. Private citizens naturally do not want to pick up the tab for trouble caused by huge power, oil, and chemical corporations, and the decrease in available federal funds has slowed the process.

Other regions of the world have lagged behind even these inadequate measures. In the late 1990s few of the tens of thousands of contaminated and hazardous toxic sites identified in Europe were subjected to any form of remediation. Recently the European Union has considered comprehensive legislation known as REACH (registration, evaluation, and authorization of chemicals) that would cast a regulatory net over a broad range of commonly used substances. By contrast, U.S. laws and regulations set standards for new chemicals brought to the market but do not cover ones in products already in use: paints, toys, and fabrics. REACH is more ambitious than any regulatory program now in place anywhere, but it has encountered debate and setbacks. For example, the first vote on REACH in the European Parliament in November 2005 called for phasing out the most hazardous chemicals but allowed continued information gaps on thousands of others. On December 13, 2006, the EU Parliament voted to support a REACH regulation—planned for implementation by June 2007—over a period of eleven years, of 30,000 chemicals.

In Asia the number of seriously hazardous sites is difficult even to estimate. Despite the proclamations and initiatives made by various governments, some of the world's worst toxic sites remain. The Union Carbide pesticide plant near the heart of the city of Bhopal, India, is now abandoned following the horrible disaster in 1984, when a water leak into a holding tank of deadly chemicals caused a violent phase change and created a noxious cloud that engulfed nearby shantytowns, killed 3,000 people, and caused serious illness in 150,000 to 600,000 more. The urban infrastructure of Bhopal collapsed, and people were trampled to death trying to escape the death cloud. It is estimated that in the end as many as 15,000 people died as a direct result of this accident.

The killing fields remain in Bhopal, even though the hazardous industrial activity is no more. Continual surveys of the ground under the plant and various dump sites show dangerously high levels of mercury, nickel, lead, copper, and other heavy metals, as well as organic contaminants including chlorinated benzenes. There is evidence that these toxins have permeated the groundwater and affected drinking water. Citizens of the region as well as international environmental organizations like Greenpeace are protesting Union Carbide's failure to compensate the survivors of the 1984 disaster and to rid the site once and for all of its dangerous pollutants. Its intransigency puts the health of more than 1.4 million people at serious risk.

Some surveys locate five of the world's ten most polluted places in Asia.

These include the nuclear disaster site of Chernobyl in Ukraine, the chemical complex at Dzerzhinsk, Russia, and the enormous outflow of tannery wastes and associated chemicals near the city of Ranipet, India. Even sites that didn't make this dubious list represent huge problems. Japan has the biggest landfill complex of all, a dump west of Tokyo contaminated with dioxins and heavy metals. In the beautiful valley surrounding the city of Kathmandu, Nepal, are several stockpiles of obsolete, imported, or donated pesticides, a cumulative mass of seventy tons with high levels of DDT, lindane, and other hazardous chemicals. Nepal, wedged between India and China, the world's two most dynamically changing economies, is still an underdeveloped country with most of its rural communities depending on agriculture for a livelihood. Roads are often unreliable in the world's most mountainous terrain, as is the electricity and communications infrastructure. Much of the country can be reached in a reasonable time only by chartered plane, as I saw when I traveled there some years ago. I was also impressed with the massive garbage piles at the dump sites around Kathmandu.

These problems are common in Africa and South America too, of course. Their big cities and industries also register on the worst toxic sites list. A mining complex near the city of Kabwe in Zambia releases lead and cadmium impurities that potentially endanger 250,000 people, and a metallic smelter near the town of La Oroya, Peru, contributes severe amounts of soil contaminants as well as sulfur dioxide and other harmful gases. But the biggest factor in soil pollution in Africa and South America is one that also afflicts Nepal: developing countries without the resources to purchase new, expensive, and less poisonous pesticides have been compelled to use "hand-me-downs" from wealthy nations whose discarded pesticides, judged to be horrendously harmful, are nonetheless still available and cheap. With outrageous hypocrisy some of these rejected products are characterized as "aid" to the countries they afflict.

My childhood in smog-ridden Los Angeles is not likely to have helped increase my evolutionary fitness or, for that matter, my life span. Despite new laws to control car exhausts and clean the air, Los Angeles and other big American cities still have unsafe air. The situation is hardly better—in fact, it is often worse—elsewhere. On my recent visits to China both government officials and citizens on the streets told me that air pollution was their country's biggest problem, an opinion well supported by recent data on air quality. Beijing has nearly 8 million people and 2.5 million automobiles, many of them poorly outfitted to control emissions. The city is in a basin that restricts air circulation except when dust-laden winds from the Gobi Desert blow in and mix with the photochemicals and particulate matter of the city's fouled atmosphere. Recent satellite data indicate that Beijing may now have the worst smog of any capital in the world, a fact I can at least anecdotally support by reporting my experience of riding in a cab with a broken window on one of Beijing's clogged expressways.

When it comes to experiencing photochemical smog on such a massive scale, the Chinese are relative newcomers. Some years ago certain Mexican politicians announced that Mexico City was no longer the world's smoggiest; that dubious distinction apparently had been passed to Tokyo. Now it seems Beijing and other megacities in China are in contention. There are others. A spectacular flight over the Andes from Argentina ends in a long dive down a huge mountain chasm into a bowl of rusty smog that envelops Santiago, Chile, a sprawling city like Los Angeles with a comparable Mediterranean climate and comparable levels of air pollution. It has been my life's dream to enjoy the panorama of the Himalayas. But when I eventually got to Nepal, I was devastated to find the acrid atmosphere of smoke from burning wood and trash pits and smog from factories and cars trapped in its narrow mountain valleys. Any view of the world's most magical and imperious peaks was obliterated.

Air pollution is composed of a complex, variant mixture of gases, including carbon monoxide, lead, nitrogen dioxide, tropospheric (lower-atmosphere) ozone, particulate matter, and sulfur dioxide. In addition, there are chlorofluorocarbons, or CFCs, and hydrocarbons, such as methane (CH_4), the odoriferous gas released by industries, livestock, and humans. Some of these—carbon dioxide or sulfur dioxide, for instance—are released directly into the atmosphere as by-products of combustion in factories or cars. Others undergo secondary reactions that build new compounds, the most notable being the photochemical production of ozone in smog. These all are dangerous to health and safety. In the United States, thousands of cases of premature mortality and tens of thousands of emergency room visits annually are linked to air pollution. Acid rain, another by-product, has destroyed major tracts of forestland in North America and northern Eurasia.

As we might expect, air pollution is increasing in both intensity and scale, the steep rise in polluting emissions being now international, even global in extent. In the early 1980s satellite images showed emissions from fossil fuel combustion, forest fires, agriculture waste burning, and vegetable-fuel combustion mixing together to form an enormous gaseous swirl around Earth. People in both the developing and the industrial worlds were making reciprocal contributions to the huge global problem.

Pollutants in the atmosphere don't last forever. They have an atmospheric life span, a period during which they retain their chemical integrity, but eventually they are broken down, dissipated, or recombined with other compounds. If such pollutants persist for a few weeks in the atmosphere, they can drift from one continent to another. Ozone (O_3), a potent greenhouse gas that is also toxic to humans, animals, and plants, has suitable longevity for long-distance migration. Not all ozone is bad, and the distinction between "good" and "bad" ozone deserves comment. At high levels in the atmosphere, stratospheric ozone is vital to life because it forms a protective shield of molecules that repel harmful ultraviolet rays from the sun. The breakdown of this ozone layer and especially the formation of an ozone "hole" in the dry, cold atmosphere over Antarctica, first detected in the 1970s, is a critically serious development.

The source of this problem is again anthropogenic. Chlorofluorocarbons emitted from refrigerators, air conditioners, spray paints, and deodorants tend to drift up into the stratosphere, where they are exposed to the ultraviolet rays that have managed to penetrate the ozone layer. These rays catalyze a reaction in the CFC molecules, forcing them to release chlorine atoms. Ozone is effectively destroyed when a chlorine atom takes one of the three oxygen atoms in ozone to form ClO, leaving an oxygen molecule, or O_2, behind. Since there are many oxygen atoms already present in the atmosphere, the damaging effect of CFCs on ozone is profound. The free oxygen atoms can combine with ClO to form an oxygen molecule and a free chlorine atom, which is again in the right state to destroy yet more ozone.

The strict regulations coming out of the Montreal Protocol of 1987 have limited or terminated the use of CFCs and certain other ozone-destroying chemicals, an international effort that appears to be working. As of 2003, scientists studying the atmosphere above Antarctica via satellites and ground stations reported that the rate of depletion of ozone has slowed significantly.

But the depletion will go on because many developing nations have not fully banned CFCs. An international fund is now in place to help such countries comply with the protocol. In any case, since some ozone-damaging compounds can remain in the atmosphere for as long as fifty to one hundred years, the situation continues to be carefully monitored.

Ozone is also formed in the lower atmosphere, the troposphere, through several pathways, including a small portion of ozone descending from the stratospheric ozone layer. Most of this tropospheric "bad" ozone is a product of a photochemical reaction between atmospheric oxygen and nitrogen oxides, on one hand, and volatile organic compounds emitted by cars and power plants, on the other. The atmospheric lifetime of tropospheric ozone is about one or two weeks in the summer, one or two months in the winter. Since the wafting of air pollutants from one hemisphere to another takes about a month, ozone can make global inroads in the winter. Recent analyses show that ozone levels have doubled over mid- and high-latitude North America, Eurasia, and the Pacific Ocean since the 1860s. Because it can migrate, and has, highly concentrated ozone over remote areas of East Asia seriously threatens cropland, forests, and other natural habitats. Another pollutant, carbon monoxide (CO), most of which comes from automobile emissions and biomass burning, also "lives" long enough in the atmosphere—one to two months on average—to be transported from continent to continent. It not only is toxic in high concentrations but badly affects the atmosphere's oxidizing capacity.

Other long-distance travelers, the nitrogen oxides (NO_x), show striking differences depending on where they come from. Nitrogen oxides are usually emitted as nitric oxide (NO), which is oxidized by ozone to form nitrogen dioxide (NO_2), the air pollutant with perhaps the worst effect on human health. Nitrogen dioxide also combines with hydrocarbons to form photochemical smog and, along with sulfur dioxide, is an important constituent of acid rain. Burning gasoline, coal, and other fuels produces nitrogen oxides. The modern controls of such emissions in the United States has kept nitrogen oxide amounts nearly level since the 1980s, and in Europe, where more stringent emission controls have been applied, especially to industrial centers, the trend is actually significantly downward. In contrast, in Asia a jolting curve goes steeply upward. Asian emissions of NO_x in the 1970s were only a minor fraction of the global output, but for at least a decade its emissions of nitrogen oxides have surpassed those from Europe and North America. Given the enthusiasm for producing and buying automobiles in huge nations like China and India, it is not likely that this acceleration of Asian nitrogen oxide pollution will diminish within the next few decades.

In the future the Southern Hemisphere will contribute more and more to the decline of global air quality. Until now its deliberate burning of forest and agricultural fields was its major source of air pollution, but with the growth of economies, the surge of industrialization, and the concomitant increase in polluting emissions that will now change.

Though pollution has been a deleterious side effect of human activity for a long time, its scale and intensity today make it a modern environmental problem. Nobody seems to like pollution, not even those responsible for it, but many industries oppose clean air, water, and soil laws because the legislation would likely cost them billions of dollars. Published lists and websites scrutinize the industries that have been slow to act or are downright uncooperative. American companies are required to issue information on their pollution impact in the form of toxic release inventories (TRIs). The worst offenders, releasing the most toxic chemicals, are not surprisingly mines extracting various ores for metals, including copper. Mines have had this dubious distinction throughout the long history of human pollution extending back before ancient Greece and Rome. That is why some of the most polluted areas in the United States are not near major cities but are the empty northwestern Arctic of Alaska, the lonely region of Humboldt, Nevada, and the small desert towns of Pinal and Gila, Arizona. The most populous region high in the rankings of chemical pollution is Salt Lake City, Utah, a metropolis surrounded by mining country.

Cleaning up mines is not easy, given that mining companies are insulated by weak laws and desultory monitoring. Their operations are not always in areas where many people live, so voters do not regularly witness their destructive activities. Some of these companies have become rather cynical; one major mining polluter in North America is named U.S. Ecology Idaho Inc. The Environmental Protection Agency has a revised rule that puts limitations on the amount of information required of industries in its TRI reports. Both the House and the Senate have issued bipartisan letters of concern about this change.

Some anthropologists have argued that over history human society has gradually coalesced from tribes to chiefdoms to nations. They predict that a world state, a single, self-governing entity for all of Earth's people, may not emerge for a few more centuries. In the meantime, we have reached the point where we are bound together not only by our interdependency for food and goods but also by our deleterious environmental impact on one another. We have become a world state of pollution.

Efforts to control pollution have the commendable goal of stamping out their harmful effects on human health and human environments. But we must understand that some of these pollutants can also profoundly transform the *global* environment. We have seen how gases emitted by factories and cars concentrate in the upper atmosphere and form an insulating blanket, bouncing heat back to Earth's surface. Carbon dioxide (CO_2), one of the most abundant gases in our atmosphere, is second only to water vapor as the most effective of these insulating greenhouse gases. (Water vapor accounts for 60 percent and CO_2 for about 26 percent of the greenhouse effect.) We know that carbon dioxide has been released to the atmosphere in notably high amounts during phases of intense and widespread volcanic eruption or in the aftermath of an asteroid strike. Some correlations between CO_2 peaks and dramatic, often destructive, events may be only coincidental; but others may have real causal meaning. We have not yet sorted out all the reasons why CO_2 concentrations spiked at certain times in the past, but we know that present levels in the atmosphere are unusually high, and unlike some other maxima over the last 500 million years, we know why. The astounding rise of CO_2 and other greenhouse gases in our time is a side effect of human-generated air pollution, a testament to human domination. And it is changing the thermostat of a planet.

HEAT WAVE

In January 1990 I journeyed to Antarctica on a museum trip meant for tourists. I was a relative newcomer to that great continent, having been there once only the season before, so my passing knowledge of marine mammals, fossils, geology, rocks, ice, water, and continental movement would have to suffice in my role as a lecturer on board. In the ragged, ice-swept seas of Drake Passage we went off course and hit first land at Elephant Island, 150 kilometers (90 miles) northeast of our intended destination on King George Island. Here we encountered the "Isle of the Dead," a spectral heap of gray, ashen cliffs and dirty glaciers descending to a thundering shore break. Clouds slammed hard against the peaks, bringing with them a barrage of rain and ice that hit us like a knife-edge as we stood on deck clutching the rails. A horn sounded. Zodiacs were lowered to take us ashore. The boats rose and descended the entire height of the ship's hull on giant swells, and our disembarkation was eventually abandoned. A mere tantalizing thousand yards away was the beach where in 1916 the great explorer Ernest Shackleton and his marooned crew huddled under their overturned boats and subsisted on seal meat. It was Shackleton who with four able men struck out in an open boat against 1,280 kilometers (800 miles) of the world's worst sea, landed on South Georgia Island in the wake of a hurricane, and had a ship rescue his men on Elephant Island. Shackleton's resourcefulness proved miraculous. Every one of his crew came back alive. Shackleton would not have been intimidated by these waves.

The aborted landing was a disappointment, but we had some compensation. Our ship circled a great spire of volcanic rock that looked like the sharpened obsidian head of a prehistoric spear. Flanking the tip of this promontory was a pale sun, emasculated by the cold and the spindrift but clearly visible through a slight thinning of the clouds. Moments later, as all astronomical calculations had precisely predicted, it experienced a full eclipse. Then the storm swept in on us for a second round, and we departed.

That was the end of the bad weather. When we got to King George Island the next day, the bottom of the world seemed to be experiencing unusually mild conditions, even for its austral summer. When, during the Antarctic summer season of 1914–1915, Shackleton launched his audacious expedition to take his ship the *Endurance* far south in order to anchor close enough for an overland crossing of Antarctica, he could have profited mightily from this kind of weather. Instead, he encountered a frozen Weddell Sea and an obstinate ice pack that crushed his ship and necessitated his epic escape. But this frigid realm was not what we encountered. We landed on rocky beaches that escaped the tendrils of glaciers, and we could feel the reflected heat of the glazed, rugged ground on our faces. The wildlife was in frenzy. Seals bellowed, and penguins carried out their shrieking, rambling, and fighting with more than customary enthusiasm. And . . . what was that? Little pink flowers (*Colobanthus quietensis*) popping out between the rocks? Sure enough, pollination was in full swing here too. Antarctica at that moment seemed far from its usual categorization as a continental-scale realm in extremis, with conditions that might be expected on a planet located somewhere between Earth and Mars.

My mother, who came as my guest on the cruise, had a long-standing familial reputation as someone who always brought sunny skies to all her visits. I idly mentioned her Proserpine powers, and this got around. She became the goddess of good weather. Passengers and crew alike made ceremonious speeches of gratitude in her honor and invited her to their dining table, as though these offerings would sustain our good fortune. Apparently the rituals worked, and a weird but delightful span of Mediterranean clime continued. We cruised into Deception Island, a huge volcanic caldera that sheltered a crescent shoreline obscured by clouds of steam from hot springs, the homes of animals, plants, and microbes yet to be identified and cataloged by science. With a few energetic passengers and members of the crew I scaled the interior wall of the volcano. What had been the year before an unbroken snow slope was now an endless ascent over piles of wet slabs of rock. When we reached the rim of the caldera, we confronted far below an open ocean so incandescent with the rays of a summer sun that you could not look directly at it. We sailed on to the U.S. research base at Palmer Station, sheltered from harsher winds by its location at the head of a narrow south-facing bay. There the warm weather had an almost shocking intensity. Walking up to the station house, I took off my parka, then stripped down to my T-shirt. Palmer is known for its "banana belt weather," but this was ridiculous.

On that trip a truism about common conversation was truer than ever: everybody talked about the weather. But what did this weather mean? Just a fluke, or a harbinger of momentous events to come? An expert meteorologist from Argentina was on board. He talked about his laboratory's measurements of the ozone hole, and he explained to people that what we were experiencing was not the same as global warming. Indeed, as he noted, the increase of CFC emissions in decreasing stratospheric ozone and enlarging the ozone hole has a slightly cooling effect on global climate. The well-educated passengers seemed to have only a vague sense of what *global warming* actually meant. The Argentine scientist, at the time correctly, cautioned against a hasty judgment that this stretch of munificent weather had anything to do with a long-term climatic shift for the continent and the planet.

Things have changed. In the March 24, 2006, issue of *Science*, a series of articles, news items, and perspectives centered on troubling observations of glacier movement in Greenland and Antarctica. In many regions of the world glaciers are retreating, a sign, most agree, of warmer times. But warming conditions might also cause glaciers to migrate faster downslope to the sea, and that would have huge consequences. A glacier like La Mer de Glace is not merely a metaphor or the name of a magnificent white tongue of Mont Blanc. Glaciers are, in actuality, rivers—rivers of ice that flow. In Antarctica and Greenland they drain the vast ice sheets at the core of these landmasses. In Greenland, the ocean is destroying the glacier's snout, and the added meltwater lubricates the glacier's underbelly, promoting its further acceleration. The faster flow of glaciers to the sea means more melting of continental ice in seawater and a resultant rise in sea level. Glaciers draining the Greenland ice sheet are speeding up alarmingly, and nothing is making up for the significant loss of ice. In Antarctica the overall gain in volume of the East Antarctic ice sheet is far short of compensating for the loss of hundreds of cubic kilometers of ice annually in the West Antarctic,

Calving of a glacier in Antarctica

which lies south of the Antarctic Peninsula, Palmer Station, and other sites that we enjoyed on that balmy austral summer cruise in 1990. The reason for this loss appears to be a warmer ocean that causes ice sheets to thin. Around Palmer Peninsula, the world's most marked regional warming of the past fifty years first created puddles of meltwater on the surface of the ice shelf. Then the meltwater shot through the ice along cracks of weakness like a hydraulic cannon, breaking up the ice shelves into smaller chunks. This was the breach in a great frozen dam that allowed glaciers to migrate much faster to the sea, reducing the thickness of the great pile of ice to the south and contributing to sea-level rise.

The spectacular changes in Antarctica are not just a matter of glacier stampede and ice meltdown. The very next, March 31, issue of *Science* had an article by J. Turner and four coauthors from the British Antarctic Survey entitled "Significant Warming of the Antarctic Winter Troposphere." The authors analyzed recently digitized conversions of measurements from radiosondes, which are radio receivers suspended from weather balloons that record temperature and other atmospheric variables. Years of accumulated radiosonde data from Russian flights were especially valuable. The data showed a significant change in the winter weather over the icy continent: a shift of 0.3° to 0.7°C per decade over the past thirty years.

These fractional temperature changes may seem small and inconsequential, but as those who study climate change and its effects know, they are not; they are alarming. The average increase in temperature per decade near the surface taken from all parts of the globe for a comparable period was just 0.11°C. Even so, a compilation of measurements for surface temperatures since the 1860s shows an increase of between 0.4° and 0.8°C with an upward trend beginning about 1910. On the basis of the progressively steeper curve for temperature increase in a little more than a century, models that account for the complexities of climate predict an increase in global mean surface temperatures by 2030 to 2050 somewhere between 1.5° and 4.5°C, depending on the region and latitude (with possibly as much as an increase of 6°C in high north latitudes). The distant future is less certain, since different models vary considerably in projections for 2080 to 2100. But even they

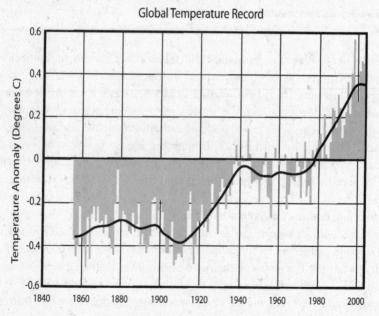

Global surface air temperature change since 1857. Vertical bars represent deviations from the 1961–90 average temperature of 14°C. The smoother dark line emphasizes thirty-year trends.

converge on a prediction that depending on location, mean annual temperatures will be 2° to 8°C higher than they are today. In both mid- and end-of-century projections, the Arctic, the high latitudes of North America and Eurasia, and selected low-latitude regions like the Sahara will experience particularly dramatic increases. Tropical rainforest regions will be affected by mean temperature increases between 2° and 3°C in as little as thirty years.

So how drastic is this single-digit increase? In recent years biologists have documented the significant effects of increases in the mean annual surface temperature of only a fraction of a degree. Peoples inhabiting small low-lying islands in the western Pacific and elsewhere have been swamped out, local populations of animals and plants extinguished, coral reefs bleached, and the range of tropically adapted species, such as malaria-carrying mosquitoes, expanded. Global mean temperatures that wobbled by only about 5° to 6°C during the later Pleistocene epoch (between eight hundred thousand and twelve thousand years ago) were enough to promote dramatic and sweeping shifts between the glacial periods, when ice sheets encrusted much of the Northern Hemisphere, and interglacial periods, when the ice sheets dramatically retreated. These oscillations were accompanied by equally dramatic fluctuations in the amount of forestland, expanding and collapsing ranges in animals, and even a few extinction events.

Do studies like the recent disclosure that Antarctic winter air is rapidly heating up suggest that progressive worldwide climate change is even more extreme than we thought it was? Turner and his coauthors were guarded in their conclusion. Antarctica is clearly warming up, but they could not definitely say why. Even though some of the available climate models predict an overall increase in global temperatures, they do not predict such extreme rates of increase at southern high latitudes. But the authors noted that warming there occurs at higher altitudes not near the sea surface. Moreover, during the winter the ocean around Antarctica is choked by sea ice—ice made directly from the freezing of salt water on the ocean's surface. Thus, the warming of the atmosphere over the southern continent is not due to a warming of the sea surface. The authors offer a scientific conclusion that is understated and at the same time subtly profound: "The observation of significant tropospheric warming at southern high latitudes, decoupled from a similar surface change, is therefore very important for those investigating natural climate variability and the possible impact of greenhouse gases."

Yet another article in *Science* was more assertive. In the May 12, 2006, issue, Richard Kerr wrote, "No doubt about it, the world is warming." He cited a report commissioned by the Bush administration whose chief editor concluded, "The evidence continues to support a substantial human impact on global temperature increase." The core of this evidence is a convergence in recordings from satellite measurements and ground stations, which show that both the lower atmosphere and the surface are warming in the way that the greenhouse climate models predicted. Given all the scientific data amassed, the mounds of scientific publications, the committee and conference reports, the media coverage, the political debate, and the popularity of Hollywood disaster scenarios like *The Day After Tomorrow*, not to mention the former vice president and presidential candidate Al Gore's documentary, *An Inconvenient Truth*, which in the winter of 2007 earned an Oscar, the conclusion of the Bush report seemed a bit anticlimactic. This is a very different tone from the one during that cruise to sunny Antarctica in 1990. Global warming is now a done deal. My Mongolian friends ascribe an especially hot week in the Gobi Desert to global warming; the phrase comes up in conversations with the superintendent of my building, with cabdrivers, and with many visitors to the museum.

Scientific consensus is now overwhelming. In February 2007, the Intergovernmental Panel on Climate Change (IPCC) issued the conclusion that global warming is "unequivocal" and humans are "very likely" (more than 90 percent likelihood) to be the cause. The last time the IPCC reported, in 2001, it assigned a conservative 60 percent likelihood to warming and stated that the link between human activity and climate change was only "likely." Even so, some scientists protested that the IPCC 2007 report was too conservative in its estimates of sea level rise because it discounts the recent disclosures on melting polar ice and sliding glaciers. In years past, many scientists regarded the IPCC results as overextended; now many are saying the opposite.

The message seems to have affected even some of the most hesitant politicians. On June 1, 2007, a week before the meeting of the Group of Eight industrialized nations, President George W. Bush announced his intention to set "a long-term global goal" for cutting greenhouse gas emissions. Specifics of the proposal were lacking, but Bush's statement at least represented a shift from resisting to accepting scientific consensus.

In 1992 the American Museum opened an exhibit on global warming.

It was well received and popular, but one might say it was almost too early. Some of our visitor surveys show that the most powerfully effective exhibits, like the one on Darwin in 2005, occur at exactly the time the public shows an overwhelming need to know, to ruminate, and to decide about that very subject. The issue of climate change did not yet have that urgency in the early 1990s. Now, obviously, it does. Many people have come to have a deep-seated conviction that the climate of the world is drastically changing. They, like Bob Dylan, are convinced that "You don't need a weatherman to know which way the wind blows." The twenty-first century may mark the first time in history that people have made the connection between a change in the weather and a transformation in the planet.

All to the good. It is important that people think about these problems. We need to do something. But what? And how fast? Moreover, what exactly *is* happening? What does an increase in temperature bode for microbes, plants, animals, us? Efforts to answer these questions are pinned on several lines of evidence: first, the fossil and geological record, which reveals the times when Earth's atmosphere heated up and cooled down and shows what happened to organisms and ecosystems when it did; second, fine-scaled prehistoric and historic records extending back thousands of years preserved in ice cores and other features, which give us a reading on climatic shifts; third, exhaustive measurements of greenhouse gases, air, water, and ground temperatures, precipitation, glacial retreat, and diverse other data, which give us diagnostics, day after day and year after year; fourth, mathematical climate models absorbing massive amounts of data and running on monster computers, which can make predictions about future climate change; and finally, numerous studies of the possible effects of recent or current climate changes on organisms, species, and ecosystems.

The grand sum of all this evidence indicates that Earth is heating up in an impressive way that is problematic for biological organisms, including us. Over the past several years many climate change issues have been clarified and even resolved, and even earlier skeptics have been satisfied that something big is indeed happening. Of course not all conclusions, models, or observations agree. It is still not clear how precisely the climate will change over how long a time. The likelihood of a major planetary trauma has given rise to a dynamic, vigorous scientific dialogue.

We need to distinguish what scientists generally recognize as climate change from climate variation. Climate change means a large-scale shift in regional, continental, or global climate extending over a short or long time. The change is directional; it does not automatically or easily reverse to initial conditions. Climate variation, by contrast, refers to rhythmic oscillations in conditions; the climate changes but then changes back to what it was— from day to day, season to season, or year to year. Thus we were warned not to ascribe too much importance to the banana belt weather of an Antarctic summer in 1990 since it was probably part of the variation normally expected from one austral summer to the next.

The climate variation can naturally oscillate over years, even millennia. The warm-water event known as the El Niño Southern Oscillation (ENSO) takes place about every three to seven years. El Niño has wide-ranging effects, including a rise in the surface temperature of much of the equatorial Pacific, a shift in winds and current patterns, and the suppression of upwelling of cold, nutrient rich water in the eastern Pacific off the coast of South America. El Niño has occurred regularly over the past several hundred years. Global warming may strengthen its effects, but it is not necessarily causing global warming.

Another climate variation is the North Atlantic Oscillation (NAO), which is driven by westerly winds over the Atlantic and incursions of cold Arctic air over Europe. There is also an Arctic Oscillation, which develops a ringlike swirl of westerly winds fanning out along the entire Northern Hemisphere. The Pacific–North American Oscillation (PNA) arises from a deeper than normal low-pressure area over the Aleutian Islands, a low-pressure area over the southeastern United States, and a higher-than-normal-pressure area over western North America. It is responsible for the rough and erratic weather in the North Pacific and other North American regions.

Finally, several large-scale oscillations in the oceans influence the climate. Perhaps the most notable of these is the Atlantic thermohaline circulation, the system that drives the famously warm current called the Gulf Stream in a northeasterly direction and gives northern Europe its unusually mild climate. A massive sinking, or downwelling, of seawater invigorates the warm currents. When sea ice forms in the frigid Arctic, it releases salt (a process known as brine rejection), which is then dissolved back into the cold water, making it denser and causing it to sink as it flows south. The sinking of this cold, deep water keeps it from mixing with the warm northward flow that is the Gulf Stream. Thus the Gulf Stream's warm water is not significantly diluted by cold water and its ameliorating effect on the European continent is essentially unimpeded. The northward-flowing currents begin to lose heat to the atmosphere only when they loop to the far North Atlantic, mixing there with cold surface water flowing from the Arctic. Similar systems of overturn are found in all oceans.

We know of other climate variations on even longer time scales. The amount of radiation reaching us from the sun is subject to variations in Earth's orbit known as Milankovitch cycles. The wobble of Earth about its rotational axis has a cycle of about twenty-two thousand years, the variation in its tilt occurs over about forty-one thousand years, and the eccentricity of its orbit (ranging from nearly circular to more elliptical) has a cycle of one to four hundred thousand years. These cycles have caused major excursions in temperature and climate, but again, they are not examples of long-term climate change.

So we need to muster evidence useful in distinguishing climate change from climate variation, even though the two phenomena are not completely separable or mutually exclusive. For example, one major trend might be for climate through time to become more variable—with, say, the summers becoming hotter and the winters cooler. This increased variation might occur even if the average annual temperature did not significantly change over the same time.

What does Earth's history tell us about climate change, and how does this bear on what is happening now? And what can we expect in the near future? Let us once again rewind time and then fast forward, but here do it in terms of changing climates. At the time of its birth 4.5 billion years ago Earth had no oxygen. Until about 3.9 billion years ago, meteorites and comets, one of which ripped off a chunk of Earth to form our orbiting moon, constantly bombarded its rocky surface. In its earliest phase Earth was bathed by only 70 percent of the current level of the sun's radiation. But knowing that fact does not allow us to reconstruct Earth's earliest climate confidently. There is much debate over whether conditions then were frigid and icy, or hot and steamy, or temperate, although the oldest rocks, 3.8 billion years old, in East Greenland, were clearly deposited in water. Doubtless, the frequent extraterrestrial impacts during this phase augmented volcanic eruptions and Earth's crust actually blew off steam; when the water vapor condensed, rained down, and filled crustal basins with oceans, the planet was ready to incubate life.

Recent studies suggest that the formation of both oceans and the continents surrounded by them may have occurred earlier than once thought, perhaps within 100 to 200 million years of the planet's formation. The primeval climate, as in the case of the primeval biota is, however, very poorly known. What little evidence we do have points to some surprising scenarios. Between about 750 and 600 million years, Earth may have experienced several global ice ages, with its crust buried under miles of ice from the equator to the poles. This early deep freeze, commonly referred to as the Snowball Earth, is suggested by ancient rocks of that age in both polar and equatorial regions that were scoured by the movements of massive glaciers. Snowball Earth may have actually been "son of Snowball Earth"; there is some evidence for such a global-scale frigid phase as long ago as 2.3 billion years.

The record of Earth's climate gets better at roughly the same time that our record of Earth's life improves. Since the beginning of the Phanerozoic in the Cambrian Period, about 500 million years ago, we have a decent picture of the major episodes of climate change, ranging from hothouse Earths to semisnowball Earths. When life moved onto land about 475 million years go, climates were probably quite mild, fostered by extensive warm tropical seas that flooded many lowland areas. There is evidence that within a few million years a south polar ice cap covered much of what is now Africa and South America, while the Northern Hemisphere probably experienced a warm, equable climate. During the Carboniferous, some 350 million years ago, when coal forests proliferated, the megacontinent of Pangea simmered in an expansive tropical climate. In time it grew cooler, a phase that culminated some 20 million years later in the great polar ice sheets of the Permian Period. That situation changed dramatically at 250 million years, the time of the great Permian/Triassic extinction event. The global mean temperature then may have been the highest it has ever been in the last 500 million years—of more than casual interest when we consider that such a hothouse climate coincided with a depletion of atmospheric oxygen and the most catastrophic mass extinction event in the history of life.

The dinosaurs, flowering plants, and pollinating insects of the succeeding Mesozoic Era thrived for the most part under warm conditions, although global temperature dipped sharply with accompanying glaciation about 130 million years ago, during the transition between the Jurassic and the Cretaceous. The Cenozoic Era that followed the great Cretaceous extinction event shows a marked climatic schizophrenia. During its first 62 million years global conditions were generally warm to temperate; the hottest conditions occurred during the Paleocene-Eocene Thermal Maximum (PETM), at 55 million years ago, when crocodiles, turtles, and broadleafed trees flourished even on landmasses at or within the polar regions. A gradual decline in global temperatures spawned the first Cenozoic ice in Antarctica, about 35 million years ago. By the beginning of the Quaternary, about 2.4 million years ago, the temperature again dropped, and glaciation in both the Northern and Southern Hemispheres increased.

We have already considered the alterations between frigid, icebound conditions during glacial periods and milder conditions during the interglacial periods. In our present postglacial age we can enjoy the warm soft breezes of a Mediterranean sea, swelter in the steamy heat of a tropical jungle, and hike in shorts and T-shirts as I did once, on a sunny day in June over Alaskan tundra not far south of the Arctic Circle. But in a broader temporal perspective, we are not—at least not yet—living in a very warm interval of Earth's history. The last 2.4 million years represent one of the planet's cooldown phases.

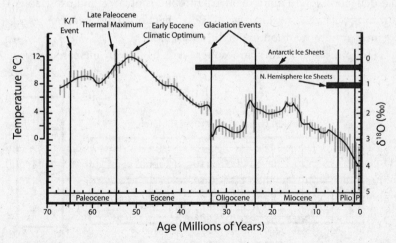

Global temperature change over the past 70 million years, with major events, including polar ice ages, indicated. Vertical bars are temperature ranges, the black line is average temperature.

Our picture of climate gets better as we approach the present. Once we reach the time of the most recent ice age, the Quaternary Period, climate shifts can be measured in thousands rather than millions of years. The last major advance of polar ice occurred during the Wisconsinian glacial interval, a pulse that ended around 12,000 years ago. There is evidence of a pronounced cold phase about 11,000 years ago known as the Younger Dryas, named after an Arctic alpine flower that was first described from sediments of this age in Denmark. The Younger Dryas persisted for a millennium, after which global average temperature increased, perhaps more than coincidentally, with the appearance of the first agricultural centers. In the broader sweep of things, we are now living in yet another interglacial stage.

Thus we might expect that the climate will eventually go through a cool intermediate stage and another fully glacial age. Extrapolating from the frequency of ice ages in the recent past, we would expect to become icebound again in about ten thousand years. Of course two exceptional circumstances could change this. An unexpected catastrophe—an asteroid impact, an orgy of volcanism, or the like—could wreak the same havoc that it did in the past and redirect the climate trend set in motion during the ice age. But it seems pointless to dwell on so unlikely and so unpredictable an event. (Believe it or not, major public money is being spent on strategies for asteroid deflection.) The more likely catastrophe is obvious and explicable: If anything delays the next ice age, it is the effect of all the greenhouse gases we are currently contributing to the atmosphere.

The physics of Earth's atmosphere tells us that an increase in green-

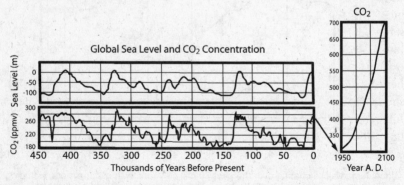

Global sea level temperatures and atmospheric CO_2 levels over the past 450,000 years, with a detail for rise in CO_2 levels since 1950 and a projection to 2100

house gases heats up the planet. The relative concentration of these gases in past times gives us key evidence for the pattern of climate change over 500 million years. We know from the geologic record, for example, that very warm periods in Earth's history are associated with higher concentrations of greenhouse gases. CO_2 is particularly sensitive and important in this regard. The correlation between the amount of CO_2 in the atmosphere and the atmospheric temperature is so tight that it gives us a handy index for making projections about the future. And even a cursory look into the past can be disarming: during the Late Paleocene–Early Eocene 55 to 60 million years ago, the time of a hothouse Earth and the highest global temperatures for at least the last 65 million years, CO_2 levels may have exceeded one thousand parts per million by volume (ppmv). As noted, this was also a time when sea levels were fifty meters (164 feet) higher than they are today.

CO_2 levels were later reduced through what one might call reverse greenhouse effects. Marine organisms fixed carbon during photosynthesis, died off in great numbers, and sank to the bottom of the ocean. Erosion and deposition of continental rocks into the shallow seas were very rapid, effectively burying carbonate minerals, the products of transformed atmospheric CO_2, under piles of sediment. The removal of CO_2 cooled things off considerably and initiated the most recent development of the Antarctic ice sheet, between 30 and 40 million years ago. CO_2 continued to decline; 3 or 4 million years ago it was at about 290 ppmv. During the subsequent ice ages CO_2 levels rose and fell, in lockstep with temperature, both being high when glaciers receded and sea levels were high, and low when ice sheets grew and sea levels were low. In the current postglacial phase the CO_2 level stayed essentially at 290 ppmv until the beginnings of the Industrial Revolution. Now it is at 380 ppmv. That means that no organism on Earth has experienced such a high level for the last 10 million years. During the earlier times the climate was significantly warmer, there were no major ice sheets on Greenland, and sea levels were several meters higher than now. Incursions of the sea covered many coastlines occupied today by some of the world's most concentrated human populations. Florida was merely a sliver of a peninsula. Even 130,000 years ago, when climates were warmer and sea levels higher than today, CO_2 levels in the atmosphere were lower. And of course 380 ppmv is not where CO_2 concentrations in the atmosphere will stop. At the current rate of human-generated CO_2 emissions, the level is likely to be double the preindustrial level of 290 ppmv, to nearly 600 ppmv by the middle of this century.

Then what next? Doubling the preindustrial CO_2 level in the atmosphere may warm the surface and the lower atmosphere (troposphere) by about 1.2°C. But other factors will affect temperature too. One estimate based on a doubling of CO_2 predicts an increase between 2.5° and 3°C. With still more uncertainties accounted for, the prediction becomes less precise, as we have seen, yielding an increase of 1.5° to 4.5°C for the mid-twenty-first century. Only 5°C is what makes the difference between the mild northern European climate of Berlin and that of Moscow (the coldest capital city in the world, with the exception of Ulan Bator, Mongolia).

My principal interest here is the effect that such warming on a global scale will have on biodiversity and ecosystems, but it will also dramatically shift many physical aspects of the planet's system. Increased heat induces more evaporation of surface moisture and also allows the atmosphere to hold more water, but the effects of this will vary. Dry areas like the Sahara will probably become even more intensively subject to drought, since the hot atmosphere will less readily give up its moisture. As heat waves become hotter, vegetation will either wither and die or be consumed by frequent wildfires. On the other hand, precipitation will occur elsewhere to balance the increased evaporation. Increased precipitation in subtropical and temperate regions could generate more storms, heavier rainfall, and flooding. Annual precipitation has increased with the warming trend over the past hundred years from 5 to 10 percent in northern Europe and other high-latitude regions. All the events that are products of the current climate regime—heat waves, cold snaps, droughts, and monsoonal rains—could become more extreme.

Or could they? One would expect that tropical cyclones and mid-latitude storms, indicators of extreme weather, would become more frequent with significant climate warming. Despite the common impression that the southeastern United States is being battered by more hurricanes with each passing year, the cyclic pattern in the last several decades suggests otherwise. An intense period of hurricane activity also occurred in the 1950s. On the other hand, there is now some debated evidence that the first true hurricane of the southern subtropical Atlantic occurred in 2004, a possible indication that the overall warming of the Atlantic is generating powerful tempests in both hemispheres. But the evidence does not yet clearly point to a significant trend, and the historical record with reliable data is too short to be a basis for predictions.

At higher latitudes the pattern is ambiguous. Storms have become more frequent in the Northern Hemisphere but less so in the Southern Hemisphere. On another scale, there is the measurement of annual extremes in temperature and precipitation. The number of very cold winter days and very hot summer days has increased in many regions. Many summer days with both high temperatures and high humidity have plagued much of the United States recently. Yet there is no apparent increase of other phenomena associated with a major climate shift: tornadoes, thunder days, and hailstorms.

Perhaps the clues to this paradox lie in the synergy between major patterns of climate variation and climate change. Data show that well-studied El Niño events have become more frequent and intense since the 1970s, further promoting drought conditions in southern Africa, Indonesia, and eastern Australia, as well as flooding in northwestern South America, East Africa, and California. The extensive spread of very warm surface temperatures associated with El Niño in its newly powerful mode promoted the bleaching and destruction of many tropical coral reefs in the 1990s.

The North Atlantic Oscillation is associated with windy, wet, and mild winters alternating with periods of very cold and dry winters in northwestern Europe. Data show that since the early 1970s the intermittent periods of cold, dry winters have diminished; the balance has tipped toward prolonged, marked periods of winter warmth and humidity. The North Atlantic Thermohaline circulation responsible for the warm Gulf Stream has varied over decades, centuries, and millennia, and we lack a clear understanding of its patterns. Nevertheless, there are indications that the returning flow of subsurface waters in the northern loop between Greenland and Scotland has slowed by about 20 percent in the past fifty years. This "stallout" in the circulation pattern could destabilize northwestern Europe's mild climate.

Though not all the repercussions of the recent, continuing rise in atmospheric and surface temperatures are clearly understood, it is important to stick with the big picture, which we *do* understand. Accumulated climate data indicate a significant climate warming on a massive scale over the past century, warming that has become even more pronounced in the past thirty years. The predicted effects range from highly compelling to ambiguous and poorly substantiated. But one thing is certain. It makes no sense to develop strategies for preserving biodiversity and securing ecosystems, as we once did, on the assumption that the climate is stable. What we know of the past tells us that climates can shift quickly and affect the entire planet in the process.

Many effects of climate transformations are irreversible, and the extinction of species ill adapted for change is one of them. It might take tens of million of years, not decades or centuries, for climate trends to reverse. We know enough to say that what is going on right now is in all likelihood a change in climate with huge implications for the ecological future of the world. And the breathlessly short timescale for this change may be unprecedented.

Paleontologists have come to recognize five major mass extinction events occurring over the last 500 million years, but these are not the only times when extinctions occurred. There are many instances in which a critical number of species went extinct simultaneously. If this had not occurred, we could not conveniently slice up eons into eras, eras into periods, periods into epochs, epochs into ages, and ages into still-smaller intervals.

What lies behind this handy calibration? At least in the case of the K/T event we have seen the evidence for the asteroid impact and the possible contributions of volcanism and climate change. The causes of other past extinction events are enigmatic, but climate change is invariably proposed as one cause, the default explanation of extinction events, in the absence of a specific, directly recorded cause like an asteroid or a series of massive volcanic eruptions. Even in those cases, one cannot rule out climate. During the Late Ordovician, 475 million years ago, while some organisms were making a go of it for the first time on land, a mass extinction event occurred in the oceans. This, like many other extinction events, has been attributed to climate change—more often with cooling than warming—but very little is known about the actual mechanisms involved. The effects of temperature change alone are not readily evident, and disruptions may have been a function of the interplay of seasonality, dissolved oxygen, energy flow, currents, and other oceanographic conditions.

In the Tertiary Period, after the K/T apocalypse, the climates grew hotter, then cooler and then frigid, at least near the opposite northern and southern edges of Earth. Climates were also affected by major tectonic

events, including the rise of the Andes and the Tibetan plateau and the closure of the Isthmus of Panama. As we have seen, the Panamanian bridge
allowed reciprocal invasions that, on balance, left the resident South American mammal fauna as the loser. In Australia changing climates in the Late
Tertiary reduced forests and induced mammal extinction. The Australian
fauna was thus set up for the next onslaught, the massive destruction of
large animals coincident with the arrival of humans some fifty thousand
years ago. New Zealand had vertebrate extinctions following the Oligocene
inundation of land areas between thirty-seven and twenty-four million years
ago. These preceded other extinction events caused by human migration
from southeastern Asia and New Guinea in the Late Holocene and, later,
by the invasion of European settlers and their domestic animals.

Scientists who try to predict the outcomes of the current climate change
search the more recent past for clues. A critical phase is the important transition from the Tertiary to the Quaternary Period, between 2.5 and 1.8 million years ago. We have already mentioned the drastic changes brought
about by the closure of the Isthmus of Panama. Such profound resculpturing of the land in combination with global cooling, reorientation of ocean
currents, and expansion of ice sheets pushed back on Earth's biota. Organisms, especially those at high latitudes or in vulnerable sea areas, had to
cope with wildly shifting environments, and they did so by expanding and
contracting their ranges. In a few cases they ultimately followed the road to
near or total extermination. Polar forests were replaced by tundra. Some
trees, such as the tall, elegant coastal redwood (*Metasequoia*), became fugi-
tives, sequestered in small corners of their former ranges, while others
(*Glyptostrobus*) went globally extinct. Marine habitats were, if anything,
even harder hit; one-fifth of all mollusk species and two-thirds of all coral
species of the Caribbean went extinct as the ice ages began. This is thought
to have been prompted by marked cooling of the sea surface by about 5° to
6°C; the resultant catastrophe somehow spared such organisms in other regions, such as the northern Pacific.

This shakeup in climate and biota was followed by the erratic climate
period known as the Pleistocene. Between 1.8 million and 12,000 years ago
the twenty Pleistocene glacial intrusions were interspersed with warmer interglacial phases. The effect of this climate schizophrenia on species extinction is not clear: many mammal extinctions of the Late Pleistocene have
been associated with it but more likely were caused by human overhunting.

Some extinction during this interval, however, cannot be blamed on humans. In southern North America, the glacial stage woodlands were populated by the spruce *Picea critchfieldi*, a tree that had disappeared by the beginning of the Holocene 12,000 years ago. Some fossil seeds in sediments in Scotland representing the last 1,000 years of the glacial stage belong to a saxifrage that is no longer with us.

Extinction events connected with the ice ages were few. By contrast, their effects on migration and range changes were dramatic and virtually ubiquitous. The flip-flopping conditions forced organisms into spasms of migration to lower latitudes or elevations in front of the ice sheets that were spreading from the poles and in the opposite direction when the glaciers receded. Expanding ice sheets killed plants at the northern edge of their ranges. The escaping plant populations moved south and reassembled into new combinations of species. We know this because records are good for large trees, and the movements they chart are significant. After the last glacial pulse and beginning with the warming trend between ten thousand and two thousand years ago in Europe, spruce and beech shifted their range boundaries northward by one to two thousand kilometers, at an impressive maximal rate of one to two kilometers a year. In these range shifts, trees and other plants had company: the long-distance travelers included snails and beetles.

These range changes are naturally most evident in areas near the glaciers. At low latitudes and in the tropics the effects are subtler. Fossil pollen shows that tropical biodiversity was not notably reduced during the Pleistocene yet also shows that tropical climates were far from stable, since they mirrored, albeit less emphatically, the great oscillations to the north and south. Just the same, the overall evenness of tropical climate is a great moderator. Studies have shown that an animal in the subtropical Andes that was forced to descend two thousand meters (6,600 feet) in elevation because of a 5°C decrease in temperature may have needed to migrate only a distance of forty to fifty kilometers.

However, one cascade effect from climate change in the tropics is particularly profound. The warming trend at the end of the last glaciation and the beginning of the Holocene caused sea levels to rise 125 meters (413 feet). Coastlines and lowland areas all over the world were affected, with flooding of the land and disruption of the biota, most severely where lowland areas were extensive, of course. The lowland tropics, notably those of southeastern Asia, experienced land loss, as we've seen, that may have caused the extinction of at least 10 percent of all species. Tropical shifts in the land biota during the Late Pleistocene and Early Holocene coincided with the disappearance of large mammals as the result of climate, human hunting, and other factors.

One would expect that the Southern Hemisphere at high latitudes, unlike the tropics, would mimic the biotic traumas of the Northern Hemisphere, but this is not the case. Even a brief glance at a globe shows us that there is more land on the top half of Earth's surface than on the bottom half. Africa and South America are wide in girth in their tropical regions but taper toward the south, while expansive Asia lies almost entirely above the equator. Australia and Antarctica, the remaining large masses of the Southern Hemisphere, are entirely encircled by oceans, and the latter is scarcely inhabited except at its edges. Because habitable land areas in southern temperate and polar regions were smaller, the moderating influences of the surrounding oceans dampened the effects of the ice ages. Here the pattern of changes in populations alternates fragmentation with expansion, rather than showing large range shifts from low to high latitudes and back again.

Despite this antipodal difference, it is virtually impossible to eliminate the connection between climate and the various range changes of organisms during the Pleistocene and the Early Holocene. In addition, the fossil record demonstrates an important mark in all the range shifts: species reacted to climate changes individualistically rather than in the aggregate. Communities of animals and plants rarely expanded or contracted their ranges in a coordinated network. Communities that were reestablished after major migration events, for example, rarely duplicated earlier ones but were instead composites assembled from the independent movement of species responding more or less rapidly to a particular climate shift. This individualistic, complex, and often unpredictable movement has much bearing on our consideration of the present and the future.

England has a traditional preoccupation with nature. Bird-watching, butterfly catching, and horticulture are thoroughly ingrained in English culture, and the English countryside is one of the most thoroughly mapped to show the distribution of creatures large and small. Much of this expresses the efforts over centuries of amateur country and town folk with an eye and a love for the great outdoors. Charles Darwin grew up in this tradition, and he demonstrated how far a fondness for beetle collecting could get you in science. Other pursuits by Darwin's nature-loving countrymen were perhaps not so monumental, but they proved to have unanticipated importance for the great environmental questions of our day.

Many of these historical observations concern what is called phenology, the study of the outward appearances, or *phena*, of organisms—their activity cycles, breeding behavior, growth, movement, and so on. In 1736, Robert Marsham, a naturalist and a fellow of the Royal Society, began making notes in his ledger on the timing of the first spring frog croaks on his country estate just north of Norwich, in Norfolk. Keeping records for this frog-croaking almanac was a consistent passion of the Marsham family over many generations and two centuries. The data collecting suddenly stopped in 1947 because the phenological committee of the Royal Society, experiencing financial exigencies and shifting priorities, suspended publication of such records. A rediscovery of the remarkable Marsham document came in 1995, when the ecologists T. H. Sparks and P. D. Carey revealed that it chronicled a very clear relationship: the first date of frog calling north of Norwich was tightly correlated with temperature. Further, more refined studies reaffirmed that frogs are cued to their croaks by seasonal shifts in temperature. The time of arrival of sexually mature common toads (*Bufo bufo*) at breeding ponds in the United Kingdom between 1980 and 1998 showed a clear correlation with mean temperatures over the forty days preceding their appearance. The rhythm of these life cycles changed as over time England grew warmer. Two frog species in their northern range in Great Britain spawned two to three weeks earlier in 1994 than in 1978.

Frogs are not peculiar in their sensitivity to weather. The current warming trend influences myriad species. It has shifted the timing of zooplankton blooms, peak insect abundance, the arrival and departure of breeding birds, and the blooming of trees. Changes are more marked at higher latitudes and altitudes. In an important review published in 2003, T. L. Root and several colleagues studied life history data in 694 species recorded between 1951 and 2001. They found that on average these species were either breeding, blooming, or doing other seasonally related activities 5.3 days earlier each decade.

It is not only the seasonal breeding behavior of organisms that climate affects. The current warming trend has set species in motion. Two-thirds of fifty-three species of butterflies studied in both North America and Europe have shifted their ranges. Fifty-nine breeding birds have taken a mean northward shift of 18.9 kilometers (11.7 miles) over twenty years. The red fox of northern Canada has expanded northward over the past thirty years while the arctic fox during the same decades has contracted its range, retreating toward the Arctic Ocean. Prior to the current warming phase, accidental transport failed to establish red foxes in the more northerly regions they now occupy. Many other species have moved upslope. Forest lowland birds have ascended Costa Rica's mountains, and some populations of Edith's checkerspot butterfly have shifted up 105 meters (347 feet) in the Sierra Nevada of California. Finally, there is evidence, some of it currently debated, that malaria-carrying mosquitoes are migrating from the tropics and subtropics to higher latitudes.

On the other hand, some species have drastically contracted their ranges, and their surviving, marginalized populations have been reduced to precariously low levels. Amphibians in the high cloud forests of montane Costa Rica have been pushed upward by warmer temperatures to less spacious, isolated elevations, a pattern also observed in chameleons on the island of Madagascar. Many stands of low-elevation pines in Florida now crown scattered hilltops. Ecological field experiments have demonstrated that shrubs have invaded habitats formerly occupied by tundra plants. Off the California coast some cold-water zooplankton and intertidal fish and invertebrates have been replaced by more southerly species. Precipitation changes have caused major shifts in the Sonoran desert flora. The extended drought of 1940 to 1970 affected the creosote bush (*Larrea*) and also shifted the boundary between low-elevation juniper and high-elevation pine by more than two kilometers (1.24 miles). Increased precipitation in recent years is shaping a new profile for the vegetation and the kinds of ants, reptiles, and rodents that make up the community.

It is clear, then, that warming climates have made many species trade places. Shifts of organisms from warm, low latitudes toward the poles are prevalent. They may not seem as significant in the long run or as traumatic as the expansion and collapse of entire populations, or species, because of habitat loss or the conversion of land for agriculture or industry, yet these weak but persistent forces may ultimately eclipse the others, since they will denude populations, destabilize communities, and drastically reorganize ecosystems.

The potential of climate change as a destabilizing force depends on the evolutionary processes at work in the organisms affected. Rapid alteration of a species's range can distend its populations and transform the nature of its genetic enrichment. A small population of hardy colonizers often establishes the leading edge of a range shift. But these colonizers may have a very low level of gene diversity, leaving them susceptible to rapid environmental changes. Meanwhile, genetic diversity may persist at the trailing edge of the migration, but these populations will start to fragment when unfavorable environmental conditions break up their preferred habitats. The populations "left behind" could become smaller and eventually even disappear.

Just how a population's genetic makeup and evolution are transformed depends on the rate at which environmental forces challenge it. If the pace of environmental change is slow enough, the species may, with a combination of movement and adaptation, maintain genetic diversity at levels that allow it to persist. If changes are drastic and rapid, the species will be thinned out at the trailing edge and highly vulnerable to habitat perturbations at the leading edge of its range. Picture a sand dune that drifts with the wind. It will normally lose sand on its windward side and pick it up on its lee side, maintaining its mass overall as it drifts along. But if there is no sand to pick up, the dune will erode on both sides, eventually fading to a lump of tiny quartz grains on the desert pediment. Some species are like drifting dunes without enough sand—namely, genetic variation—to maintain their genetic robustness in the face of drastic climate change.

After all, we must remember here that any migration is itself an adaptive response, an attempt to go where the favorable habitat is going. It is easy to recognize that an adjustment may be very challenging. Most high-latitude species now exposed to warming trends are ancient enough to have evolved under the much cooler conditions that have persisted for the last million years. The current, abrupt change in climate is a shock to many of them, and simply moving to a cooler habitat does not guarantee that the genetic composition of the migrating populations will be robust enough to sustain them.

Thus climate change today, as in the past, is likely to influence evolutionary processes in many species. As species move in response to drifting climatic conditions, they become subject to new selective pressures. How strong are these pressures? Answering that question in the real world is not easy. Models of populations, genes, and migration fail to explain a bewildering diversity of examples in nature. And observations of different species can prompt conflicting conclusions. Some genetic studies suggest that climate change has easily outrun the rate at which a given population can adjust to it; others indicate that local gene pools are preserved through numerous past climate fluctuations. There is evidence, for example, that some species, especially in mountainous and tropical regions, are buffered by the amount of genetic variation already resident in their populations. In such cases, evolution in the species as a whole will not drastically change with climate, although there may be strong effects on local populations. In extreme cases, climate change may promote intense, directional selection, leading to the collapse of populations and eventually the extinction of the species.

But can we demonstrate that species extinction resulting from climate change is occurring right now? We know that countless extinction events have occurred in the past, and a great percentage of them were likely driven by climate change, but we have the advantage of looking back over a record measured in millions of years rather than during the short lifetime of an investigator. Asteroid bashings and other such calamities aside, extinction, like evolution itself, is not generally an overnight phenomenon.

Nonetheless, we are aware of striking contemporary evidence that climate change in combination with other factors is killing off certain species. Although there is some debate over the matter, it seems that the drastic decline of coral reefs in many regions of the world resulted from a clear sequence of events. The first blow came with overfishing, beginning in the eighteenth century, on reefs in the Caribbean, the Great Barrier Reef, the Red Sea, and elsewhere, which culled predators that had kept destructive species from overrunning the coral. This was followed in the nineteenth and twentieth centuries by human activity causing pollution and the spread of disease. Now climate change making for warmer water temperatures may be delivering the coup de grace, driving some coral species toward extinction. Although the rate at which ocean surface temperatures will warm is somewhat uncertain, we can be assured that the projected increases in CO_2 and temperature over the next fifty years will result in environmental stresses not seen during the last five hundred thousand years of coral reef evolution. Even if coral reefs today were in pristine condition—and they are not—one would expect major shifts in the dominance patterns.

Such changes are not good for life in the sea in general. The devastation of many of the world's coral reefs has had and will have huge impacts on marine biodiversity. The scaffolding of the sea's richest ecosystem, like the canopy of a rainforest, harbors greater biodiversity by an order of magnitude than all other coastal systems. More than a million named species, as well as myriad species yet to be described, may inhabit about four hundred thousand square kilometers of coral reef. Darwin, in his consistent prescience, noted that coral reefs thrived under seemingly less than optimal conditions—namely, nutrient-poor waters around the equator—yet appeared to support an extremely complex food web with many positive feedbacks that recycled limited nutrients and kept the system churning along.

The concentration of today's coral reefs around the equator turns out to be a feature that may spell their doom, at least in these regions. Tropical oceans are 0.5° to 1.0°C warmer than they were one hundred years ago, and this phenomenon has led to what is commonly called coral bleaching. Corals depend on the mutualistic companionship of tiny organisms known as dinoflagellates (zooxanthellae) that live in their tissues by the millions. Dinoflagellates also carry pigments that lend a spectacular array of colors to coral. Tropical waters that heat up to a critical temperature kill dinoflagellates, removing the source of pigments and bleaching the color out of these colonies. When coral reefs turn white, they have lost their vital symbiotic dinoflagellates, and they too are destroyed. Not all species of coral are susceptible to bleaching; indeed, many species show a high tolerance to climate change. Nonetheless, warmer climates may have been the cause of mortality rates of up to 90 percent for some species of coral. Massive bleaching of tropical reefs has been recorded for the years 1979 to 2002. The bleaching and devastation expectedly intensified with the El Niño phases between 1997 and 1998, the most extreme El Niño conditions for the twentieth century. During El Niño the ocean current patterns change and the waters around the equator across the Pacific become unusually warm. Of the coral reefs throughout the world's oceans, 10 to 16 percent died off between 1997 and 1998. In the western Indian Ocean, mortality was as much as 46 percent. All it may take to accomplish this damage is a shift in water surface temperature of only 1° to 2°C; the effect varies geographically. Corals near the equator are shown to have survived water temperatures as high as 31°C, while higher-latitude corals will bleach at 26°C. In the 1980s in lagoons off Belize the coral species *Acropora cervicornis* was completely killed off by disease. It was replaced in the 1990s by another species of *Acropora*, which in turn succumbed in 1998 to higher water temperatures. Sediments in the Caribbean region record the history of local coral reefs through three thousand years but lack any sign of such massive destruction at earlier intervals.

The destruction of reefs worldwide can be plotted on a blacklist according to assessed reduction of species richness and the overall health of the reef. Heavily degraded reefs lack large animals like turtles, sharks, and groupers; the water quality is poor; and weedy soft corals and seaweed replace the unhealthy, dying, or dead large corals. What was once a rainbow of coral color becomes dark green, brown, and bleached white. When these criteria for decline are used, it is clear that major reef systems of the ironically named Virgin Islands, Moreton Bay in Australia, the Bahamas, eastern Panama, the main Hawaiian islands, the Florida Keys, the Cayman Islands, Bermuda, Belize, and the northern Red Sea are more than 50 percent degraded. Reefs off Jamaica and western Panama are worse off still, being roughly 80 percent degraded. Even the Great Barrier Reef (GBR) off Australia, the world's largest and best-managed coral reef, shows between 20 and 30 percent degradation. Australia's stiff regulations and management strategies—more than a third of the GBR has been designated a no fishing, "no take," zone—will test our ability to save coral. Another project, conducted by scientists at the American Museum of Natural History and several collaborating universities and marine labs, connects scattered reefs in the Bahamas by oceanic corridors in which fishing and human invasion are restricted, allowing the reefs to recruit organisms from the whole aquatic area rather than from local, limited sites. Unfortunately, most reefs in the world have not been or cannot be coddled in this manner. J. M. Pandolfi and many coauthors noted in a 2005 policy forum paper that U.S. coral reefs are "on the slippery slope to slime." The reefs of the Florida Keys are in very bad shape, yet despite new restrictions, only 6 percent of the Florida Keys National Sanctuary is zoned as "no take," and some experts think the zones were poorly chosen.

Coral reefs may be the most visible marine casualties of deleterious changes in the global system, but other marine creatures are massively affected as well, not directly from climate change but from the increase in atmospheric carbon dioxide. The ocean absorbs much of the carbon dioxide released from the burning of fossil fuels and the CO_2 combines with water to form carbonic acid (the weak acid found in carbonated beverages). In the process, hydrogen, carbonate, and bicarbonate ions are released into solution. The concentration of free hydrogen ions is indicated by the pH scale, with the former relating to the latter by a factor of ten. A fluid with a pH of

seven has ten times the concentration of free hydrogen ions in pristine seawater, with a pH of eight. (Fluids like vinegar, with low pH, are said to be
acidic, while fluids like seawater, with high pH, are said to be alkaline.) Ab-
sorption of CO_2 in the sea is necessary because many marine organisms depend on carbonate ions to build their shells and other hard parts out of
calcium carbonate. But when CO_2 levels in the atmosphere are high and
greater concentrations of the gas are absorbed by the oceans, the overabun-
dance of carbonic acid leads to an excess of hydrogen ions, which readily
lower the pH of seawater, a process known as acidification. Since free hydrogen ions combine with carbonate ions to form bicarbonate ions, the
overall concentration of carbonate ions is reduced. Studies now project that
the concentration of carbonate ions in the oceans could drop by a half over
this century.

This is truly bad news for the organisms that need calcium carbonate,
which will have great difficulty growing. These include the most abundant
and ecologically important organisms in the oceans: photosynthesizing
phytoplankton known as coccolithophorids, foraminifera, and small marine
snails called pteropods, all of which occur in huge amounts and are a vitally
important food for many fish and some whales. Coral reefs also have constituents that build their skeletons out of calcium carbonate, and some scientists believe that coral bleaching may result from both extreme warmth
and ocean acidification. This whole process is now being carefully scrutinized, because the projected acidification of the seas could rampantly disrupt the marine ecosystem, removing food for a great number and variety of
fish. The loss of resource for hundreds of species, including our own, that
feed on fish would be enormous.

On land, numerous important biological communities also show clearly
worrisome vulnerabilities to climate change. Organisms that live in lakes,
streams, rivers, and other bodies of freshwater are among the most endangered on Earth. Mounting evidence shows that they are especially susceptible to climate change because they cannot escape its effects, being captive
in their habitat. They are also victimized when rising waters allow the in-
flow of alien and destructive species.

This extreme vulnerability also serves an illustrative purpose. With their
limited powers of movement and escape, freshwater organisms show us
what can happen to once mobile and adaptive organisms, on land and in
the sea, as we slice up their environments and destroy their habitats with

factories, roads, cities, suburbs, and farms. We are creating new barriers for species and depriving them of the option to escape and to survive.

What kind of future does all this suggest for life in a warm climate? Perhaps, as noted, the most important insight offered by the study of the fossil record is that species respond independently to climate change, but that habitat loss greatly constrains migrations that are one important response to such change. With these two basic realities in mind we can make a few predictions.

First, substantial global warming threatens most obviously the habitats and species at high latitudes, the northern tundra and polar deserts such as those on the Arctic islands and Antarctica as well as species inhabiting high Alpine or montane habitats at middle to low latitudes. Susceptible species include some familiar ones, musk oxen, polar bears, and seals that breed on Arctic Sea ice, as well as a great diversity of small flowering plants confined to the fleeting habitats on the threatened list.

But we should remember that species reactions to climate change are idiosyncratic. Although mountaintop species are at high risk if they have a low tolerance to heat, mountainous terrain offers complex habitats with their own microclimates, which tend to protect more effectively from wholesale climate changes than flatlands do. And though global warming will be more marked at high latitudes, biodiversity might be most affected in the tropics, where species have very narrow niches and a low tolerance for environmental changes, including daily and seasonal temperatures shifts. Unfortunately, we have very few studies of the effects of climate change in tropical hotspots, where the most biodiversity is at risk.

Much of the anticipated disruption comes in the form of range shifts and not necessarily extinctions. But the human destruction of natural habitats, and the consequent lack of places in which displaced species can find refuge, certainly help forge a connection between the two. Computer models that simulate the effects of climate change have been designed to take into account many factors, including the reproductive capacity of a species, its capabilities in dispersal, and its climate tolerances, in combination with projected estimates for temperature, precipitation, and other climate changes. They allow us to predict both the range shifts and the ultimate viability of the species under scrutiny. These models have been used in case studies of plants in a region very likely to be affected, the far southern edge of the African continent. As noted before, the results predict that 25 percent of the *Protea* flower species in the Fynbos hotspot of southern

Africa will be extinct by mid-century. Projections like these, vital to an understanding of what is in store for ecosystems worldwide, are unfortunately all too few. We need more data in order to identify the likely victims as well as survivors of the current environmental changes.

One candidate for survival is of course *Homo sapiens*, a species that is able to respond to environmental disruption. Like certain other species, humans can move to new habitable areas when their original habitats are destroyed. Unlike other species, humans have been able to convert unsuitable habitats to livable ones with their heaters, air conditioners, insulated homes, and shipped-in water and food. But humans, like other species, have limits. Native Alaskans and Aleuts living on islands in the far north Arctic are worried about the rising sea level and the denudation of the shore that has removed the land around their homes and disrupted their coastal food sources. Inhabitants of many small low-lying Pacific islands have been similarly threatened. An entire village on the island of Tegua, in Vanuatu, was forced to move to higher ground. Kiribati, an archipelago nation of thirty-three coral atolls barely six feet above sea-level, is starting to drown. The president of the country warned that an anticipated sea level rise of only nineteen inches could force a mass exodus of many of Kiribati's 92,500 people to New Zealand and Australia.

In the developed world, Venice, New Orleans, and many other coastal communities are living on borrowed time, as global temperatures and sea levels rise. Venice is sinking through a combination of movements in the underlying continental crust as well as a rise in sea level. Because of high tides and high prices, the resident population has decreased from 150,000 in the 1950s to 60,000. Within decades, habitation could be hugely impractical, if not impossible. The roar of Hurricane Katrina showed the vulnerability of coastal lowlands around New Orleans, and what if, as some suggest, hurricanes proliferate in the future? As a recent government report stated, such events coupled with sea-level rise "portend serious losses of life and property." So too goes Florida. In the southeastern United States alone rising waters might displace millions of people. We, like the arctic fox or the Fynbos flora, would be a species on the move.

It is difficult to think of a topic more relevant to the connection of the past, present, and future than the effect of climate change on the success or failure of species. But not all the lessons the past teaches are straightforward. As we have seen, Pleistocene climates did not cause widespread extinction, nor did the warming of the Early Holocene. And proponents of the theory of human hunting as the primary cause now effectively dispute the once popular idea that large mammals were brought down by climate change during the ice ages. Still, species extinction in the future might be different. Holocene warming did not raise the temperature range established over the previous 100,000 years, whereas now the predicted near-term warming will far exceed these limits, especially for high latitudes. And if species are lost, recovery of our modern ecosystems in any form familiar to us will hardly be imminent. The restoration of plants, marine creatures, and large mammals after a mass extinction like the K/T event takes from 3 to 15 million years. This is a span of time that exceeds by nearly a hundredfold the time of our own species's existence on Earth.

As I've said, everybody now talks about the weather. Is there something we can do about it? The answer is yes, we can, because we have overwhelming evidence for the connection between our own fouling of the air and the rise in global temperature now threatening life on Earth and the livability of the human habitat. Despite this, many people, including those of power and influence, continue to resist the notion of reducing our use of fossil fuels and moving to renewable, less problematic sources of energy. The United States, Russia, and Australia rejected the Kyoto Protocol largely by taking the position that there was great scientific uncertainty over this issue. (That the protocol itself needed further modification should have encouraged rather than dissuaded the United States and other industrialized nations from participating in the dialogue and reaching a resolution.) The recent shift to acceptance and action by some formerly reluctant political and industrial leaders is encouraging. Still, I often encounter individuals otherwise very open to the discoveries of science who deny that humans could disrupt the balance of a planet, the very atmosphere and the climate it controls, in such an enormous way. They say such an idea is astounding, beyond belief. But science has eventually convinced us before of the unbelievable.

FUTURE WORLD

When I first went to China in 1990, the country had already been open to the West for more than a decade. The capital city of Beijing was bristling with cranes and the early skeletal frameworks of skyscrapers. Yet there was still evidence of tradition. In the sultry evening, bicyclists in thin white cotton shirts and dresses floated through the streets like fireflies. In the early morning, crowds of people of all ages gathered in small parks and pathways for their tai chi. The city was becoming explosively modern, but there was still the low hum of serenity.

Now Beijing has changed the way my childhood hometown of Los Angeles did years before I was born. I have been in China at least twenty times since 1990, and each visit is another epiphany, another eye-opening demonstration of the awesome power of high-rise, high-tech humanity. The new buildings of glass and polished metal are impressive, though often architecturally repugnant, and the new flying ramps and roadways are wonders of modern construction. But polluted haze enshrouds the buildings and is slowly suffocating the people below. There are far fewer bicyclists; indeed, this mode of transportation has become a dangerous anachronism in the heavy traffic. People still gather for tai chi as the roar of the traffic rises with the sun, but they are mainly elderly. The younger generations need that extra time for the morning commute. Many people in the crowded streets still don traditional white cotton masks, and one can't be sure whether this is meant to filter rather ineffectively the polluted air or to keep strangers from catching the respiratory illnesses that sweep through Beijing. Doubtless the two are connected. Recent estimates show that as many as 300,000 people a year in China die prematurely from respiratory diseases.

Spurred on by a need to improve air quality before the 2008 Olympic Games, Beijing has implemented programs to replace coal for cooking and heating fuel with natural gas and liquid petroleum gas and to increase the use of electric public vehicles, but it is still one of the world's worst cities for air pollution. (According to a June 2006 report by the World Bank, sixteen of the world's twenty most polluted cities are in China. A 2007 version of this report removed some "socially sensitive" passages on health hazards, allegedy under pressure from the Chinese government.) And just south of Beijing is the catastrophe of Baiyangdian Lake, as we have seen. In ten years China has lurched from the past to a dangerous present. But the sheer momentum of its metamorphosis involving 1.3 billion people, points to a future I'm not sure I want to be part of.

What is that future? Surely it is not summed up in the smog of Beijing or the world's other exploding cities. Forests are disappearing, but after all, they still occupy 30 percent of the land surface. Some coral reefs, such as those in northwestern Hawaii, are almost untouched, and others might be rejuvenated. Clean water still runs in some mountain streams with their unique assemblages of tiny fish and insect larvae. Will not the future of the planet be more like a mosaic? Won't it strike a balance between conversion to total human domination in some places and other areas left inviolable, free zones where the life wrought over eons still breathes unadulterated? Or will nature be reduced to tiny sanctuaries isolated in the midst of vast acreage controlled by humanity? Unfortunately, much of the world already looks like that. What of the rest? What is likely to happen, and what does it bode for the planet and for us?

Prognostication is a dangerous business in science. Yet science is by definition a predictive enterprise, so why not foretell the future? After all, what good is all this documentation of extinction in the topsy-turvy world of the past unless it gives us some useful indication of what's in store for us? The easiest prediction to make is that the natural world will continue to change dramatically. Three premises underscore our certainty about this. The first is that the changes we are witnessing today are in many ways unparalleled not only in the entire length of human history but in the many millions of years preceding that history. The second is that facets of the current events were nonetheless foreshadowed by certain events in the past, from which we can learn something. The third premise is that our environmental and evolutionary future will be altered and redefined by any major change occurring in the vital services already provided—nutrient recycling, productivity, CO_2 sequestering—by the present ecosystems and their diverse species. What has occurred, is occurring, and will occur over the next few decades has the power to transform a 100-million-year old ecosystem.

This transformation has two key ingredients. Take 6 billion human consumers, predicted to increase to 8.9 billion by 2050, and blend them with what remains of Earth's natural habitats, and you get a rather volatile mix made all the more unstable because humanity has not made a commitment to designing a different, sustainable way of using natural resources. Serving this exploding population, and making only slow and erratic progress in converting the performance of these services into more efficient and environmentally less damaging behavior, has put us on this collision course with nature.

Scientists call the warming years after the last ice age some 12,000 years ago, which saw the beginnings of agriculture and the emergence of civilization, the Holocene, or "wholly new," epoch. The Nobel laureate Paul J. Crutzen, developing a thought first offered by scientists like Antonio Stoppani, V. I. Vernadsky, and Pierre Teilhard de Chardin, has coined a new term, *Anthropocene*, for the epoch that began in the late eighteenth century, the signal event being James Watt's invention of the steam engine in 1784, which marked a change in the planet as profound as that caused by the great paroxysm at the end of the Cretaceous and the beginning of the Tertiary 65 million years ago.

The Anthropocene is easy to characterize, as we have seen, by human activity that has nearly halved Earth's forestland, consumed its limited freshwater reserves at a rate that outstrips human population growth, wiped out many of its terrestrial and marine organisms, and infused its atmosphere with enough pollutants to change its climate. Every year thousands, perhaps tens of thousands of species are going extinct. Perhaps a third or a half will be extinct by the mid-twenty-first century. We are at the beginning of the sixth great extinction event.

Some people think all this talk about the sixth extinction event is simple hyperbole, fueled by hysteria rather than by real science. They usually issue the following challenge: it may have been nice to have all these species, but the world is changing in response to human need. We have to give some things up. It might as well be a stone fly, a spider, a pond shrimp, a South African *protea* flower, even a rare antelope in a forest in Vietnam, and a gazelle on the plains of the Gobi Desert. The disappearance of these creatures, great and small, is regrettable, but we cannot compromise an ever-expanding population and global economy, whose collapse would leave billions to starve. Moreover, tree farms and other restored green spaces perform many environmental services, including the sustainable provision of a few products, the regulation of water flow, and carbon sequestration.

They sometimes link this argument with an often ignored fact about the modern ecosystem: humans have been exterminating its components for forty thousand years. Much of what we find beneficial and appealing— "natural"—has suffered at least some measure of historical degradation. What we have come to call natural habitats in our time were already grossly modified by humans centuries ago. In New Zealand, for example, flightless birds were reduced in a few centuries from thirty-eight to nine species. David Steadman, who has chronicled the extinction and precipitous decline of birds on Pacific islands, puts it well: ". . . the biodiversity crisis is over. People won: native plants and animals lost." So, the argument goes, this is an inevitable state of affairs, and we should not expend huge resources and energy to protect habitats simply to sustain all their biodiversity.

This two-part dismissal warrants vigorous response. All science consists of connecting theories with evidence, and scientists know they must refute a theory if an observation contradicts it. Scientists are not supposed to claim that they are offering profound truths, yet the following is as close to facts as anything scientists know of, and I label them accordingly:

Important fact: What we are experiencing is not just the "normal" rate of extinction, the background rate. The current extinction rate is soon to be as much as ten thousand times faster than the background rate. The projected mid-century loss of 30 to 50 percent of species is perhaps not as big a deal as the 90 percent loss at the end of the Permian, 250 million years ago, but it approaches the Cretaceous extinction event 65 million years ago and surpasses in magnitude many extinction events that mark the boundaries of time over the last 500 million years. Also, the current assault on Earth's biota is not a matter of epochs, millennia, or even centuries, but of decades, a mark too small to jot down on a room-size Earth calendar. Remember that the error range for dating the Cretaceous extinction event at 65 million years is, at its most precise, between 50,000 and 100,000 years, nearly the entire evolutionary history of *Homo sapiens*!

Second fact: Although humans have assaulted the ecosystems of the world for more than forty thousand years, there is no scientific indication that their indefinite exploitation or abuse of the environment will ensure livable conditions in the future. With the recent acceleration of abuses, we are approaching, indeed may have already crossed, the threshold to catastrophe for much of life on this planet. Again, what is at stake is not a century, or forty thousand years, but 100 million years of modern living.

Third fact: Extinction is irreversible. Species that die out will never come back. While humans might accomplish the Herculean feat of "restoring" destroyed grassland, forest, or rivers, they will never resurrect any extinct species that once lived in those habitats. Michael Soule has said, "Death is one thing, an end to birth is something else." And the impact of the extinction events such as those we are now experiencing reverberates through many other species that interact with those that have been lost, perhaps for one that carries on for millions of years.

Fourth fact: This irreversibility of species extinction and its reverberating effect clearly impede ecosystem recovery. The past provides us with useful evidence. The "very speedy" post-Cretaceous recovery of some species, among them ferns and some marine invertebrates and microorganisms, took hundreds of thousands of years, hundreds of times longer than recorded human history and nearly as long as the brief span of our species's history. After the Cretaceous event, key species in the ecosystem, including large plant-eating vertebrates, took even longer, in some cases millions of years. In anticipating recovery from the current biodiversity crisis, we can take little comfort from what the fossil record shows.

Fifth fact: The loss of species is not an event unto itself; it is one event that inevitably leads to others that can threaten ecosystem collapse. We have learned that the interconnectedness and complexity of ecosystems maximize the impact of species loss. For example, spiders, which are not humans' most cherished organism but are the most important invertebrate predator on insects in many habitats, are unusually slow and dogged in apprehending prey, and they are particularly susceptible to plant and soil toxins and other products of human intervention. The endangerment and ultimately the extinction of spider species in many areas, a trend now being explicitly documented by my colleagues at the American Museum of Natural History, will allow the increase of swarms of crop-destroying, disease-carrying insects. These multitudes, if unchecked, could wipe out other species and would not be welcomed by humans.

Sixth fact: The diversity of species is directly related to the sustenance of human life, not to mention our health, our pleasure, and our happiness. An analysis of African ecosystems and human population distribution shows that greater biodiversity is directly associated with areas of greater human population. The two are closely correlated; people need biodiversity. We use thousands of species of plants and other organisms for food, pharmaceuticals, and raw materials. The modern land ecosystem depends on plant pollination by insects. We have seen that bees, including many species of wild bees, pollinate a majority of the world's crop species and are thus directly or indirectly responsible for much of the food we produce for ourselves.

Gauging the expected level of extinctions by mid-century is a first big step in prognostication. But we can expect other biological effects, first-order effects, as the experts Norman Myers and Andrew H. Knoll call them. These include massive losses of population even in species that do survive, accelerated invasions of alien species, progressive depletion and homogenization of biotic communities, global reduction in biomass, and the severe reduction, if not virtual elimination, of some biomes such as tropical forests and coral reefs.

All these first-order effects have obvious evolutionary consequences: collapse of species ranges, disruption of gene flow, depletion of gene reservoirs, and the increase in exchange of species between different areas and ecosystems. A biota that is so severely manipulated, as Myers and Knoll point out, is likely to experience many new evolutionary shocks.

We might expect an outburst of new species as old ones go extinct and their adaptive zones or niches are vacated. This is the pattern we see in the fossil record following major extinction events. What new species might these be? We have some clues from what we have learned about the uncontrolled evolutionary success of invasive species like zebra mussels and water hyacinths, species that thrive in degraded, human-dominated ecosystems.

Species that reproduce explosively and invest little time and resource in the prolonged development of offspring might be favored. These are quick to spread and less sensitive to vicissitudes in climate and other conditions, indeed resilient to natural catastrophes, such as intensive storms or droughts. They can afford to lose offspring under stress because they produce so many. The laws of chance give them a chance. We might have an Earth dominated by pest and weed ecology.

Our principal locus of biodiversity, the tropics, will no longer be the key place for the replenishment of species. The rapid deforestation in the tropics and the difficulty of restoring those regions to levels that generated real biological wealth ensure that this great source for Earth's evolutionary prospects will be severely diminished. During hundreds of millions of years the evolution of life on land saw rampant diversification and species flow from the warm, even-tempered climates of the tropical regions. When tropical conditions extended from the equator to the poles, the diversity of some groups in North America and Eurasia, such as mammals, was at an all-time high. The evolutionary potential of the planet's biota will be greatly reduced when the wellspring for this diversity is so massively damaged.

Species will decrease not only in numbers but in the range of their structures, functions, and adaptations. This aspect of biodiversity, which scientists call biodisparity, explains why there are mammals as big as a 100-ton blue whale or as small as a 0.13-ounce pygmy shrew no bigger than your little finger. Biodisparity allows new opportunities for yet more species and increases the range of evolutionary potentials not only for species but for the major groups to which they belong.

The twenty-first century may mark the evolutionary dead end of large vertebrates. As we have seen, much of the devastation that humans have wrought over the past forty thousand years has been unusually focused on big animals. The survivors of this onslaught now hang on in confined, degraded habitats, with small, isolated populations that maintain only a meager portion of their once enriched gene variation. We may have already deprived them of the genetic potentials for evolutionary change and adjustment they accumulated over millions of years. Despite recent conservation efforts, even some of the largest protected areas might be too small to provide a matrix for such evolutionary change. Particularly vulnerable are bears (especially polar bears), elephants, rhinoceroses, apes, and big cats. With strict implementation of international conservation regulations, whales in the open ocean may fare better despite the disruption of food chains in the ocean caused by overfishing, pollution, and global warming.

As evolution has always demonstrated, novelties will invariably emerge, and not all of them will be welcomed by humans. Small mammals (e.g., rats) and insects (e.g., cockroaches) will likely thrive in human-dominated ecosystems and develop evolutionary resistance to any chemical or other means of disposing or controlling their numbers. One source of novelty has already been demonstrated: disease-carrying insects from the tropics that spread to high-latitude regions, where their status as invaders makes them strong enough to dominate resident species and resist human defenses against them. The bacteria and viruses that travel with insects and other disease vectors are the greatest threat here, since they can reproduce stupendously, evolve explosively, and kill rampantly. The HIV epidemic is an important reason why the revised estimate of global human populations by mid-century is somewhat lower than originally predicted. We are also watching to see if other novelties, such as the virus responsible for avian flu, will evolve to a form readily transferred from human to human. With all our ecological tampering, one can predict with some confidence that such prospects will be part of our future.

As the last point shows, evolution and the biota intersect with all the trends in global human society—changes in health, wealth, distribution of resources, trade, government, and societal prerogatives. We have seen how a perfect ecological storm is brewing in West Africa because of the complex interplay between overfishing by both African and European nations offshore and the periodic devastation of wildlife on land for bushmeat. We have also learned that human population densities in Africa are higher where biodiversity is higher, suggesting that biodiversity is itself a better index for resource and comparative power than we realized. Large-scale migration of humans can be understood as one outcome of a situation in which there is a disparity in resources, especially one in which impoverished communities seek safe, sustainable conditions elsewhere. But nations that become target destinations have pushed and will push back. Such pressures lead to conflict. Many important areas rich in biodiversity lie on international borders, especially tropical rainforests. Some straddle the boundaries of Brazil, Colombia, and Peru, of Vietnam, China, and Laos, and of many countries in western and central Africa. These nations have not always maintained the most peaceful of relations. History shows that people have made war over gold, oil, and water; they may do so over biodiversity.

This is not a very uplifting prospect, but neither I nor other scientists who have studied the modern ecosystem as it has evolved over the past 100 million years can see a way of opting for less alarming scenarios. I can understand that none of us necessarily wants to hear this message. But we must guard against denial and passivity, which will lull us into ignoring the carnage that is engulfing life on this planet. In *Monty Python and the Holy Grail*, Sir Lancelot, on an ill-conceived mission of rescue, goes on a rampage at a wedding, slaughtering many of the guests. Standing amid the corpses and the moaning wounded, the royal father of the bride entreats the remaining survivors: "This is supposed to be a happy occasion. Let us not bicker and argue about who killed who." The absurd proclamation reminds me of the exuberant, Panglossian statements of world leaders who praise our thriving economies and our beautiful Earth, with nary an expression of concern about the dark side of the future and nary a finger pointed to the source of the darkness, ourselves.

When the signs of the times became clear more than a decade ago, many of us suspected that some of the doom and gloom might be over-wrought. We convinced ourselves that the negative trends would be mitigated by complex factors we would come to understand or that conservation awareness and action would improve matters. One area of skepticism concerned climate change, for evidence seemed too sketchy to allow a clear look into the future. Unfortunately, in this case science has been prescient. Those early-warning signals have been, if anything, reinforced. This is not like the stories we read as children about people of Earth facing an oncoming comet, an alien invasion, or a lethal shift in the composition of the atmosphere, trying to beat back or adapt to the onslaught, even leaving the planet. What we are dealing with is not science fiction. It is the reality of here and now. It is science fact.

Still, human ingenuity, commitment, and shared responsibility have great potential for ensuring more positive prospects. History has demonstrated our capacity for improving situations; think of those Romans who centuries ago constructed an elaborate, superbly functional sewage system in their polluted city. Do current positive efforts indicate a clear possibility that we can at least slow biological decline and environmental degradation? I shall say yes, but with a caveat: our recent efforts, however meritorious, must become more intensive and global; they must be implemented from the top down and given higher priority and more investment. We know which recent and current efforts to conserve water, restore forests, ban hunting of threatened wildlife, manage fisheries, arrest the invasion of alien species, curb pollution, and control the release of greenhouse gases are successful. They should continue and proliferate.

Let us consider other progressive actions that should be taken. The IUCN Red List of 2006 documented a few reversals in disturbing trends, cases in which the status of species was improved from "Critically Endangered" to "Endangered," from "Endangered" to "Threatened," or from "Nearly Threatened" to "Least Concern." Among such species were the

European white-tailed eagle (*Haliaeetus albicilla*), the seabird Abbott's booby (*Papasula abbotti*), the 300-kilogram Mekong catfish (*Pangasianodon gigas*) of Southeast Asia, and the Indian vulture (*Gyps indicus*), which had suffered a 97 percent population crash and was listed in 2002 as "Critically Endangered." (The veterinary drug that accidentally poisoned the vultures, diclofenac, is now banned in India, and a noninjurious substitute has been used.) Some bird species, such as Rodrigues fody (*Foudia flavicans*), Seychelles warbler (*Acrocephalus sechellensis*), Seychelles magpie-robin (*Copsychus sechellarum*), and black robin (*Petroica traversi*), were once categorized as candidates for imminent extinction but have been successfully brought back from the edge. Many other species, such as the humphead wrasse (*Cheilinus undulatus*) and saiga antelope (*Saiga tatarica*), are also the subject of concerted conservation campaigns; other plants and animals highlighted in previous Red Lists are now the focus of conservation, and their status should improve. These are good but unfortunately tentative steps to a comprehensive global-level effort for which we all should work.

Most disconcerting is the potential loss of key species in ecosystems that we have simply neglected, the lost little creatures, including, of course, insects. Those dwindling populations of butterflies on the British Isles and dragonflies in Sri Lanka are small indications of a rampant loss of vital species whose evolutionary and ecological health has not been assessed. The scientifically based conservation agenda needs to embrace all of diversity beyond the familiar flora and fauna that occupy most of our attention.

Meanwhile, the creatures that humans do recognize and cherish have yet to run the gauntlet successfully. The precarious situation for the few tigers scattered about Asia will not improve until we enforce, rather than simply proclaim, the interdiction of the trade in traditional medicines derived from their body parts. The reserves established for such large animals are laudable, but most of them are not big enough to support populations sizable enough to secure a healthy evolutionary future for the animals. Creating corridors lacing through human-dominated regions to connect wildlife reserves and nature preserves is a welcome idea, and it has caught on in South Africa, in the areas around the coral reefs of the Bahamas, in the borderland between Vietnam and Laos, and the Rocky Mountain corridor stretching from Alaska to Colorado. These causeways for wild species are and will be controversial; they disrupt the master plans for housing tracts and highways and are often inconveniences for the people near them; they also curtail activities, including hunting and fishing, that many believe are their birthright or, in some places, their means of survival. But connecting reserves with these lifelines is a genuine necessity to keep the otherwise dismembered parts of our ecosystem functioning.

Obviously, these necessary actions concern land use. If land is given over wholly to industry, agriculture, or human habitation and, worse, is "used up" and left barren, polluted, toxic, or otherwise unproductive, the planet as we know it will not survive. But by now we know land use that sustains ecosystems is possible and that it confers a triple reward to the environment, society, and the economy. There are the coffee farms I've mentioned, planted near forests and benefiting from wild pollinators, which increase coffee yields. There is New York City's purchase for a billion dollars of watersheds in the Catskill Mountains that naturally purify water, instead of building a filtration plant that would have cost six to eight billion dollars plus annual operating costs of three hundred million. Forests in the Yangtze River watershed that help control the discharge of river water allow for efficiencies in hydroelectric supply equal to forty million kilowatt-hours per year. Safe habitats for birds control caterpillar pests and raise production in European apple orchards. There are plans for reflective roofing, green space, and increased shade to cool urban "heat islands" that will lower the energy costs of a city like Sacramento by twenty-six million dollars per year and reduce peak ozone concentrations by 6.5 percent. The use of natural species, such as larval-feeding fish, to control mosquitoes in China has minimized the need for toxic chemicals, reduced malaria outbreaks, and simultaneously stocked rice paddies with edible fish.

All these are good starts, but we have a long way to go. Jonathan Foley and other scientists have argued that a more global strategy of land use must distinguish among three options. Option 1, to preserve natural ecosystems, would maximize all the benefits we expect from ecosystems—water filtration and flow, biodiversity and habitat health, mediation of infectious diseases, good air quality, forest production, and carbon sequestration—but of course does not provide food crops. Option 2, to develop intensive croplands, is the converse and fails to provide the ecosystem services under Option 1. Option 3, to develop croplands mixing agriculture with natural components, would provide crop foods and restored ecosystem services. Though its productive capacity will not equal that under Option 2, it does provide a good deal of the other aspects of the natural habitat we value. We need to shift away from a strategy that is on the whole still directly converting Option 1 into Option 2 landscapes. It is Option 3 we should aim for. The fate for the remainder of what we call natural ecosystems will depend on our adopting it and, at the same time, securing as much natural habitat as possible from *any* kind of conversion. With a global human population that will increase by 2.6 billion by 2050—100 million more than the total population of the planet in 1950—some of these natural habitats will invariably be lost. To lesson the blow, Option 3 land conversion should be widely adapted to serve both humans and the nature we are part of. Catskill and Yangtze watersheds, organic coffee farms, bird-friendly apple orchards, and urban green spaces point the way.

This brings us to the challenge of improving the prospects for the planet on the largest, most integrative scale. Climate change itself can be good for some organisms, as we've seen. Under normal conditions many species can adapt over time, are able to migrate and expand their ranges. But the extremely rapid climate changes that are predicted, in combination with habitat fragmentation, preclude this kind of flexibility and adaptation. The steps taken to deal with this problem have been limited and insufficient. Emissions of polluting gases such as nitrogen oxides have leveled off in North America and even declined in Europe, but air pollution, including CO_2 emissions, remains much too high. To make matters worse, countries in Asia, especially China and India, are making huge contributions to air pollution and climate change with their increasing emissions of pollutants, including carbon dioxide and nitrogen oxides. These were unprecedented just a few years ago.

What can be done then? Containing this accelerating planetary transformation will take a truly mammoth effort that goes beyond a drastic reduction in the use of substances producing greenhouse gases. Two prominent environmental scientists, Thomas Lovejoy and Lee Hannah, have well stated the nature of the problem and the possible solutions. The current global yearly use of fossil fuels is about six gigatons of carbon, which will have to grow by a factor of two or three *even if* we use fuel more efficiently, make major changes in our lifestyle, and substitute renewable energy sources for traditional ones. Remember the current concentration of CO_2 in the atmosphere is 380 parts per million per volume, a higher concentration than at any time in the last 10 million years. The last time it was this high much of the coastal land surface that now is home to millions of humans was under water. So we must decrease the use of carbon fuels very soon just to stay within a "stabilized" range of 450 to 750 ppmv; if we don't, carbon levels could increase by a factor of ten, a disastrous prospect for life on Earth. The future should be carbon-neutral, and that requires the immediate adoption of three major strategies:

First, we must both effect greater fuel efficiency and sequester excess carbon—concentrate and store it in safe places—by natural, biological means. If all the cars in the world were converted to hybrid power, that would effect an efficiency savings of 50 percent or more. The reforestation of natural ecosystems, as in the Option 1 approach I noted above, or the development of tree plantations and cropland enriched with biodiversity, Option 3, can absorb significant amounts of carbon.

Yet these monumental steps are merely short-term fixes. We must go beyond the natural sequestration of fossil fuel CO_2, and we must start making a transition to renewable energy sources, both programs requiring a great deal of money and technology. There are proposals to pump CO_2 into abandoned oil wells or saline aquifers or to dispose of it in the deep ocean. But ocean disposal would doubtless disrupt marine life, and there would always be the insidious problem of leakage back to the atmosphere. This might take the form of a gigantic burst of CO_2, like the catastrophic cloud of gas released from Lake Nyos in Cameroon in 1986 that killed more than seventeen hundred humans who lived nearby. Some paleontologists hypothesize that gigantic releases of CO_2 in the ocean may have been related to mass extinction events like the one occurring at the end of the Permian Period. Mineral sequestration is potentially the best alternative, since CO_2 chemically reacts with rocks such as serpentine and peridotite; it is estimated that mineral carbonates could store fifty thousand gigatons of carbon for hundreds of thousand of years, a storage vault bigger than any requirements we might have for a carbon-neutral future. The technology for such sequestration has yet to be developed, much less tested. But sequestration must be pursued as a viable remedy, with a focus on technologies that are safe and efficient.

A third strategy would require us to switch entirely to renewable energy sources such as solar and wind power. A comprehensive solution that minimizes or virtually eliminates our dependency on fossil fuels is a must if we are to stabilize the current conditions. Many renewable energy technolo-

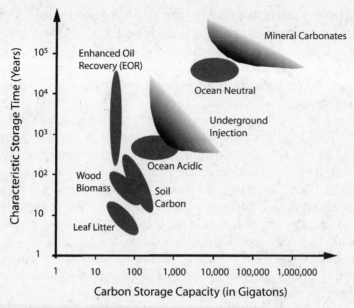

Getting rid of carbon: storage capacity and time of different reservoirs. Shaded ellipses indicate the range in both capacity and duration of storage. For example, leaf litter has a
storage capacity ranging from about twenty to one hundred gigatons of carbon for a duration ranging between five and twelve years. Ellipses with unclosed edges indicate uncertainty concerning upper limits of storage capacity and duration. Note that an acidic ocean
(with a lower pH) absorbing an excess of carbon dioxide will have much less capacity than
a neutral ocean where carbon dioxide uptake is in balance with carbon dioxide release.

gies are benign at local levels but have problems when applied at a larger
scale. Already there are conflicting political problems in using wind power
in the form of expansive "forests" of wind turbines, which require large sec-
tors of land or ocean and may interfere with other environmental prerogatives, such as assuring safe passage for migrating birds.

But these steps, however difficult and complex, must be taken. Climate
change, in combination with habitat fragmentation and other human-
induced factors, leaves the deepest mark on our modern ecosystem. Bleaching coral reefs and withering plants in South Africa are just a taste of what is
in store for the biota. Any measure of success depends not only on international cooperation but also on the leadership of those nations and economies
most empowered to do something. Unfortunately, such leadership is dismay-

ingly erratic or absent. It is no secret that the United States and a few other powerful nations refused to sign on to the Kyoto Protocol. Moreover, countries like China and India undergoing rapid development were exempted from signing on. Given the huge contributions those two nations are now making to the environmental degradation of the planet, such an exemption was outrageously shortsighted.

As of 2007, the holdout nations have yet to take major steps. Many U.S. leaders, including President Bush, now more openly acknowledge the problem but do not explicitly accept the measures recommended by the Kyoto Protocol, stating that countries like China and India are unjustifiably exempted. China, for its part, did issue in June 2007 a strategy to improve energy efficiency and reduce greenhouse gases, but it rejected mandatory caps on emissions, which it thought might constrain the country's explosive economy. Its leaders also continued to maintain that compliance with the Kyoto Protocol was the responsibility of those countries that produced the majority of greenhouse gases since the industrial revolution. Unfortunately, that rationale, based on prior history, will soon be irrelevant: China is projected to surpass the United States by 2009 or 2010 as the world's biggest source of greenhouse gases.

When it comes to implementation, many administrations still apply a strategy that ignores the problem because of the scientific debate concerning aspects of the predictions for change. Any technical disagreement is an excuse to dispense with the whole matter. Appropriate funding for research of environmental urgency is hence marginalized by continued subsidies of massive and wasteful agriculture and tax breaks for industries tied to traditional, rapidly depleting, and polluting energy sources. In order to bolster its case to constrain environmental regulation, certain federally supported projects, including oil drilling, pipelines, and power plants, the U.S. government in 2005 enlisted David Legates, the head of the Center for Climatic Research at the University of Delaware, Newark, and a well-known skeptic of the scientific evidence for global warming. Legates claimed that data for the rise in temperature over the twentieth century were biased by measurements taken near unnaturally warm urban centers, that Greenland was actually cooling, coral bleaching was a beneficial response to environmental change, and droughts during the twentieth century were relatively benign. Scientific work published at that time and certainly since then has thoroughly debunked these claims. The recorded rise in temperature over the last century, for example, clearly takes into account any bias from the urban heat effect. That Greenland is cooling is outrageously wishful thinking, given the striking new data on the melting ice cap and sliding glaciers. Coral bleaching has been directly linked to the extinction of species. As for droughts being benign, tell that to Mongolian herdsmen or, worse, the millions of suffering people in sub-Saharan Africa.

Following the release in early 2007 of the IPCC report, and at least a recognition by President Bush and other once resistant world leaders that climate change was a problem in need of a solution, the tide may be turning. Yet Bush was among those who rejected aggressive programs for the control of emissions offered at the June 2007 Group of Eight conference. The lack of action for more than a decade since the warning signs emerged has seriously narrowed our window of opportunity.

This skepticism and indifference to the wreckage around us and the delay in response to the problem are the source of some dire warnings by influential people. In a rare public appearance in June 2006 at a news conference in Hong Kong, the cosmologist Stephen Hawking, perhaps the best-known scientist alive today, emphasized that the world as we know it is facing imminent catastrophe in the form of global warming, nuclear war, or genetically engineered viruses. He called upon society to plan for a serious expedition to settle a hospitable planet in a distant star system sometime this century. I am not one to decry the importance of space travel and colonization as part of our future, but I hope that the awesome investment required for such an enterprise isn't substituted for the investments we can and must make here on Earth to improve the situation. I would rather stand and fight than cut and run. And the examples of progress related above show that, with due commitment, we can fight successfully.

What is the future for Terra? Will it resemble the desolate slag heaps around those toxic swamps in New Jersey or a pristine rainforest in the Truong Son Mountains of Vietnam? In reality neither extreme is a good predictive metaphor. The human population would crash in the unproductive, unsafe environments of a globe completely encrusted with slag heaps and engulfed by poisonous air. That kind of unmitigated disaster might shock even Malthus himself. Conversely, there are only a paltry number of the national and international tracts of land usually called pristine. Indeed, what does the word *pristine* mean in a world ecosystem dominated and transformed by humans, who can reach virtually everywhere? We have gone way past any point where we might be able to restore the world to that pure, mysterious, and unblemished condition that curious, exploratory humans, then only a billion strong, encountered just two centuries ago.

But perhaps our dazzling evolutionary success points to a bright future. Given our successes, why isn't it reasonable to assume that *Homo sapiens* has the capacity to clean up the water, air, soil, stop depleting the now dangerously diminished and unevenly distributed products of long-dead biota— oil and coal—and put ecosystems back on the right track? Do we have the will to turn back the Four Horsemen at the gates of the city? Is our instinct to do so a real one or simply another example of our vanity and hubris? Unfortunately, for all our progress in plotting the diversity and distribution of species, our refinements of ecology, and our exhaustive recording of life's ups and downs in the fossil record, we can't predict the future of the human condition.

I have tried to show, on one hand, the marvelous intricacy and sustainability of life and living processes and, on the other, their susceptibility to unexpected shocks that leave Earth utterly and irrevocably changed. Perhaps this simple lesson from the past is enough to make us skeptical about our ability to reverse the insidious forces in play. Evolutionary success is not an endowment; it is a legacy, and we may or may not continue to fulfill it. As Darwin maintained, there is no optimal goal or guarantee of everlasting success. As evolution goes, much of the future may be out of our hands.

Still, we find ourselves in an extraordinary moment, a species with a profound knowledge of its past, with a huge ambition for change, and with a disturbing ability to radically transform the water, land, air, and life of an entire planet. By scientific consensus, we are our own destroyers, our Vishnus and asteroids. But science tells us we also can be our own benefactors, as seen in the work of a few, whose efforts should be an inspiration to all of us. One of these is the discoverer of the saola, the Vietnamese ecologist Do Tuoc, who still hacks his way through the dense forests of the Truong Son Mountains to catch sight of one of his beloved creatures. He works tirelessly to educate villagers who share the saolas' habitat to protect rather than kill them. Do Tuoc is undaunted by the destructive forces in play; since my visit to Vietnam in 2002 the fate of the rare, elusive saola has become even more uncertain. In 1992, when Tuoc first discovered them, the saolas were thought to number as many as one thousand; now the most optimistic counts hover around two hundred. But the work of Do Tuoc and a few others has finally resulted in a government action plan for saola conservation that includes a long-delayed ban on hunting this animal. There are also plans to clone a captive saola and eventually reintroduce individuals bred in captivity to the wild, a project, like many other such cloning attempts, that has such a low chance for success it has been called a genetic Hail Mary. Meanwhile, Do Tuoc would prefer to fight for the saola on its own home turf. It is said of him, "Optimists about the saola's fate are about as rare as the animal itself, but Tuoc is one of them."

There may indeed be reasons for other optimists. Some nations have actually reduced the levels of some of the gases that pollute the atmosphere and raise the global thermostat. And when it comes to saving species, I have related success stories in bringing back the European white-tailed eagle, Abbott's booby, the Indian vulture, and others from the point of near extermination. At the same time, these successes should fuel, not allay, our concerns for the future. They demonstrate how much work is required to save a few among many others that need our attention. Had only those efforts been applied to some of the species already wiped out in the brief span of our history! The passenger pigeons that once flew above North America in flocks of two billion strong were "as beautiful as they were numerous." Henry David Thoreau described them as "a dry slate color, like weather-stained wood . . . a more subdued and earthly blue than the sky . . ." As a paleontologist examining the fossils I've discovered I often wonder about a world of the past neither I nor anyone else can ever see. The vision of a nineteenth-century prairie shadowed by those countless birds is now as gone as a scene in a Cretaceous forest. The battle for the passenger pigeon is lost, the battle for the European white-tailed eagle may be turning in our favor, but the battle for Terra has just begun.

NOTES AND REFERENCES

PROLOGUE: THE HYENA

ix "On a summer morning, just like summer mornings for hundreds of millennia, the Sahara turned its scorched earth toward the sun." The age of the Sahara is controversial, as shown in a May 2006 forum in *Science* 312:1138–39. Estimates range from 86,000 to 2.5 million years before present.

ix–x "We were a small group led by the paleontologist Dr. Maureen O'Leary . . ." Results of these expeditions are in O'Leary, M. A., et al., 2006, "Malian Paenungulata (Mammalia: Placentalia): New African afrotheres from the Early Eocene," *Journal of Vertebrate Paleontology* 26:981–88.

x "Over the years I had worked in many hot places . . ." These expeditions are described in Novacek, M., 2002, *Time Traveler*, Farrar, Straus and Giroux, New York.

x ". . . in the Darfur region of Sudan, hundreds of thousands of people were being driven from their homes and arable lands." One of numerous references on the genocide in the Sudan is Millard, B. J., and Collins, R. O., 2006, *Darfur: The Long Road to Disaster*, Markus Wiener, Princeton, NJ.

x "The Sahara is part of a global environmental disaster, desertification . . ." United Nations Environment Programme (UNEP), 1992, *World Atlas of Desertification*, Edward Arnold, UK; Batterbury, S.P.J., and Warren, A., 2001, "Desertification," in Smelser, N., and Baltes, P. (eds.), *International Encyclopædia of the Social and Behavioral Sciences*, Elsevier Press, Amsterdam, pp. 3526–29.

x–xi ". . . of all natural resources, the most tenuous . . . freshwater ranks number one." Dudgeon, D., et al., 2005, "Freshwater biodiversity: importance, threats, status and conservation challenges," *Biological Reviews* 81:163–82.

xi ". . . harsh conditions faced by 2-million-year-old human populations in Africa." Early humans as prey are described in Hart, D., and Sussman, R. W., 2005, *Man the Hunted: Primates, Predators, and Human Evolution*, Westview Press, Boulder, CO.

xi ". . . human predation on Africa's wildlife seemed oddly well managed." Martin, P. S., 1984, "Prehistoric overkill: the global model," in Martin, P. S., and Klein, R. G. (eds.), *Quaternary Extinctions: A Prehistoric Revolution*, University of Arizona Press, Tucson, pp. 354–403.

xi "African animals . . . evolved into forms more wary of humans . . ." Ibid.

xi "During a single year, 1911, one safari company killed 700 to 800 lions." Herne, B., 1999, *White Hunters*, Henry Holt, New York, pp. 78–86.

xi ". . . poachers have killed thousands of African animals . . ." Ogutu, J. O., Dublin, H. T., 2002, "Demography of lions in relation to prey and habitat in the Maasai Mara National Reserve, Kenya," *African Journal of Ecology* 40:120–29; Ogutu, J. O., Bhola, N., and Reid, R., 2005, "The effects of pastoralism and protection on the density and distribution of carnivores and their prey in the Mara ecosystem of Kenya," *Journal of Zoology* 265:281–93.

xi ". . . numbers of elephants in Virunga National Park . . ." "Elephants, large mammals recover in Africa's oldest NP," www.terradaily.com/reports/Elephants_Large_Mammals _Recover_from_Poaching_in_Africas_Oldest_NP_999.html.

xi "Poachers had relentlessly culled the elephant populations . . ." Ibid.

xii "Even today wild animals kill hundreds of humans." "Human-animal conflict," World Wildlife Fund Report, www.panda.org/about_wwf/what_we_do/species/problems /human_animal_conflict/index.cfm; 2005 "Human wildlife conflict manual," Wildlife Management Series, WWF-World Wildlife Fund for Nature and South African Regional Programme Office (SAPRO), http://assets.panda.org/download/human _wildlife_conflict.pdf.

xii "In many cases, these deaths simply result from unhappy accidents . . ." "Human-animal conflict."

xii "Since 1990 in Tanzania lions have killed . . ." Packer, C., et al., 2005, "Lion attacks on humans in Tanzania: understanding the timing and distribution of attacks on rural communities will help to prevent them," *Nature* 436:927–28.

xii "A 2005 World Wildlife Fund dispatch . . ." www.worldwildlife.org/expeditions /mozambique/dispatch8.htm.

xiv "Ecologists often point out . . ." Vitousek, P. M., et al., 1997, "Human domination of earth's ecosystems," *Science* 277:494–99; Western, D., 2001, "Human-modified ecosystems and future evolution," *Proceedings of the National Academy of Sciences* 98:5458–65.

xiv ". . . humans became the primate group to watch." The huge literature on human evolution includes such overviews as Diamond, J. M., 1992, *The Third Chimpanzee: The Evolution and Future of the Human Animal*, HarperCollins, London; Tattersall, I., 1995, *The Fossil Trail: How We Know What We Think We Know About Human Evolution*, Oxford University Press, Oxford, UK; Tattersall, I., 1995, *The Last Neanderthal: The Rise, Success and Mysterious Extinction of Our Closest Human Relatives*, Macmillan, Basingstoke Hampshire, UK; Stringer, C., and Andrews, P., 2005, *The Complete World of Human Evolution*, Thames and Hudson, London.

xiv ". . . primitive, humanlike species lived out desperate lives . . ." Hart and Sussman, *Man the Hunted*.

xiv "They decimated large animals . . ." A useful compendium on the causes of extinction of large animals during the past hundred thousand years that emphasizes the likely impact of humans is MacPhee, R. D. E. (ed.), 1999, *Extinctions in Near Time, Causes, Contexts, and Consequences*, Kluwer Academic, Plenum Publishers, New York. Also see Solow, A. R., Roberts, D. L., and Robbirt, K. M., 2006, "On the Pleistocene extinctions of Alaskan mammoths and horses," *Proceedings of the National Academy of Sciences* 103:7351–53.

xiv "The advent of agriculture . . ." Diamond, J., 1999, *Guns, Germs, and Steel: The Fates of Human Societies*, W. W. Norton, New York; Diamond, J. M., 2002, "Evolution, consequences and future of plant and animal domestication," *Nature* 418:700–707; Bellwood, P., 2004, *First Farmers: The Origins of Agricultural Societies*, Blackwell Publishing, Oxford, UK.

xiv ". . . not until 1800 that global human population reached 1 billion." Cohen, J. E., 2003, "Human population: the next half-century," *Science* 302:1172–75.

xiv–xv "Antarctica . . . was not seen by humans until 1820." Huntford, R., 1985, *The Last Place on Earth*, Pan Books, London; Headland, R. K., 1990, *Chronological List of Antarctic Expeditions and Related Historical Events*, Cambridge University Press, Cambridge, UK; Landis, M. J., 2003, *Antarctica: Exploring the Extreme: 400 Years of Adventure*, Chicago Review Press, Chicago.

xv "A well-known European map . . ." Melish, J., 1820, *Map of the World from the Latest Discoveries*, John Melish, Philadelphia.

xv "Not until well along in the nineteenth century did Western scientists . . . name the Sumatran rhino . . ." The first date of the naming and description of the mammals mentioned can be found in McKenna, M. C., and Bell, S. K., 1997, *Classification of Mammals Above the Species Level*, Columbia University Press, New York. Also see Wilson, D. E., and Reeder, D. M. (eds.), 2005, *Mammal Species of the World: A Taxonomic and Geographic Reference*, 3rd ed. Johns Hopkins University Press, Baltimore.

xv ". . . the passenger pigeon . . . numbered an estimated 3 to 5 billion birds . . ." Schorger, A. W., 1955, *The Passenger Pigeon: Its Natural History and Extinction*, University of Wisconsin Press, Madison.

xv ". . . the change in terms of the human population explosion." Cohen, "Human population."

xv "This requires a boost of 50 to 60 percent in food production . . ." Myers, N., 1998, "Population dynamics and food security," in Johnson, S. R. (ed.), *Food Security: New Solutions for the 21st Century*, Iowa State University Press, Ames, pp. 185–220.

xvi "5 billion tons of topsoil have been removed . . ." UN Food and Agricultural Organization (FAO), 1999, *State of the World's Forests*, FAO, Rome.

xvi ". . . 4.3 million square kilometers of cropland . . . have been abandoned . . ." Ibid.

xvi "From the advent of agriculture eleven thousand years ago . . ." Diamond, *Guns, Germs, and Steel*.

xvi "Today's forests occupy only about 30 percent of the total land surface" Mygatt, E., 2006, "World's forests continue to shrink," Earth Policy Institute, www.earth-policy.org /Indicators/Forest/index.htm; FAO, *Global Forest Resources Assessment 2005*, Rome: 2006, www.fao.org/forestry/site/32038/en.

xvi "Deforestation is carried out at a global annual rate of 7.3 million hectares . . ." Mygatt, E., 2006.

xvi "The situation for tropical rainforest . . . is particularly distressing." Ibid.

xvii ". . . (Jared Diamond has called a very similar list the Evil Quartet) . . ." Diamond cites habitat loss, species introduction, extinction cascades, and resource overexploitation as the evil four. See Diamond, J. M., 1984, "'Normal' extinctions of isolated populations," in Nitecki, M. H. (ed.), *Extinctions*, University of Chicago Press, Chicago, pp. 191–246. Here I replace extinction cascades with pollution, as the former pertains to all factors and seems a second-order consequence of the others. My listing reflects the consensus in a meeting of biodiversity specialists sponsored by the National Academy of Sciences and cited in a colloquium report by Novacek, M. J., and Cleland, E. E., 2001, "The current biodiversity extinction event: scenarios for mitigation and recovery," *Proceedings of the National Academy of Sciences* 981:5466–70.

xvii "Fisheries have erased more than 25 percent of the productivity . . ." Crutzen, P. J., 2002, "Geology of mankind," *Nature* 415: 23.

xvii "Invasive species have wiped out . . ." Lowe, S., et al., 2000, *100 of the World's Worst Invasive Alien Species: A Selection from the Global Invasive Species Database*, Invasive Species Specialist Group (ISSG) of the World Conservation Union (IUCN)'s Species Survival Commission (SSC), pp. 1–12, www.issg.org/booklet.pdf.

xvii ". . . coral reef systems . . . have been . . . degraded . . ." Pandolfi, J. M., et al., 2005, "Are U.S. coral reefs on the slippery slope to slime?," *Science* 307:1725–26.

xvii "The sixteenfold increase in energy use . . ." Crutzen, "Geology of mankind."

xvii ". . . air pollution became global in extent . . ." Akimoto, H., 2003, "Global air quality and pollution," *Science* 302:1716–19.

xvii–xviii ". . . an ultragreenhouse atmosphere . . ." An important status report is a special section, "Climate change, breaking the ice," in the March 24, 2006, issue of *Science* 311:1669–1785. A comprehensive recent volume is Lovejoy, T. E., and Hannah, L. (eds.), 2005, *Climate Change and Biodiversity*, Yale University Press, New Haven, CT.

xviii "CO_2 levels in the atmosphere have not been so high . . ." Ibid. Also see Kennedy, D., and Hanson, B., 2006, "Ice and history," *Science* 311:1673.

xviii "Models. . . predict an increase in global mean surface temperatures . . ." Kennedy and Hanson.

xviii "Adapting to this profound climatic change will be enormously difficult for many species . . ." Ibid. and Allen, J. D., Palmer, M., and Poff, N. L., 2005, "Climate change and freshwater ecosystems," in Lovejoy and Hannah, *Climate Change and Biodiversity*, pp. 274–90.

xviii "Species, the fundamental units in biology . . ." The relevance of the species to the assessment of biodiversity and the biodiversity crisis is described by Wilson, E. O., 2003, "The encyclopedia of life," *Trends in Ecology and Evolution*, 18:77–80; Novacek, M. J. (ed.), 2001, *The Biodiversity Crisis: Losing What Counts*, American Museum of Natural History and New Press, New York (distributed by W. W. Norton, New York); Novacek, M. J., and Wheeler, Q. C. (eds.), 1992, *Extinction and Phylogeny*, Columbia University Press, New York; Novacek, M. J., 1992, "The meaning of systematics and the biodiversity crisis," in Eldredge, N. (ed.), *Systematics, Ecology, and the Biodiversity Crisis*, Columbia University Press, New York, pp. 101–108; Wilson, E. O., 1992, *The Diversity of Life*, Belknap Press of Harvard University, Cambridge, MA.

xviii ". . . extinguished or endangered large animals likely resulted from hunting . . ." MacPhee, *Extinctions in Near Time*.

xix ". . . every year thousands, perhaps tens of thousands of species are going extinct. This means that 30 percent and perhaps as much as 50 percent of all species might be extinct by the middle of this century." The original and convergent works for this well-known projection are Wilson, *The Diversity of Life*, Lawton, J. H., and May, R. M., 1995, *Extinction Rates*, Oxford University Press, Oxford, UK; Pimm, S. L. , Russell, G. J., Gittleman, J. L., and Brooks, T. M., 1995, "The future of biodiversity," *Science* 269:347–50; Myers, N., 1996, "Two key challenges for biodiversity: discontinuities and synergisms," *Biodiversity Conservation* 5:1025–34.

xix "The present-day diversity of life—some 1.75 million named species . . ." The number.

of named species may lie between 1.5 and 1.8 million with no exact count available. Here I opt for the 1.75 million, a tally cited in Harrison, I., Laverty, M., and Sterling, E., 2004, "Species diversity," *Connexions*, http://cnx.org/content/m12174/1.3/. See also Wilson, "The encyclopedia of life."

xx "Indeed, even scientists have long emphasized the radical transformation of ecosystems . . ." Kirchner, J. W., and Weil., A., 2000, "Delayed biological recovery from extinctions throughout the fossil record," *Nature* 404:177–80; Erwin, D. H., 2001, "Lessons from the past: biotic recoveries from mass extinctions," *Proceedings of the National Academy of Sciences*, 98:5399–403.

xx "It indeed long preceded the 7-million-year-old appearance of humanlike species . . ." Stringer and Andrews, *The Complete World of Human Evolution*.

xx ". . . before *Tyrannosaurus rex* stalked the Earth . . ." Benton, M., 1993. "Life and time," in Gould, S. J. (ed.), 1993, *The Book of Life*, W. W. Norton, New York, pp. 22–37; Novacek, M. J., 1996, *Dinosaurs of the Flaming Cliffs*, Anchor/Doubleday, New York.

xxi ". . . from 475 million years ago, when creatures from the sea invaded the land . . ." Edwards, E., 2001, "Early land plants," in Briggs, D.E.G., and Crowther, P. R. (eds.), *Palaeobiology II*, Blackwell Publishing, Oxford, UK, pp. 63–66.

xxi ". . . when the dark pine forests and fern gardens . . . dominated the landscape . . ." Friis, E. M., Pedersen, K. R., and Crane, P. R., 2001, "Origin and radiation of angiosperms," in Briggs and Crowther, *Palaeobiology II*.

xxi ". . . 65 million years ago, a catastrophe . . ." Alvarez, L. W., 1983, "Experimental evidence that an asteroid impact led to the extinction of many species 65 million years ago," *Proceedings of the National Academy of Science* 80:627–42; Stanley, S. M., 1987, *Extinctions*, Scientific American Books, New York; Archibald, J. D., 1996, *Dinosaur Extinction at the End of an Era: What the Fossils Say*, Columbia University Press, New York; Novacek, *Dinosaurs of the Flaming Cliffs*; Dingus, L., and Rowe, T., 1998, *The Mistaken Extinction: Dinosaur Evolution and the Origin of Birds*, T. H. Freeman, New York. For additional references, see chapter 13.

xxii "Many of the classic and powerful disclosures . . ." Leopold, A., 1949, *A Sand County Almanac and Sketches Here and There*, Oxford University Press, New York; Carson, R., 1962, *Silent Spring*, Houghton Mifflin, Boston; Silko, L. M., 1977, *Ceremony*, Viking, New York; Finch, R., and Elder, J. (eds.), 1990, *The Norton Book of Nature Writing*, W. W. Norton, New York.

xxii ". . . the UN Convention on Biological Diversity in 1992 . . ." The official text and related information can be found on www.biodiv.org/convention/articles.asp.

xxii ". . . the Kyoto Conference in 1997 . . ." The Kyoto Protocol is an amendment to the United Nations Framework Convention on Climate Change. The full text can be found on http://unfccc.int/resource/docs/convkp/kpeng.html. For a recent analysis of the protocol and a scientific rejoinder to the objections to it raised by officials from the United States and other countries, see Watson, R. T., 2005, "Emissions reductions and alternative futures," in Lovejoy and Hannah, *Climate Change and Biodiversity*, pp. 375–95.

xxiii "Wilson was the first to publish the word *biodiversity* . . ." Wilson, E. O. (ed.), and Peter, F. M. (associate ed.), 1988, *Biodiversity*, National Academy Press, Washington, D.C. The conference was the National Forum on Biological Diversity, organized by the National Research Council (NRC), which was held in 1986.

xxiii "Wilson himself wrote a book . . ." Wilson, *The Diversity of Life*.

xxiii "Subsequently there have been many books, including one that I edited . . ." Novacek, *The Biodiversity Crisis*; Eldredge, *Systematics, Ecology, and the Biodiversity Crisis*; Eldredge, N., 2000, *Life in the Balance: Humanity and the Biodiversity Crisis*, Princeton University Press, Princeton, NJ; Mooney, H. A., and Hobbs, R. J., 2000, *Invasive Species in a Changing World*, Island Press, Washington, D.C.; Peters, R. L., and Lovejoy, T. E., 1992, *Global Warming and Biological Diversity*, Yale University Press, London; Lovejoy and Hannah, *Climate Change and Biodiversity*; Heywood, V. H. (executive ed.), and Watson, R. T. (chair), 1995, *Global Biodiversity Assessment*, Cambridge University Press, Cambridge, UK, and New York; Wilson, E. O., 2002, *The Future of Life*, Knopf, New York.

xxiii ". . . dovetailed with the physical transformation of the planet . . ." Overpeck, J., Cole, J., and Bartlein, P., 2005, "A 'paleoperspective' on climate variability and change," in Lovejoy and Hannah, *Climate Change and Biodiversity*, pp. 91–108; Falkowski, P. G., et al., 2005, "The rise of oxygen over the past 205 million years and the evolution of large placental mammals," *Science* 309:2202–4.

1: A CREATURE IN THE FOREST

3 "This is the mysterious saola . . ." Robichaud, W. G., 1998, "Physical and behavioral description of a captive saola, *Pseudoryx nghetinhensis*," *Journal of Mammalogy* 79:394–405; Timmins, R. J., et al., 2005, "*Pseudoryx nghetinhensis*," 2006 IUCN Red List of Threatened Species, SSC, www.iucnredlist.org/.

3 "Most biologists, like George Schaller . . ." Schaller, G., 1998, "On the trail of a new species," *International Wildlife* 28:36–44.

3–4 ". . . the saola was first discovered . . ." Dung, V. V., et al., 1993, "A new species of living bovid from Vietnam," *Nature* 363:443–45.

4 "In 1812 the great naturalist Baron Georges Cuvier . . ." See Schaller, "On the trail of a new species."

5 ". . . a diverse order of mammals known as artiodactyls . . ." McKenna and Bell, *Classification of Mammals*.

5 ". . . new evidence from DNA and anatomical features . . ." Grauer, D., and Higgins, D., 1994, "Molecular evidence for the inclusion of cetaceans within the order Artiodactyla," *Molecular Biology and Evolution* 11:357–64; Gatesy, J. C., et al., 1996, "Evidence from milk casein genes that cetaceans are close relatives of hippopotamid artiodactyls," *Molecular Biology and Evolution* 13:954–63; Gatesy, J., 1997, "More support for a Cetacea/Hippopotamidae clade: the blood clotting protein gene g-fibrinogen," *Molecular Biology and Evolution* 14:537–43.

5 "Some students of saola anatomy have assigned it to the tribe Caprini . . ." Thomas, H., 1999, "Cranial anatomy and phylogenetic relationships of a new bovine (*Pseudoryx nghetinhensis*) discovered in the Vietnamese Annamite mountain range," *Mammalia* 58:453–81.

5 ". . . but recent studies based on DNA put it within the tribe Bovini . . ." Hassanin, A., and Douzery, E.J.P., 1999, "Evolutionary affinities of the enigmatic saola (*Pseudoryx nghetinhensis*) in the context of the molecular phylogeny of Bovidae," *Proceedings of the Royal Society of London B: Biological Sciences* 266:893–900; Gatesy, J., and Arctander, P., 2000, "Hidden morphological support for the phylogenetic placement of

Pseudoryx nghetinhensis with bovine bovids: a combined analysis of gross anatomical evidence and DNA sequences from five genes," *Systematic Biology* 49:515–38.

5 ". . . researchers have accumulated only a small collection of twenty partial specimens . . ." Schaller, "On the trail of a new species."

5 "twenty-one saolas were killed and three were taken alive . . ." Ibid.

6 ". . . the first major U.S. exhibition on Vietnamese culture . . ." Nguyen Van Huy and Kendall, L., 2003, *Vietnam: Journeys of Body, Mind, and Spirit*, University of California Press, Berkeley. The exhibit is featured on the American Museum of Natural History website www.amnh.org/exhibitions/vietnam/.

6 ". . . Vietnam has more than fifty ethnic groups . . ." Dang Nghiem Van, Chu Thai Son, and Luu Hung, 2000, *Ethnic Minorities of Vietnam*, Gioi Publishers, Hanoi.

6 ". . . Eleanor wrote . . ." Sterling, E. J., Hurley, M. M., and Le Duc Minh, 2006, *Vietnam: A Natural History*, Yale University Press, New Haven, CT.

6–7 ". . . Vietnam ranked second only to Thailand in exports of wood . . ." Ibid., p. 333.

7 ". . . lost a staggering 51 percent of its primary forests . . ." FAO, 2006, *Global Forest Resources Assessment 2005* (Rome), www.fao.org/forestry/site/ 32038/en.

7 ". . . the blunt edge of the wedge into the forest." Sterling et al., *Vietnam*.

7 ". . . relocated almost five million people . . ." Ibid., p. 334.

8 "A special gift to Vietnam is its cycad flora." Nguyen Tien Hiep and Phan Ke Loc, 1999, "The cycads of Vietnam," in Chen, C. J. (ed.), *Biology and Conservation of Cycads: Proceedings of the Fourth International Conference on Cycad Biology, Panzhihua, Sichuan, China, May 1–5, 1996*, International Academic Publishers, Beijing, pp. 24–32.

8 ". . . the Mesozoic Era . . ." See International Commission on Stratigraphy (ICS), 2005, *International Stratigraphic Chart*, International Union of Geological Sciences, www.stratigraphy.org/chus.pdf.

9 "The 2006 Red List . . ." 2006 IUCN Red List.

9 ". . . orchids, as many as 897 named species . . ." Averyanov, L. P., and Averyanov, A. L., 2003, *Updated Checklist of the Orchids of Vietnam*, Vietnam National University Publishing House, Hanoi.

9 "Charles Darwin was obsessed with orchids . . ." Darwin, C. R., 1862, *On the Various Contrivances by Which British and Foreign Orchids are Fertilised by Insects, and on the Good Effects of Intercrossing*, John Murray, London.

9 ". . . slipper orchids (genus *Paphiopedilum*) are in particular danger." Ministry of Science, Technology, and the Environment, 1996, *Red Data Book of Vietnam, vol. 2. Plants*, Science and Technics Publishing House, Hanoi.

11 "Hoan Kiem Lake means 'Lake of the Returned Sword' . . ." Nguyen Vinh Phuc, 2006, *Hanoi Streets of the Old Quarter and Around Hoan Kiem Lake*, Gioi Publishers, Hanoi.

11 "Shanghai softshell turtle, *Rafetus swinhoei* . . ." Farkas, B., and Webb, R. G., 2003, "*Rafetus leloii Hà Dinh Dúc, 2000*—an invalid species of softshell turtle from Hoan Kiem Lake, Hanoi, Vietnam (Reptilia, Testudines, Trionychidae)," *Zoologische Abhandlungen* (Dresden) 53:107–12; Meylan, P. A., and Webb, R. G., 1988, "*Rafetus swinhoei (Gray)* 1873, a valid species of living soft-shelled turtle (family Trionychidae) from China," *Journal of Herpetology* 22:118–19.

12 ". . . Hanoi has invested serious money in dredging centuries of accumulated sewage-infused mud . . ." http://vietnamnews.vnanet.vn/2004-01/10/Stories/02.htm.

12 ". . . Cham Art Museum . . ." Guillon, E., 2006, *Cham Art—Treasures from the Da Nang Museum, Vietnam*, River Books, Bangkok.

12 "U. S. armed forces had sprayed Vietnam's countryside with more than 20 million gallons (76 million liters) of Agents Orange, White . . ." Sterling et al., *Vietnam*.

13 "As Eleanor has noted in her recent book . . ." Ibid.

14 ". . . more than fifteen hundred species of plants serve as food for humans." Roubik, D. W., 1995, "Pollination of cultivated plants in the tropics," in Food Agricultural Organization, Rome, *Plants for a Future*, www.pfaf.org/leaflets/intro2.php.

14 "Snake wine is a problematic product." Sterling et al., *Vietnam*.

14 "Perhaps the most emblematic of victims in this regard is the tiger (*Panthera tigris*)." Ibid. and Sinha, V. R., 2003, *The Vanishing Tiger: Wild Tigers, Co-predators and Prey Species*, Salamander, London; Ellis, R., 2005, *Tiger Bone and Rhino Horn: The Destruction of Wildlife for Traditional Chinese Medicine*, Island Press/Shearwater Books, Washington, D.C.

14 "The Environmental Investigation Agency (EIA) . . . estimated that at least one tiger a day was being killed for Chinese medicine." The agency is an international campaigning organization committed to investigating and exposing environmental crime, www.eia-international.org/.

14 ". . . nineteen hundred kilograms of tiger bone . . . were exported to Japan . . ." *Tiger Crisis* website, www.tigerfdn.com/Tigerworld/W5X1.html.

14–15 ". . . tiger penis soup . . ." Highley, K., 1993, "A market for tiger products on Taiwan, a survey," *Earthtrust*, Taiwan, www.earthtrust.org/tiger.html.

15 ". . . Convention on International Trade in Endangered Species (CITES) . . ." www.cites.org/.

15 "In 1999 China did respond to international appeals . . ." TRAFFIC and WWF Briefing, 2002, "Conservation of tigers and other Asian big cats," www.traffic.org/cop12/ABC_CoP12.pdf.

15 "Hunting of tigers . . . is . . . prohibited in all countries . . ." www.worldwildlife.org/trade/faqs_tiger.cfm; www.cites.org/.

15 ". . . a workshop in Beijing . . ." TRAFFIC, April 24, 2006, Beijing, "TRAFFIC works with Chinese leaders to promote the conservation of Traditional Chinese Medicine," www.traffic.org/.

15 ". . . Red List still categorizes *Panthera tigris* as endangered . . ." 2006 IUCN Red List.

17 "Fossil deposits are the only evidence we have that these spectacular big mammals . . ." Discussed in Novacek, *Time Traveler*.

17 "Vietnam has an arresting number of unique species . . ." Sterling et al., *Vietnam*.

17 "A warming trend at the end of the last glaciation some twelve thousand years ago caused sea levels to rise 125 meters." Bush, M. B., and Hooghiemstra, H., 2005, "Tropical biotic responses to climate change," in Lovejoy and Hannah, *Climate Change and Biodiversity*, pp. 125–37.

17 ". . . classic ecological principles offered by Robert MacArthur and E. O. Wilson . . ." MacArthur, R. H., and Wilson, E. O., 1967, *The Theory of Island Biogeography*, Princeton University Press, Princeton, NJ.

18 "Mammals bigger than fifty kilograms . . . suddenly disappeared." Bush and Hooghiemstra, "Tropical biotic responses to climate change."

18 "Recent studies of the Amazonian ecosystem . . ." Andressen, E., 1999, "Seed dispersal by monkeys and the fate of dispersal seeds in a Peruvian rain forest," *Biotropica* 31:145–58; Silman, M. R., Terborgh, J. T., and Kiltie, R. A., 2003, "Population regulation of a rainforest dominant tree by a major seed predator," *Ecology* 84:431–38.

18 "The big mammals . . . were victims of extinction events . . ." MacPhee, *Extinctions in Near Time.*

18 ". . . the overkill theory . . ." Ibid.

18 "Extermination of large mammals in recent centuries has been much more intensive than for mammals in general." Alroy, J., 1999, "Putting North America's end-Pleistocene megafaunal extinction in context; large scale analyses of spatial patterns, extinction rates, and size distributions," in MacPhee, *Extinctions in Near Time,* pp. 105–43.

19 ". . . the most arresting examples is the green turtle . . ." Jackson, J.B.C., et al., 2001, "Historical overfishing and the recent collapse of coastal ecosystems," *Science* 293:629–37.

19 "The largest nest sites today . . ." Ibid.

19–20 ". . . thirty-two index sites worldwide have shown a 48 to 65 percent decline . . ." Ibid. and 2006 IUCN Red List.

20 "A recent sighting and photograph of a solitary right whale . . ." National Oceanic and Atmospheric Administration (NOAA) News Online (story 2467), 2005, www.noaanews.noaa.gov/stories2005/s2467.htm. Also see Martin, A. R., and Walker, F. J., 1997, "Sighting of a right whale (*Eubalaena glacialis*) with calf off s.w. Portugal," *Marine Mammal Science* 13:139.

20 ". . . we continue to devastate marine life for sustenance." Pauly, D., et al., 1998, "Fishing down marine food webs," *Science* 279:860–63; Pauly, D., et al., 2003, "The future for fisheries," *Science* 302:1359–61.

20 ". . . conference at the American Museum of Natural History held in 2002 . . ." "Sustaining Seascapes: The Science and Policy of Marine Resource Management," Center of Biodiversity and Conservation Spring Symposium, 2002, http://research.amnh.org /biodiversity/symposia/archives/seascapes/index.html.

20 "Daniel Pauly, a well-known fisheries scientist . . ." Malakoff, D., 2002, "Daniel Pauly profile: going to the edge to protect the sea," *Science* 296:458–61.

20 ". . . life on Earth all the way back to its beginning 3.5 billion years ago." The rocks of the Coonterunah Group of Australia and the Theespruit Formation of southern Africa, dated at 3.52 billion years before present, offer the oldest compelling evidence for life on Earth. Buick, R., 2001, "Life in the Archean," in Briggs and Crowther, *Palaeobiology II,* pp. 13–21.

20 ". . . but 99.999 percent of all life that ever existed is extinct." Novacek, M. J., and Wheeler, Q. C. (eds.), 1992, *Extinction and Phylogeny,* Columbia University Press, New York.

21 ". . . wiped out when a piece of rock . . . collided with Earth . . ." Alvarez, "Experimental evidence."

21 ". . . the current biodiversity crisis—what many scientists are calling the sixth mass extinction . . ." Leakey, R., and Lewin, R., 1995, *The Sixth Extinction: Patterns of Life and the Future of Humankind,* Anchor/Doubleday, New York; also many references to the sixth extinction in Novacek, *The Biodiversity Crisis;* Eldredge, *Systematics;* Eldredge, *Life in the Balance.*

21 "The recovery of resilient organisms . . . shows a particular pattern . . ." Kirchner and Weil, "Delayed biological recovery"; Erwin, "Lessons from the past."

2: LUSH LIFE

22 "... a spider barely twitches ..." Reports of spiders near the summit of Mount Everest go back to early expeditions. See Younghusband, Sir F., 1926, *The Epic of Mount Everest*, new ed., Pan Books, London.

22 "A few thousand feet lower in the Himalayas, life is easier to observe." Key references on the distribution of animals at high altitudes in the Himalayas are Pocock, R. I., 1939, *The Fauna of British India Including Ceylon and Burma Mammalia*, vol. I. *Primates and Carnivora*, and 1941, *The Fauna of British India Including Ceylon and Burma Mammalia*, vol. II. Taylor & Francis, London. A summary for the general reader was issued as Pocock, R. I., 2002, "Animal life at high altitudes," http://wondersbook4.myanimalcenter.com/313.shtml.

22–23 "Spiders are known to float on balloonlike webs ..." Brunet, B., 1998, *The Silken Web*, Reed New Holland, Sydney.

23 "The diverse life ... may have even tipped into the upper atmosphere." Wainwright M., et al., 2002, "Microorganisms cultured from stratospheric air samples obtained at 41km," *FEMS Microbiology Letters* 218:161–65.

23 "More than a mile below the sea surface, vents spew out sulfur-laden boiling water ..." Crabtree, R. H., 1997, "Where smokers rule," *Science* 276:222.

23 "... life as we think we know it ..." A summary of the diversity of different biological groups is in Raven, P. H., "What have we lost, what are we losing?" in Novacek, *The Biodiversity Crisis*, pp. 58–61; Harrison, I., Laverty, M., and Sterling, E., 2004, "Species Diversity," *Connexions*, http://cnx.org/content/m12174/1.3/.

24 "A possible estimate of 10 to 20 million or more species ..." Ibid. and Wilson, "The encyclopedia of life."

24 "... comprises more than 5,000 different living species ..." Ibid. and Wilson, D. E., and Reeder, D. M. (eds.), 2005, *Mammal Species of the World. A Taxonomic and Geographic Reference*, 3rd ed. Johns Hopkins University Press, Baltimore.

24 "... 300,000 species of plants; about 240,000 of these are the angiosperms ..." Ibid.

25 "... more than 940,000 named arthropod species ..." Ibid.

25 "The American Museum of Natural History ... has a collection of 32 million specimens ..." Descriptions of the museum's scientific division and its collections can be found at www.amnh.org/science/index.php.

26 "... for example, dung beetles—are responsible for such vital services ..." Fincher, G. T., 1981, "The potential value of dung beetles in pasture ecosystems," *Journal of the Georgia Entomological Society* 16:301–16.

26 "But the most spectacular function served by insects is pollination ..." Proctor, M., Yeo, P., and Lack, A., 1996, *The Natural History of Pollination*, Timber Press, Portland, OR.

26 "... 337 insect species ... routinely visit carrot flowers ..." Ibid.

26 "... ratio of plant to insect diversity ..." Myers, N. , 2005, "Death and taxas," *Nature* 435:566.

27 "... Edward Tyson ..." see Boorstin, D. J., 1983, *The Discoverers*, Random House, New York.

28 "Aristotle... loved to collect things." Aristotle, *Historia animalium*. For analysis of Aristotle's classification, see Gregory, W. K., 1910, "The orders of mammals," *Bulletin of the American Museum of Natural History* 27:1–524; Ross, D., 1995, *Aristotle*, 6th ed., Rutledge, London; Adler, M. J., 1978, *Aristotle for Everybody*, Macmillan, New York.

29 "Charles Darwin himself came to his great synthetic insight . . ." The voluminous literature on the life and works of Darwin is listed on various websites. A good compilation can be found on The C. Warren Irvin, Jr., Collection of Charles Darwin and Darwiniana www.sc.edu/library/spcoll/nathist/darwin/darwin.html. Many of the comments in this section are also taken from the text of the American Museum of Natural History Exhibit *Darwin*, curated by Niles Eldredge and cowritten by Laurie Halderman; see www.amnh.org/exhibitions/darwin/. An excellent recent collection of essays on Darwin and evolution is in *Natural History* magazine, November 2005, vol. 114, no. 9. A classic biography is Desmond, A., and Moore, J., 1991, *Darwin*, Michael Joseph, London.

30 ". . . taxonomy . . . systematics . . . binomial system and the formal classification . . ." The binomial system is called the Linnaean system after the great eighteenth-century taxonomist Carolus Linnaeus, Linnaeus, C., 1735, *Systema naturae, sive regna tria naturae systematice proposita per classes ordines, genera & species*, Fol. Lugduni Batavorum. A traditional treatment of systematics and taxonomy is in Mayr, E., 1942, *Systematics and the Origin of Species*, Columbia University Press, New York. More modern treatments are Hennig, W., 1966, *Phylogenetic Systematics*, University of Illinois Press, Urbana; Eldredge, N., and Cracraft, J., 1980, *Phylogenetic Patterns and the Evolutionary Process*, Columbia University Press, New York; Nelson, G., and Platnick, N. I., 1981, *Systematics and the Biogeography—Cladistics and Vicariance*, Columbia University Press, New York; Queiroz, K. de, and Donoghue, M. J., 1988, "Phylogenetic systematics and the species problem," *Cladistics* 4:317–38; Nixon, K. C., and Wheeler, Q. D., 1992, "Extinction and the origin of species," in Novacek and Wheeler, *Extinction and Phylogeny*, pp. 119–43; Gaffney, G., Dingus, L., and Smith, M., 1995, "Why cladistics?" *Natural History* 104:33–35.

30 "A recent study by David Fleck . . ." Fleck, D. W., Voss, R. S., and Simmons, N. B., 2002, "Undifferentiated taxa and sublexical categorization: an example from Matses classification of bats," *Journal of Ethnobiology* 22:61–102.

31 ". . . the same Aristotle . . ." *Historia animalium*.

31 ". . . classifications of John Ray . . ." Ray, J., 1686–1704, *Historia plantarum*; 1693, *Synopsis methodica animalium quadrupedum et serpentini generis*; Raven, C. E., 1942, *John Ray: Naturalist: His Life and Works*, Cambridge University Press, Cambridge, UK. See also Gregory, "The orders of mammals . . ."

32 ". . . myriad separate acts of God." Boorstin, *The Discoverers*.

32 ". . . other visionaries of the late eighteenth and early nineteenth centuries . . ." Ibid.

32 "The twentieth-century exploration of the genome . . ." Keller, E. F., 2001, *The Century of the Gene*, Harvard University Press, Cambridge, MA.

32 "Biologists have indulged in endless debate about the nature of species . . ." Harrison et al., "Species diversity."

33 ". . . the biological species concept . . ." Mayr, *Systematics and the Origin of Species*.

33 "In *Histoire naturelle* . . . Buffon recognized . . ." Buffon, G.L.L., 1749–1788, *Histoire naturelle, générale et particulière*, vol. 1–36, L'Imprimerie Royale, Paris.

33 "We have observed directly the breeding habits . . . of only a paltry few of the 1.75 million named organisms." Wilson, "The encyclopedia of life."

33 "Many entomologists use such features in their taxonomy." A good example is Schuh, R. T., and Slater, J. A., 1995, *True Bugs of the World (Hemiptera: Heteroptera): Classification and Natural History*, Cornell University Press, Ithaca, NY.

34 ". . . sibling species . . ." For a general review, see Knowlton, N., 1993, "Sibling species in the sea," *Annual Review of Ecology and Systematics* 24:189–216.

34 ". . . simply to revisions in our recognition of species . . ." Cracraft, J. A., 1992, "The species of the birds-of-paradise (Paradisaeidae): applying the phylogenetic species concept to a complex pattern of diversification," *Cladistics* 8:1–43.

35 "Edward O. Wilson, Peter Raven, and others described both the audacious mission . . ." Raven, P. H., and Wilson, E. O., 1992, "The fifty-year plan for biodiversity surveys," *Science* 258:1099–1100.

35 ". . . staggering counts of true diversity." Erwin, T. L., 1988, "The tropical forest canopy: the heart of biotic diversity," in Wilson, E. O. (ed.), *Biodiversity*, National Academy Press, Washington, D.C., pp. 123–29.

35 ". . . as many as 30 or maybe even 50 million species." Ibid.

35 "Current estimates . . ." Heywood, V. H., and Watson, R. T., 1995, *Global Biodiversity Assessment*, Cambridge University Press, Cambridge, UK.

35 "About 100,000 fungi have been described and named . . ." Harrison et al., "Species diversity."

36 "Four out of every five animals on Earth may be nematodes." Wilson, "The encyclopedia."

36 "Bacteria and . . . archeans . . ." Ibid.

36 "Wilson characterized as the black hole of systematic biology . . ." Ibid.

36 ". . . *Prochlorococcus*, possibly the most abundant organism on the planet . . ." Ibid.

36 ". . . six thousand species of bacteria are recognized . . ." Ibid.

37 "Venter and his coworkers estimated at least eighteen hundred different microbial species in . . . their samples." Venter, J. C., et al., 2004, "Environmental genome shotgun sequencing of the Sargasso Sea," *Science* 304:66–74.

37 ". . . seven thousand different species of bacteria . . . in a few grams of soil." Curtis, T. P., and Sloan, W. T., 2005, "Exploring microbial diversity—a vast below," *Science* 309:1331–33.

37 "In a paper published in 2005, Jason Gans and coauthors . . ." Gans, J., Wolinsky, M., and Dunbar, J., 2005, "Computational improvements reveal great bacterial diversity and high metal toxicity in soil," *Science* 309:1387–90.

38 "There are about six thousand taxonomists worldwide." Wilson, "The encyclopedia."

38 ". . . botanist Carl Peter Thunberg . . ." Thunberg, C. P., 1784, *Flora japonica*, Leipzig; 1794–1800, Prodromus plantarum Capensium, Uppsala. See also Ford, B. J., 2000, "Eighteenth century scientific publishing," in Hunter, A. (ed.), *Scientific Books, Libraries and Collectors*, Aldershot and Vermont.

38 "Lee Herman . . . published . . . a catalog of . . . rove beetles." Herman, L. H., 2001, "Catalogue of the Staphylinidae (Insecta, Coleoptera): 1758 to the end of the second millennium," *Bulletin of the American Museum of Natural History* 265:1–4218.

39 "Several websites are able to network the efforts of bioexplorers . . ." Edwards, M., and Morse, D. R., 1995, "The potential for computer-aided identification in biodiversity research," *Trends in Ecology and Evolution* 10:153–58. For some examples, see www.morphobank.org/; www.barcodinglife.org/views/login.php.

40 ". . . E. O. Wilson has promoted the idea of a new encyclopedia of life . . ." Wilson, "The encyclopedia."

3: EPHEMERAL LIFE

41 "The 4.6-billion-year history of Earth . . ." Most dates hover between 4.5 and 4.6 billion years with a recent formulation of 4.570 billion years. Dalrymple, G. B., 1991, *The Age of the Earth*, Stanford University Press, Stanford, CA. Also see United States Geological Survey (USGS) website on geologic time, http://pubs.usgs.gov/gip/geotime/age.html.

41 ". . . a calendar with designated time intervals." USGS website.

41 ". . . famous examples of endurance . . ." Eldredge, N., and Stanley, S. M. (eds.), 1984, *Living Fossils*, Springer Verlag, Berlin.

41–42 ". . . on average, species probably endure for about 1 or 2 million years." Stanley, S. M., 1979, *Macroevolution: Pattern and Process*, W. H. Freeman, San Francisco.

42 "In 1860 the paleontologist John Phillips . . ." Phillips, J., 1860, *Life on the Earth: Its Origin and Succession*. Macmillan and Co., Cambridge and London (rep. ed., 1980, Arno Press, New York).

43 ". . . geological calendar has been much refined." An engaging, highly readable account of the formulation of the Earth calendar is Berry, W.B.N., 1968, *Growth of a Prehistoric Time Scale*, W. H. Freeman, San Francisco.

43 ". . . calendar has been recalibrated in million-year dates that differ notably from those used by Phillips." Ibid.

43 ". . . tens of thousands of species every year may now be going extinct . . ." Wilson, *The Diversity of Life*; Lawton and May, *Extinction Rates*; Pimm et al., 1995, "The future of biodiversity."

43 ". . . one hundred to one thousand times the background rate." Ibid.

43 ". . . the Los Alamos team . . ." Gans et al., 2005, "Computational improvements."

44 ". . . degradation of soils worldwide . . ." McNeill, J. R., and Winiwarter, V., 2004, "Breaking the sod: humankind, history, and soil," *Science* 304:1627–29.

45 "Biodiversity is not distributed evenly . . ." Myers, N., 1988, "Threatened biotas: 'hot spots' in tropical forests," *Environmentalist* 8:187–208; Myers, N. et al., 2000, "Biodiversity hotspots for conservation priorities," *Nature* 403:853–58; Prendergast, J. R., et al., 1993, "Rare species, the coincidence of diversity hotspots and conservation strategies," *Nature* 365:335–37; Sala, O. E., et al., 2000, "Global biodiversity scenarios for the year 2100," *Science* 287:1770–74.

45 ". . . 'hotspots' . . ." Ibid.

45 "Some organisms have very narrow temperature tolerances . . ." Lovejoy and Hannah, *Climate Change and Biodiversity*.

45 ". . . frogs . . . are showing considerable decreases . . ." Stuart, S. N., et al., 2004, "Status and trends of amphibian declines and extinctions worldwide," *Science* 306:1783–86.

46 ". . . Edith's checkerspot butterfly . . ." Parmesan, C., 1996, "Climate and species' range," *Nature* 382:765–66.

46 ". . . 2,000 species of birds on many Pacific Islands went extinct . . ." Steadman, D. W., 1995, "Prehistoric extinctions of Pacific island birds: biodiversity meets zooarchaeology," *Science* 267:1123–31; Steadman, D. W., 2006, *Extinction and Biogeography of Tropical Pacific Birds*, University of Chicago Press, Chicago.

46 ". . . Pacific has 289, or 24 percent, of the world's globally threatened birds . . ." BirdLife International, 2000, *Threatened Birds of the World*, Lynx Edicions, Barcelona and Cambridge, UK, www.birdlife.org/regional/pacific/overview.html.

46 "... Hope is the thing with feathers..." Dickinson, E. , 1924, *The Complete Poems of Emily Dickinson,* Little, Brown, Boston.

46 "In 2006, of 40,177 species assessed..." 2006 IUCN Red List.

46 "... goitered gazelle..." Mallon, D., 2005, *"Gazella subgutturosa,"* ibid.

47 "... a distressing decline in big fish..." Morey, G., et al., 2006, *"Squatina squatina,"* ibid. See also Pauly et al., 1998, "Fishing down marine food webs"; Jackson et al., "Historical Overfishing and the Recent Collapse of Coastal Ecosystems."

47 "Freshwater ecosystems are especially vulnerable..." Dudgeon et al., "Freshwater biodiversity: importance, threats, status and conservation challenges."

48 "... freshwater ecosystems are in far worse shape..." 2006 IUCN Red List.

48 "The Mediterranean region..." Ibid. and Myers et al., "Biodiversity hotspots."

48 "Several plant species are facing extinction..." An example is the pallid squill, a handsome flower on Cyprus; see Kadis, C. and Christodoulou, C. S., 2006, *"Scilla morrisii,"* 2006 IUCN Red List.

48 "... study... on centers of imminent extinction..." Ricketts, T. H., et al. (twenty-nine coauthors), 2005, "Pinpointing and preventing imminent extinctions," *Proceedings of the National Academy of Sciences* 102:18497–501.

49 "... Fynbos biome..." Cowling, R. M., 1992, *The Ecology of the Fynbos: Nutrients, Fires and Diversity,* Oxford University Press, Cape Town.

49 "... biologists... used a computer simulation model..." Midgley, G. F., and Miller, D., 2005, "Modeling species range shifts in two biodiversity hotspots," in Lovejoy and Hannah, *Climate Change and Biodiversity,* pp. 229–31.

51 "J. R. Malcolm... used a computer model..." Malcolm J. R., et al., 2006, "Global warming and extinctions of endemic species from biodiversity hotspots," *Conservation Biology* 20:538–48.

51 "... butterflies on the British Isles..." Thomas, J. A., et al., 2004, "Comparative losses of British butterflies, birds, and plants and the global extinction crisis," *Science* 303:1879–81.

52 "... of the 564 dragonfly and damselfly species... 1 in 3 (174) are threatened..." Boudot, J.-P., 2005, *"Macromia splendens,"* 2006 IUCN Red List.

52 "... a single river dolphin..." Guo, J., 2006, "River dolphins down for the count and probably out," *Science* 314:1860.

4: ELEPHANTS, DUNG BEETLES, AND ECOSYSTEMS

53 "Although census data for the African elephant... from earlier decades are uneven..." A historical summary and the current status of the African elephant are in Blanc, J. J., et al., 2003, "African elephant status report 2002: an update from the African elephant database," IUCN/SSC, African Elephant Specialist Group, Gland, Switzerland, and Cambridge, UK. Also see African Elephant Specialist Group, 2004, *"Loxodonta africana,"* in 2006 IUCN Red List.

53 "... Convention on International Trade in Endangered Species (CITES)..." Ibid.

53 "... international ban on ivory..." Ibid.

53 "... resurgence of elephant populations in many parts of Africa." Ibid. Note that the African elephant is still listed as vulnerable in the IUCN 2006 Red List because of the pressures of human population growth, culling, and poaching.

54 ". . . unusual and very problematic pressure on the African elephant's food sources . . ." The data here are given in Sukumar, R., 2003, *The Living Elephants: Ecology, Behaviour, and Conservation*, Oxford University Press, New York. See also Owen-Smith, R. N., 1988, *Megaherbivores: The Influence of Very Large Body Size on Ecology*, Cambridge University Press, Cambridge, UK.

54 ". . . promoting energy flow in an ecosystem." A comprehensive treatment of ecosystems may be found in Cotgreave, P., and Forsyth, I., 2002, *Introductory Ecology*, Blackwell Publishing, Oxford, UK. College-level websites on this topic are plentiful; a good one is provided by Professor Dave McShaffrey at Marietta College: www.marietta.edu /~biol/102/ecosystem.html.

54 ". . . trophic interaction . . ." Ibid.

54 ". . . does not mean that energy is recovered." Ibid.

55 ". . . the sun . . . bombards Earth with about 175 peta watts . . ." Karl, T. R., and Trenbeth, K. E., 2005, "What is climate change?" in Lovejoy and Hannah, *Climate Change and Biodiversity*, pp. 15–28.

55 ". . . photosynthesizers . . . consumers . . . decomposers . . ." Cotgreave and Forsyth, *Introductory Ecology*, and www.marietta.edu/~biol/102/ecosystem.html.

55 ". . . food chain . . . food webs . . ." Ibid.

56 ". . . principle of efficiency when it comes to energy transfer." Ibid.

57 ". . . trophic roles of giant sauropod dinosaurs . . ." Taggart, R. E., and Cross, A. T., 1997, "The relationship between land plant diversity and productivity and patterns of dinosaur herbivory," *Dinofest, International Proceedings* 1:403–16.

57 ". . . carnivores are more specialized still." Nowak, R. M., 1991, *Walker's Mammals of the World*, vols. I and II, Johns Hopkins University Press, Baltimore.

57 "Predation . . ." Cotgreave and Forsyth, *Introductory Ecology*.

57–58 "The predaceous spider . . ." Hillyard, P.,1994, *The Book of the Spider*, Random House, New York.

58 ". . . Moche civilization . . ." Popson, C. P., 2002, "Grim rites of the Moche," *Archaeology* 55 (2), www.archaeology.org/0203/abstracts/moche.html; Bourget, S., 2006, *Sex, Death, and Sacrifice in Moche Religion and Visual Culture*, University of Texas Press, Austin.

58 ". . . parasite . . ." Matthews, B. E., 1998, *An Introduction to Parasitology*, Cambridge University Press, Cambridge, UK.

58 "Competition . . ." Cotgreave and Forsyth, *Introductory Ecology*.

58 "Fifty million years ago in North America and Eurasia . . ." Janis, C. M., 1993,"Tertiary mammal evolution in the context of changing climates, vegetation, and tectonic events," *Annual Reviews of Ecology and Systematics* 24:467–500.

59 "An ecosystem is a unit of nature . . ." The relationship between biodiversity and ecosystems is described in Purvis, A., and Hector, A., 2000, "Getting a measure of biodiversity," *Nature* 405:212–19; Gaston, K. J., 2000, "Global patterns in biodiversity," *Nature* 405:220–27.

59 "But an ecosystem is not simply the sum of all its species parts." Purvis and Hector, 2000.

59–60 "But how stable are ecosystems in the real world?" An excellent review article on this issue is McCann, K. S., 2000, "The diversity-stability debate," *Nature* 405:228–33.

60 "But Robert May in 1973 published provocative research . . ." May, R. M., 1973, *Stability and Complexity in Modern Ecosystems*, Princeton University Press, Princeton, NJ.

60 "I eagerly devoured May's book . . ." A paper based on my graduate research that I authored is highly influenced by May's theories: Lawson, D. A., and Novacek, M. J., 1981, "Structure and change in three Eocene invertebrate (primarily molluscan) communities from nearshore marine environments," in Boucot, A., and Berry, W.B.N. (eds.), *Communities of the Past: Proceedings of the Symposium, Paleontology Convention of North America II*, Dowden, Hutchinson, and Ross, Stroudsburg, PA.

60 ". . . many weak interactions mute . . . a few strong producer-consumer interactions." MacArthur, R. H., 1955, "Fluctuations of animal populations and a measure of community stability," *Ecology* 36:533–36.

60 ". . . drastic changes . . . can occur with even the removal or addition of a single species." See also many papers cited in McCann, "The diversity-stability debate."

60–61 "In a classic study R. T. Paine experimentally removed . . ." Paine, R. T., 1984, "Ecological determinism in the competition for space," *Ecology* 65:1339–48.

61 "Ancient Egyptians took to this image; they revered dung beetles . . ." Andrews, C., 1994, *Amulets of Ancient Egypt*, University of Texas Press, Austin.

61 "Dung beetles . . ." My description of the fascinating biology of this group is drawn from Halffter, G., and Matthews, E. G., 1966, *The Natural History of Dung Beetles: Of the Subfamily Scarabaeinae*, Folia Entomologica Mexicana Xalapa, Veracruz, México; Fincher, G. T., 1981, "The potential value of dung beetles in pasture ecosystems," *Journal of the Georgia Entomological Society* 16:301–16; Fincher, G. T., and Morgan, P. B., 1990, "Flies affecting livestock and poultry," in Habeck et al. (eds.), *Classical Biological Control in the Southern United States*, Southern Cooperative Series Bulletin No. 355, p.152; Hanski, I., and Cambefort, Y., 1991, *Dung Beetle Ecology*, Princeton University Press, Princeton, NJ; Thomas, M., 2001, "Dung beetle benefits in the pasture ecosystem," ATTRA, http://attra.ncat.org/attra-pub/dungbeetle.html.

61 "The average cow releases . . . about three hundred liters of methane a day." Gillespie, A., 2003, "The problem of methane in international environmental law," *Review of European Community & International Environmental Law* 12(3): 321.

62 ". . . dung slurpie . . ." Thomas, M., 2001, "Dung beetle benefits."

64 "The impressive power of the industrious dung beetles . . ." Thomas, "Dung beetle benefits"; Losey, J. E., and Vaughan, M., 2006, "The economic value of ecological services provided by insects," *BioScience* 56:311–23, www.biosciencemag.org.

64 "Nitrogen. . . to be 'fixed' . . ." Cotgreave and Forsyth, *Introductory Ecology* and www.marietta.edu/~biol/102/ecosystem.html.

65 ". . . United States alone averts $380 million in annual losses . . ." Losey and Vaughan, "The economic value."

65 ". . . red tide . . ." Halwell, B., 2006, "Nitrogen: too much of a good thing?" *World Watch* 19(1): 7; Van Dolah, F. M., 2000, "Marine algal toxins: origins, health effects, and their increased occurrence," *Environmental Health Perspectives* 108(suppl.1): 133–41; Sellner, K. G., Doucette, G. J., and Kirkpatrick, G. J., 2003, "Harmful algal blooms: causes, impacts and detection," *Journal of Industrial Microbiology and Biotechnology* 30:383–406.

65 "*Plankton* is actually a grab bag term . . ." Australian Museum, 2002, *Beyond the Reef: What Is Plankton?*, www.amonline.net.au/exhibitions/beyond/index.htm.

66 ". . . to the nitrogen cycle . . ." Cotgreave and Forsyth, *Introductory Ecology*, and www.marietta.edu/~biol/102/ecosystem.html.

67 ". . . denitrification." Ibid.

67 "... Fritz Haber ... devised a scheme to short-circuit the nitrogen cycle ..." Stoltzenberg, D., 2005, *Fritz Haber: Chemist, Nobel Laureate, German, Jew: A Biography*, Chemical Heritage Foundation, Philadelphia.

68 "This nutrient overload is ... more serious in the atmosphere and on land." Vitousek, P. M., et al., 1997, "Human alteration of the global nitrogen cycle: causes and consequences," *Issues in Ecology* 1:1–17.

68 " In northern California, diverse plants ... are being crowded out ..." Tilman, D., and Lehman, C., 2001, "Human-caused environmental change: impacts on plant diversity and evolution," *Proceedings of the National Academy of Sciences* 98:5433–40.

69 "... organisms require other elements besides nitrogen ..." Cotgreave and Forsyth, *Introductory Ecology*, and www.marietta.edu/~biol/102/ecosystem.html.

69 "All these items have their own cycles." Ibid.

69 "... carbon ... carbon cycle ..." Ibid.

71 "... our own species is making major contributions ..." Ibid. and Lovejoy and Hannah, *Climate Change and Biodiversity*.

71 "Charles Keeling ..." Keeling, C. D., 1960, "The concentration and isotopic abundance of carbon dioxide in the atmosphere," *Tellus* 12:200–203; Keeling, C. D., Chin, J.F.S., and Whorf, T. P., 1996, "Increased activity of northern vegetation inferred from atmospheric CO_2 measurements," *Nature* 382:146–49; Keeling, C. D., and Whorf, T. P., 2005, "Atmospheric CO_2 records from sites in the SIO air sampling network," in *Trends: A Compendium of Data on Global Change. Carbon Dioxide.* Information Analysis Center, Oak Ridge National Laboratory, U.S. Department of Energy, Oak Ridge, TN.

72 "Using historical records and measurements in ice cores ..." Etheridge, D. M., et al., 1996, "Natural and anthropogenic changes in atmospheric CO_2 over the last 1000 years from air in Antarctic ice and firn," *Journal of Geophysical Research* 101:4115–28; Smith, H. J., Wahlen, M., and Mastroianni, D., 1997, "The CO_2 concentration of air trapped in GISP2 ice from the Last Glacial Maximum-Holocene transition," *Geophysical Research Letters* 24 (1): 1–4; Overpeck, Cole, and Bartlein, "A 'paleoperspective' on climate variability and change"; Kennedy and Hanson, "Ice and history."

72–73 "... greenhouse gases ..." Kiehl, J. T., and Trenberth, K. E., 1997, "Earth's annual global mean energy budget," *Bulletin of the American Meteorological Society* 78:197–208.

73 "... the increase in Earth's average atmospheric temperature." Houghton, J. T., et al. (eds.), 2001, *Climate Change 2001: The Scientific Basis Contribution of Working Group I to the Third Assessment Report of the Intergovernmental Panel on Climate Change (IPCC)*, Cambridge University Press, Cambridge, UK.

73 "Sea-level ... would flood many coastal cities ..." Kennedy and Hanson, "Ice and history."

73 "... other phases of Earth's history ..." Overpeck et al. "A 'paleoperspective.'"

74 "When photosynthesizing bacteria emerged ... some 2 billion years ago ..." Gould, S. J. (ed.), 1993, *The Book of Life*; Stanley, S. M., 1999, *Earth System History*, W. H. Freeman, New York.

5: EVOLUTION—LIFE THROUGH A NEW LENS

75 "... new theories ... must be testable ..." This explicit requirement for science, one that I and many other scientists accept, largely stems from the work of the philosopher Karl Popper. Popper, K., 1934, *Logik der Forschung*, Springer, Vienna. An expanded English version was published in 1959 as *The Logic of Scientific Discovery*.

75 "... intelligent design ..." Two very well-reasoned statements against the notion that intelligent design "theory" is science are National Academy of Sciences, 1999, *Science and Creationism: A View from the National Academy of Sciences*, 2nd ed., and John E. Jones III, December 20, 2005, Ruling, Kitzmiller v. Dover Area School District, Case No. 04cv2688. 400 F.Supp.2d 707 (M.D. Pa. 2005).

76 "... Hurricane Katrina in 2005 was one of history's most avoidable disasters." Fischetti, M., 2001, "Drowning New Orleans." *Scientific American*, October 2001; Warrick, J., and Grunwald, M., "Investigators link levee failures to design flaws." *Washington Post*, October 24, 2005.

76 "... the theory of evolution ..." Among the best of the current texts are Futuyma, D. J., 1998, *Evolutionary Biology*, 3rd ed., Sinauer Associates, Sunderland, MA; Futuyma, D. J., 2005, *Evolution*, Sinauer Associates, Sunderland, MA. For a more general treatment, see Carl Zimmer, 2001, *Evolution: The Triumph of an Idea*, HarperCollins, London.

76 "... Thomas Robert Malthus ..." Malthus, T. R., 1798, "*An Essay on the Principle of Population, as It Affects the Future Improvement of Society with Remarks on the Speculations of Mr. Godwin, M. Condorcet, and Other Writers*," J. Johnson, London.

76–77 "... Darwin ... Wallace ..." A history of the convergent discoveries of Darwin and Wallace is in Boorstin, *The Discoverers*. Also see *The C. Warren Irvin, Jr., Collection*, www.sc.edu/library/spcoll/nathist/darwin/darwin.html; the exhibit *Darwin*, www.amnh.org/exhibitions/darwin/; *Natural History*, November 2005, 114 (9); Desmond and Moore, *Darwin*.

77 "... what some have called their 'dangerous idea' ..." Dennett, D., 1995, *Darwin's Dangerous Idea*, Simon & Schuster, New York.

77 "... Comte de Buffon ... Jean-Baptiste Lamarck ... Étienne Geoffroy Saint-Hilaire ... Charles Lyell ..." A classic account of the development of evolutionary thought is Greene, J., 1959, *The Death of Adam*, University of Iowa Press, Ames. Also excellent is Sloan, P., 2005, "Evolution," in Zalta, E. N. (ed.), *The Stanford Encyclopedia of Philosophy*, http://plato.stanford.edu/archives/sum2005/entries/evolution/.

77 "The word *evolution* was first used by ... Saint-Hilaire ..." Saint Hilaire, É. G., no date, "Lectures on zoology (evolution)," Étienne Geoffroy Saint Hilaire Collection, American Philosophical Society, File Box 2, pp. 1–12. His major published work is Saint-Hilaire, É. G., 1818, *Philosophie anatomique*, Paris.

77 "Charles Lyell ... claimed that species had appeared and disappeared ..." Lyell, C., 1830, *Principles of Geology*, vols. 1–3, John Murray, London; electronic version, www.esp.org/books/lyell/principles/facsimile/title3.html.

77–78 "... Erasmus Darwin ..." Darwin, E., 1794–1796, *Zoonomia, or, The Laws of Organic Life*, 2 vols., printed for J. Johnson, London.

78 "His *Beagle* notebooks ..." *Darwin*, www.amnh.org/exhibitions/darwin/; Eldredge, N., 2005, *Darwin: Discovering the Tree of Life*, W. W. Norton, New York.

78 "He read 'for amusement' ..." Darwin, www.amnh.org/exhibitions/darwin.

78 "Darwin . . . kept them a secret for sixteen more years." Ibid.

79 "On July 1, 1858, three papers, two by Darwin and one by Wallace . . ." Ibid., and Boorstin, *The Discoverers*.

79 "Thomas Bell . . . declared with little prescience . . ." Ibid.

79 "Darwin had published his . . . book . . ." Darwin, C. R., 1859, *On the Origin of Species by Means of Natural Selection, or the Preservation of Favoured Races in the Struggle for Life*, 1st ed., John Murray, London. The first edition is i–x, 1–502 pages; later editions had added sections.

79 "It became a huge seller . . ." Boorstin, *The Discoverers*.

79 "His fame was soon hugely eclipsed by Darwin's . . ." Ibid.

80 "An Austrian monk, Gregor Mendel . . ." Mendel, G. "Versuche über Pflanzen-Hybriden," *Verhandlungen des naturforschenden Vereines, Abhandlungen, Brünn* 4, pp. 3–47 (1866). (Translation:."Experiments on plant hybridization," *Proceedings of the Natural History Society of Brunn*.) For a biography, see R. M., Henig, 2000, *Monk in the Garden: The Lost and Found Genius of Gregor Mendel, the Father of Genetics*, Houghton Mifflin, New York.

80 ". . . de Vries . . ." Vries, H. de, 1889, 1910, *Intracellular Pangenesis*, trans. C. S. Gager, Open Court, Chicago; Darden, L., 1976, "Reasoning in Scientific Change: Charles Darwin, Hugo de Vries, and the Discovery of Segregation," *Studies in History and Philosophy of Science* 7:127–69.

80 ". . . culminated in the late 1950s with the disclosure of . . . DNA." Watson, J. D., and Crick, F.H.C., 1953, "A structure for deoxyribose nucleic acid," *Nature* 171:737–38; Watson, J. D., *The Double Helix: A Personal Account of the Discovery of the Structure of DNA*, Norton Critical Edition, New York; Judson, H. F., 1996, *The Eighth Day of Creation: Makers of the Revolution in Biology*, Cold Spring Harbor Laboratory Press, New York; Watson, J. D., et al., 2003, *Molecular Biology of the Gene*, 5th ed., Benjamin Cummings, New York.

80 "As this century begins we have the complete map of all 30,000 genes . . ." Venter, J., et al., 2001,"The sequence of the human genome," *Science* 291:1304–51; Wolfsberg, T., McEntyre, J., and Schuler, G., 2001, "Guide to the draft human genome," *Nature* 409:824–26; Ridley, M., 1999, *Genome: The Autobiography of a Species in 23 Chapters*, Fourth Estate, London; Human Genome Project Information Website, www.ornl.gov /sci/techresources/Human_Genome/home.shtml.

80 ". . . Jenny the orangutan . . ." www.amnh.org/exhibitions/darwin.

81 ". . . 3 billion DNA nucleotides, of which 96 percent are exactly alike." The publication of the chimp genome in 2005 reduced the level of similarity from the 99 percent figure common in the literature. Chimpanzee Sequencing and Analysis Consortium, 2005, "Initial sequence of the chimpanzee genome and comparison with the human genome," *Nature* 437:69–87; Cheng, Z., Ventura, M., et al., 2005, "A genome-wide comparison of recent chimpanzee and human segmental duplications," *Nature* 437:88–93. An excellent nontechnical summary is Powell, A., 2005, "Chimp genome effort sheds light on human evolution," *Harvard University Gazette*, www.hno .harvard.edu/gazette/2005/09.15/11-chimp.html/

81 ". . . Darwin drew sticklike trees with branches . . ." The exhibit *Darwin* website, www.amnh.org/exhibitions/darwin.

81 "Konrad von Gesner's monumental . . ." Gesner, K., 1551–58, *Historia animalium libri I–IV. Cum iconibus. Lib. I. De quadrupedibus viviparis*, C. Froschauer, Zurich.

81 "... hodag ..." www.museumofhoaxes.com/photos/hodag.html.

82 "... the pycnogonids, or sea spiders ..." Dunlop, J. A., and Arango, C. P., 2005, "Pycnogonid affinities: a review," *Journal of Zoological Systematics and Evolutionary Research* 43:8–21.

82 "At some basic level all life-forms are alike." The concept of basic similarity and relation to features shared by descent (homology) is the subject of many works. Early notions were presented by Saint-Hilaire, *Philosophie anatomique*; see also Patterson, C., 1982, "Morphological characters and homology," in Joysey, K. A., and Friday, A. E. (eds.), *Problems in Phylogenetic Reconstruction*, Academic Press, London. A recent treatment that incorporates genomics, evolution, and development is Caroll, S. B., 2005, *Endless Forms Most Beautiful: The New Science of Evo Devo and the Making of the Animal Kingdom*, W. W. Norton, New York.

83 "This branching structure is a phylogeny ..." Hennig, W., 1966, *Phylogenetic Systematics*; Eldredge and Cracraft, *Phylogenetic Patterns and the Evolutionary Process*; Nelson and Platnick, *Systematics and the Biogeography—Cladistics and Vicariance*; Novacek and Wheeler, *Extinction and Phylogeny*; Gaffney, G., Dingus, L., and Smith, M., 1995, "Why cladistics?" *Natural History* 104 (6): 33–35.

83 "... the tree of life." Ibid.

83–84 "... cladistics." Cracraft, J., and Donoghue, M. J., 2004, *Assembling the Tree of Life*, Oxford University Press, New York.

84 "... branching of the tree of life at several levels ..." Ibid.

84 "... DNA ..." Watson et al., *Molecular Biology of the Gene*; Clayton, J. (ed.), 2003, *50 Years of DNA*, Macmillan, Palgrave, UK. For an excellent website, see Cold Spring Harbor Laboratory, www.dnai.org/.

86 "...information in the average genome ..." Gregory, T. R. (ed.), 2005, *The Evolution of the Genome*, Elsevier, Amsterdam; Saccone, C., and Pesole, G., 2003, *Handbook of Comparative Genomics*, John Wiley & Sons, New York; Sea Urchin Genome Sequencing Consortium, 2006, "The genome of the sea urchin *Strongylocentrotus purpuratus*," *Science* 314:941–52, also updated websites: www.nslij-genetics.org/seq/; www.genome newsnetwork.org/resources/sequenced_genomes/genome_guide_p3.shtml.

87 "... fossil record ... is in some critical cases good enough ..." Novacek, M. J., 1992, "Fossils, topologies, missing data, and the higher level phylogeny of eutherian mammals," *Systematic Biology* 41:58–73; Gauthier, J., Kluge, A. G., and Rowe, T., 1988, "Amniote phylogeny and the importance of fossils," *Cladistics* 4:105–209; Donoghue, M., et al., 1989, "The importance of fossils in phylogeny reconstruction," *Annual Reviews of Ecology and Systematics* 20:431–60; Novacek, M. J., and Wheeler, Q. C. (eds.), 1992, *Extinction and Phylogeny*, Columbia University Press, New York.

89 "... influenza type A virus ..." The Writing Committee of the World Health Organization (WHO), 2005, "Consultation on Human Influenza A/H5," *New England Journal of Medicine* 353:1374–85. See also Webster, R. G., and Walker, E. J., 2003, "The world is teetering on the edge of a pandemic that could kill a large fraction of the human population," *American Scientist* 91 (2): 122.

90 "... 318 reported cases, including 192 human deaths." World Health Organization, Epidemic and Pandemic Alert and Response, www.who.int/csr/disease/avian _influenza/country/cases_table_2007_07_11/en/index.html.

90 "... mutations are ... *deleterious* ..." Futuyma, *Evolutionary Biology*.

91 "... special exhibit on Darwin ..." www.amnh.org/exhibitions/darwin.

91 "... roughly half the American public rejects the theory ..." CBS News Poll, October 23, 2005, "Majority Reject Evolution"; www.cbsnews.com/stories/2005/10/22/opinion/polls/main965223.shtml.

91 "... Judge John E. Jones ruled ..." Kitzmiller v. Dover Area School District, Case No. 04cv2688. 400 F.Supp.2d 707 (M.D. Pa. 2005).

92 "Degraded habitats seem to select for novel organisms ..." Myers, N., and Knoll, A. H., 2001, "The biotic crisis and the future of evolution," *Proceedings of the National Academy of Sciences* 98:5389–92.

6: ANCIENT GROUND

95 "Let us replay time at fast forward." The highlights in this passage are covered in many texts, among them Gould, S. J., *The Book of Life*; Dalrymple, *The Age of the Earth*; Fortey, R., 1997, *Life: An Unauthorized Biography*, HarperCollins, London; Gee, H., 1999, *In Search of Deep Time: Beyond the Fossil Record to a New History of Life*, Free Press, Simon & Schuster, New York; Briggs and Crowther, *Palaeobiology II*; Eldredge, N. (ed.), 2002, *Life on Earth: An Encyclopedia of Biodiversity, Ecology, and Evolution*, vols. 1 and 2, American Museum of Natural History and ABC-CLIO Inc., Santa Barbara; Fortey, R., 2004, *Earth: An Intimate History*, Knopf, New York.

96 "... terrestrial life—is ... only about 475 million years old." Edwards, E., 2001, "Early land plants," in Briggs and Crowther, *Palaeobiology II*.

96 "Geologists have formally named these subdivisions ..." Palmer, A. R., and Geissman, J., 1999, *The Geological Society of America (GSA) Geologic Time Scale*, www.geosociety.org/science/timescale/timescl.htm; International Commission on Stratigraphy (ICS), 2005, International Stratigraphic Chart, International Union of Geological Sciences, www.stratigraphy.org/chus.pdf.

96 "... rocks and a smattering of microscopic fossils so ancient ..." Fortey, *Life*; Fortey, *Earth*.

96 "These are the eons when oceans first formed, when life was born." Ibid.

96 "... Phanerozoic Eon ..." Ibid.

97 "... mass extinction events first recognized by John Phillips ..." Phillips, *Life on the Earth: Its Origin and Succession*; Taylor, P. D. (ed.), 2004, *Extinctions in the History of Life*, Cambridge University Press, Cambridge, UK.

97 "... Paleozoic ... Permian ... Mesozoic ... Cretaceous ... Cenozoic ..." The origins of the names and the historical development of the timescale are described in Berry, *Growth of a Prehistoric Time Scale*.

98 "... radioactivity in rocks." Faure, G., 1986, *Principles of Isotope Geology*, 2nd ed., John Wiley and Sons, New York. See also Benton, M., 1993, "Life and time," in Gould, *The Book of Life*, pp. 22–37; Novacek, *Dinosaurs of the Flaming Cliffs*.

98 "Carbon 14 ..." The classic paper on the subject is Libby, W. F., 1961, "Radiocarbon dating," *Science* 133:621–29.

98–99 "... paleomagnetism ... paleomagnetics ..." Cox, A. (ed.), 1973, *Plate Tectonics and Geomagnetic Reversals*, W. H. Freeman, San Francisco; Seki, M., and Ito, K., 1993, "A phase-transition model for geomagnetic polarity reversals," *Journal of Geomagnetism and Geoelectricity* 45:79–88; Cande, S. C., and Kent, D. V., 1995, "Revised calibration of the geomagnetic polarity time scale for the Late Cretaceous and Cenozoic," *Journal of Geophysical Research* 100:6093–95; Mathez, 2001, *Earth Inside and Out*.

99 ". . . magnetic lines of force around Earth . . ." Mathez, *Earth Inside and Out.*

99 ". . . Earth's magnetic field has done flip-flops . . ." Ibid.

99 ". . . shifts in the currents of molten material . . ." Ibid.

100 ". . . the field reverses about once every 1 to 5 million years . . ." Ibid.

100 ". . . plate tectonic theory . . ." The revolution in Earth sciences is the subject of a massive literature. Good general references are Miller, R., 1983, *Continents in Collision*, Time-Life Books, Alexandria, VA; and Mathez, *Earth*. Classic volumes containing a good sampling of some of the original technical and semitechnical articles are Cloud, P. (ed.), 1970, *Adventures in Earth History*, W. H. Freeman, San Francisco, and Cox, A. (ed.), 1973, *Plate Tectonics and Geomagnetic Reversals*, W. H. Freeman, San Francisco. A well-written modern treatment is in Fortey, *Earth: An Intimate History.*

101 "The first life on land . . ." Edwards, "Early land plants."

102 ". . . the interplay of marine and terrestrial realms . . ." Visbeck, M., 2001, "The ocean's role in climate," in Mathez, *Earth: Inside and Out*, pp. 140–45.

102 "In cores drilled in 475-million-year-old rocks . . ." Edwards, "Early land plants."

102 ". . . virtually microscopic objects . . ." Ibid.

102 ". . . common to all life . . . is water . . ." Martin, W., and Russell, M. J., 2002, "On the origins of cells: a hypothesis for the evolutionary transitions from abiotic geochemistry to chemoautotrophic prokaryotes, and from prokaryotes to nucleated cells," *Philosophical Transactions of the Royal Society B: Biological Sciences* 358:59–85; Hazen, R. M., 2005, *Genesis: The Scientific Quest for Life's Origins*, Joseph Henry Press, Washington, D.C.

102 ". . . encountered evidence of the presence of water in the rocks on Mars." Greeley, R., 1987, "Release of juvenile water on Mars: estimated amounts and timing associated with volcanism," *Science* 236:1653–54.

102 ". . . liverworts . . ." Kenrick, P., and Crane, P. R., 1997, *The Origin and Early Diversification of Land Plants: A Cladistic Study*, Smithsonian Institution Press, Washington, D.C.; Crandall-Stotler, B., and Stotler, R. E., 2000, "Morphology and classification of the Marchantiophyta," in Shaw, A. J., and Goffinet, B. (eds.), *Bryophyte Biology*, Cambridge University Press, Cambridge, UK, pp. 21–70.

103 ". . . just some small bifurcating structures . . ." Gray, J., 1985, "The microfossil record of early land plants; advances in understanding of early terrestrialization, 1970–1984," *Philosophical Transactions of the Royal Society of London B: Biological Sciences* 309:167–95; Edwards, "Early land plants"; Scheckler, S. E., 2001, "Afforestation—the First Forests," in Briggs and Crowther, *Palaeobiology II*, pp. 67–71.

103 "Club mosses . . ." Ibid.

103 ". . . plants came to look like plants . . ." Ibid.

103 ". . . Rhynie Chert . . ." Taylor, T. N., Remy, W., and Hass, H., 1994, "Allomyces in the Devonian," *Nature* 367:601. Taylor, T. N., and Taylor, E. L., 2000, "The Rhynie chert ecosystem: a model for understanding fungal interactions," in Bacon, C. W., and White, J. F. Jr. (eds.), *Microbial Endophytes*, Marcel Dekker, New York, pp. 31–47; Kerp, H., 2002. "The Rhynie Chert—the oldest and most completely preserved terrestrial ecosystem," in Dernbach, U., and Tidwell, D. W. (eds.), *Secrets of Petrified Plants*, D'Oro-Verlag, Heppenheim Germany, pp. 22–27.

104 ". . . Rhynie Chert itself has for nearly a century yielded only some tiny springtails and mites . . ." Ibid. and Tillyard, R. J., 1928, "Some remarks on the Devonian fossil insects from the Rhynie chert beds, Old Red Sandstone," *Transactions of the Entomological So-*

ciety of London 76:65–71; Whalley, P., and Jarzembowski, E. A., 1981, "A new assessment of *Rhyniella*, the earliest known insect, from the Devonian of Rhynie, Scotland," *Nature* 291:317.

104 ". . . the discovery of the oldest-known insect . . ." Engel, M. S., and Grimaldi, D. A., 2004, "New light shed on the oldest insect," *Nature* 427:627–30.

104 "This two-jointed jaw . . ." Ibid.

105 ". . . an emergence of advanced flying insects . . ." Ibid.

105 "The known vertebrates today comprise only about forty-five thousand species . . ." Tudge, C., 2000, *The Variety of Life*, Oxford University Press, Oxford, UK.

106 "Vertebrates go back nearly 500 million years . . ." Carroll, R. L., 1988, *Vertebrate Paleontology and Evolution*, W. H. Freeman, New York.

106 ". . . the first tetrapods . . ." Ibid. and a review of recent discoveries by Clack, J. A., 2006, "From fins to limbs," *Natural History*, July–August 2006:36–41.

106–107 "*Ichthyostega . . . Eusthenopteran . . Acanthostega . . .*" Clack, "From fins to limbs."

106–107 ". . . *Tiktaalik* . . ." Shubin, N. H., Daeschler, E. B., and Jenkins, F. A., 2006, "The pectoral fin of *Tiktaalik roseae* and the origin of the tetrapod limb," *Nature* 440:764–71; Daeschler, E. B., Shubin, N. H., and Jenkins, F. A., 2006, "A Devonian tetrapod-like fish and the evolution of the tetrapod body plan," *Nature* 440:757–63.

109 "'. . . *Tiktaalik* is a true intermediate form . . .'" Clack, "From fins to limbs," p. 38.

109 "John Noble Wilford, a journalist at *The New York Times* . . ." "Fossil called missing link from sea to land animals," *New York Times*, April 6, 2006, http://select .nytimes.com/.

110 "He would have loved the fishapod." Novacek, M. J., 2006, "Darwin would have loved it: what his theory predicted and why it matters," *Time*, April 17, 2006.

110 "That ecosystem had changed dramatically over millions of years, and it would change again." Gould, *The Book of Life*; Fortey, *Life*.

111 ". . . Earth fostered a great hothouse of plants . . ." Fortey, *Life*.

111 "The decay of these swamps . . ." Ibid.

111 "These species were likely under stress . . ." Shubin et al., "The pectoral fin of *Tiktaalik rosea*"; Daeschler et al., "A Devonian tetrapod-like fish."

111 "The arctic fox . . . Fynbos plants . . ." Lovejoy and Hannah, *Climate Change and Biodiversity*.

7: IMPERIAL COLLAPSE

112 ". . . coal country . . ." International Energy Annual 2004 World Estimated Recoverable Coal, May–July, 2006, www.eia.doe.gov/emeu/iea/res.html; *Key World Energy Statistics*, International Energy Agency, 2006 edition, www.iea.org/.

112 ". . . 1 billion tons . . ." Ibid.

112 ". . . coal crown has passed to China . . ." Ibid.

112 ". . . enough coal in the world to last about three hundred years . . ." Ibid.

113 "The products of carbonized extinct life . . ." DiMichele, W. A., Pfefferkorn, H. W., and Gastaldo, R. A., 2001, "Response of Late Carboniferous and Early Permian plant communities to climate change," *Annual Review of Earth and Planetary Sciences* 29:461–87; DiMichele, W. A., 2001, "Carboniferous coal swamp forests," in Briggs and Crowther, *Palaeobiology II*, pp. 79–82.

113 "... the seed ferns ..." Ibid. and Scheckler, S. E., 2001, "Afforestation—the first forests," in Briggs and Crowther, *Palaeobiology II*, pp. 67–71.

113 "... gymnosperms ... angiosperms ..." For clear characterizations, see Tudge, *The Variety of Life*; Ingrouille, M., and Eddie, B., 2006, *Plants: Diversity and Evolution*, Cambridge University Press, Cambridge, UK.

114 "... life cycles ... sporophyte ... gametophyte ..." Ibid.

114 "The coal forests took root ..." DiMichele, "Carboniferous coal swamp forests"; DiMichele et al., "Response of Late Carboniferous and Early Permian plant communities"; DiMichele, W. A., and Phillips, T. L., 2002, "The ecology of Paleozoic ferns," *Review of Palaeobotany and Palynology* 119:143–59.

114 "... Pangea ..." Fortey, *Earth: An Intimate History*.

114 "... calamites ..." DiMichele, "Carboniferous coal swamp."

114–15 "... the Carboniferous Period ..." Palmer, A. R., and Geissman, J., 1999, GSA *Geologic Time Scale*, ICS, 2005, International Stratigraphic Chart; Berry, *Growth of a Prehistoric Time Scale*.

115 "At the beginning of the Carboniferous, atmospheric CO_2 levels were very high ..." Data for the temperature profile of this interval are from Berner, R. A., and Kothavala, Z., 2001, "Geocarb III: a revised model of atmospheric CO_2 over Phanerozoic time," *American Journal of Science* 301:182–204; Scotese, C. R., 2002, Paleomap Project, www.scotese.com/climate.htm.

115 "Some have suggested that it was due to the very proliferation of land plants ..." Crowley, T. J., and Berner, R. A., 2001, "Enhanced: CO_2 and climate change," *Science* 292:870–72.

115 "... Pangean megacontinent drifted into the hot zone ..." Scotese, Paleomap Project.

115 "... the concentration of atmospheric gases ..." Analysis is complex. A technical treatment is given in Berner and Kothavala, "Geocarb III." Derivation of oxygen curves from measuring carbon isotopes is explained in Falkowski et al., "The rise of oxygen over the past 205 million years and the evolution of large placental mammals."

116–17 "... isotopes are important indicators ..." Falkowski et al., "The rise of oxygen."

117 "Climates ... have been controlled by a multitude of internal interactive components ..." Karl, T. R., and Trenberth, K. E., 2005, "What is climate change?" in Lovejoy and Hannah, *Climate Change*, pp. 15–28.

117 "Climates of the past can be reconstructed ..." Overpeck et al., "A 'paleoperspective,'" ibid., pp. 91–108.

118 "... air pollution ... during Roman times ..." Hong, S., et al., 1996, "History of ancient copper smelting recorded in Roman and medieval times recorded in Greenland ice," *Science* 272:246–49; Nriagu, J. O., 1996, "History of global metal pollution," *Science* 272:223.

118 "... warm, sultry, oxygen-rich climate ..." Berner, R. A., and Canfield, D. E., 1989, "A new model for atmospheric oxygen over Phanerozoic time," *American Journal of Science* 294:56–91; Berner, R. A., and Kothavala, Z., 2001, "Geocarb III"; Scotese, 2002, Paleomap Project.

118 "... abode of a diversity of tetrapods ..." Carroll, R. L., 1988. *Vertebrate Paleontology and Evolution*, W. H. Freeman, New York.

118 "Forests ... teemed with ... giant dragonflies ..." Briggs, D. E. G., 1985, "Gigantism in Palaeozoic arthropods," *Special Papers in Palaeontology* 33:157; Dudley, R., 1998, "Atmospheric oxygen, giant Paleozoic insects and the evolution of aerial locomotor

performance," *Journal of Experimental Biology* 201:1043–50; Fountain, H., 2004, "When giants had wings and six legs," *The New York Times*, February 3, 2004, http://select.nytimes.com/.

118 ". . . *Meganeuropsis* . . ." Fountain, "When giants had wings and six legs."

118 "What accounts for all this gigantism?" Ibid.

119 "But are these explanations . . . even necessary?" Ibid.

120 "Colder, more seasonal climates . . ." DiMichele, "Carboniferous coal swamp."

120 ". . . new organisms decisively replaced the archaic ones." Ibid.

120 "For the first time beetles . . . attaining a staggering . . . diversity." Labandeira, C., and Sepkoski, J. J., Jr., 1993, "Insect diversity in the fossil record," *Science*, 261:310–15. A monumental work on insect paleontology and history is Grimaldi, D., and Engel, M. S., 2005, *The Evolution of Insects*, Cambridge University Press, Cambridge, UK.

120 ". . . the amniotes." More detailed information on the phylogeny of the amniotes is provided in Gauthier, J., Kluge, A. G., and Rowe, T., 1988, "Amniote phylogeny and the importance of fossils," *Cladistics* 4:105–209; Carroll, *Vertebrate Paleontology*; Novacek, *Dinosaurs of the Flaming Cliffs*; Dingus and Rowe, *The Mistaken Extinction*; Hopson, J. A., 2001, "Origin of mammals," in Briggs and Crowther, *Palaeobiology II*, pp. 88–94.

120–21 ". . . *Dimetrodon* . . . *Edaphosaurus* . . . synapsids . . ." Carroll, *Vertebrate Paleontology*.

121 ". . . the greatest mass extinction event." Erwin, D. H., 1993, *The Great Paleozoic Crisis*, Columbia University Press, New York. Also see Gould, *The Book of Life*; Schopf, *Major Events in the History of Life*; Erwin, "Lessons from the past."

121 ". . . 85 to 90 percent of all Permian marine species did not live . . ." Erwin, *The Great Paleozoic Crisis*.

121 "The record for land plants also shows marked devastation . . ." Ibid.

122 ". . . various views of experts . . ." Bowring, S. A., Erwin, D. H., and Isozaki, Y., 1999, "The tempo of mass extinction and recovery: the end-Permian example," *Proceedings of the National Academy of Sciences* 96 (16): 8827–28.

123 ". . . the extinction event could have taken longer than a hundred thousand years." Jin, Y. G., et al., 2000, "Pattern of marine mass extinction near the Permian-Triassic Boundary in South China," *Science* 289:432–36.

123 ". . . crater discovered under western Australia . . ." Becker, L., et al., 2004, "Bedout: A possible end-Permian impact crater offshore of northwestern Australia," *Science* 304:1469–76.

124 ". . . paleontologists have identified some rather stressful conditions . . ." Huey, R. B., and Ward, P. D., 2005, "Hypoxia, global warming, and terrestrial Late Permian extinctions," *Science* 308:398–401.

124 ". . . Carboniferous was a time of high oxygen levels . . ." Berner and Canfield, "A new model for atmospheric oxygen"; Ibid. and Berner, R. A., 1997, "The rise of plants and their effect on weathering and atmospheric CO_2," *Science* 276:544–46; Berner and Kothavala, "Geocarb III"; Scotese, Paleomap Project.

124 ". . . by the Late Permian they were only at about 15 to 18 percent . . ." Scotese, Paleomap Project.

124 ". . . Early Triassic, an all-time low." Ibid.

124–25 ". . . 'death zone' above twenty-five thousand feet . . ." Coburn, B., 1997, *Everest: Mountain Without Mercy*, National Geographic Society, Washington, D.C.

125 "... climate that was truly awful for large animals." Huey and Ward, "Hypoxia."

125 "... further reduced habitat ..." Ibid.

125 "... therapsids, with skull features ... more efficient respiration ..." Ibid.

125 "By the Late Triassic ... oxygen levels were on the rise again." Ibid. and Berner and Canfield, "A new model"; Berner, "The rise of plants"; Berner and Kothavala, "Geocarb III"; Scotese, Paleomap Project.

125 "... extinction in the Permian seas was worse ..." Erwin, *The Great Paleozoic Crisis*; Erwin, P., "Lessons from the past."

126 "... our own relentless drive to fish out the sea ..." Pauly et al., "Fishing down marine food webs."

126 "Among the major groups of marine species wiped out ..." Erwin, *The Great Paleozoic Crisis*; Erwin, "Lessons from the past"; Wignall, P. E., 2001, "End-Permian extinctions," in Briggs and Crowther, *Palaeobiology II*, pp. 226–29.

126 "... reef systems ... in the Permian ..." Wood, R. A., 2001, "Evolution of reefs," in Briggs and Crowther, *Palaeobiology II*, pp. 57-62.

126 "... Permian reefs were ... vulnerable ... in warm, tropical waters." Wignall, "End-Permian extinctions."

126 "... dramatic effects ... on today's coral-dominated reefs." Pandolfi et al., *Science* 307:1725–26; Huegh-Guldberg, O., 2005, "Climate change and marine ecosystems," in Lovejoy and Hannah, *Climate Change*, pp. 256–73.

126 "Some elements ... began to reemerge ... But the overwhelming majority of Permian species were not among them." Erwin, "Lessons from the past."

126 "... present-day reef off Belize ..." Huegh-Guldberg, "Climate change and marine ecosystems."

8: THE DINOSAURS OF MIDDLE EARTH

127 "... Triassic Period ..." ICS, 2005, International Stratigraphic Chart.

127 "The boundaries ... as Phillips recognized ..." Phillips, *Life on Earth*.

127 "... ginkgos ... cycads ... conifers ..." Crepet, W. L., 1974, "Investigations of North American cycadeoids: The reproductive biology of *Cycadeoidea*," *Palaeontographica* Abt. B 148:144–69; Miller, C. N., 1977, "Mesozoic conifers," *Botanical Review* 43:217–80; Miller, C. N., 1988, "The origin of modern conifer families," in Beck, C. B. (ed.), *Origin and Evolution of Gymnosperms*, Columbia University Press, New York; Crane, P. R., Herendeen, P., and Friis, E. M., 2004, "Fossils and plant phylogeny," *American Journal of Botany* 91:1683–99.

127 "The pollen ... is spread by wind ..." Ibid. and Ingrouille and Eddie, *Plants*.

128 "Patterns of extinction and renewal for insects ..." Labandeira, C. C., 1998, "Early history of arthropod and vascular plant associations," *Annual Review of Ecology and Planetary Sciences* 26:329–77; Labandeira, C. C., 2001, "The rise and diversification of insects," in Briggs and Crowther, *Palaeobiology II*, pp. 82–88.

129 "... an enrichment of hemipteroid insects ... and holometabolous ones ..." Labandeira, "The rise and diversification of insects."

129 "... flies (Diptera) ... and the caddis flies (Trichoptera)." Ibid.

129 "... mutual dependence between cycads and weevils ..." Ibid. and Norstog, K. J., and Fawcett, P.K.S., 1989, "Insect-cycad symbiosis and its relation to the pollination of *Za-*

mia furfuracea (Zamiaceae) by *Rhopalotria mollis* (Curculionidae)," *American Journal of Botany* 76:1380–94.

129 "Dinosaurs . . ." Weishampel, D. B., Dodson, P., and Osmolska, H. (eds.), 1990, *The Dinosauria*, University of California Press, Berkeley; Norell, M. A., Gaffney, E. S., and Dingus, L., 1995, *Discovering Dinosaurs*, Knopf, New York; Currie, P. J., and Padian, K. (eds.), 1997, *Encyclopedia of Dinosaurs*, Academic Press, San Diego.

129 "The first dinosaurs . . ." Novas, F. E., 1997, "Herrerasauridae," in Currie and Padian, *Encyclopedia of Dinosaurs*, pp. 303–11; Padian, K., 1997, "Origin of dinosaurs," ibid., pp. 481–86.

129 ". . . the long-necked dinosaur *Coelophysis* . . ." Schwartz, H. L., and Gillette, D. D., 1994, "Geology and taphonomy of the *Coelophysis* quarry, Upper Triassic Chinle Formation, Ghost Ranch, New Mexico," *Journal of Paleontology* 68:1118–30; Colbert, E. H., 1995, *The Little Dinosaurs of Ghost Ranch*, Columbia University Press, New York; Hutchinson, J. R., and Padian, K., 1997, "Coelurosauria," in Curry and Padian, *Encyclopedia of Dinosaurs*, pp. 129–33.

130 "marked by another mass extinction event . . ." Benton, M. J., 1986, "More than one event in the late Triassic extinction," *Nature* 321:857–61; Simms, M. J., and Ruffell, A. H., 1990, "Climatic and biotic change in the Late Triassic," *Journal of Geological Society London* 147:321–27; Benton, M. J., 1997, "II extinction, Triassic," in Curry and Padian, *Encyclopedia of Dinosaurs*, pp. 230–36.

130 ". . . long-necked sauropods . . ." McIntosh, J. S., 1990, "Sauropoda," in Weishampel et al., *The Dinosauria*, pp. 345–401; McIntosh, J. S., 1997, "Sauropoda," in Curry and Padian, *Encyclopedia of Dinosaurs*, pp. 654–58.

130 "Adults may have ranged between 20 and 70 tons." McIntosh, "Sauropoda."

130 ". . . sauropods ate plants." Ibid.

130 ". . . these giants were depicted as lethargic, semiamphibious creatures . . ." An engaging review of the history of dinosaur paleontology is Colbert, E. H., 1968, *Men and Dinosaurs*, E. P. Dutton, New York.

130 "By the 1970s paleontologists had taken a swipe at this scenario . . ." Bakker, R. T., 1971, "The ecology of the brontosaurs," *Nature* 229:172–74.; Bakker, R. T., 1986, *The Dinosaur Heresies*, William Morrow, New York.

130 ". . . dinosaur tails were held high . . ." Bakker, *The Dinosaur Heresies*.

131 ". . . metabolism . . . endotherms . . . ectotherms . . ." Desmond, A. J., 1975, *The Hot-Blooded Dinosaurs: A Revolution in Paleontology*, Blond and Briggs, London; Coombs, W. P., 1990, "Dinosaur paleobiology: behavior patterns in dinosaurs," in Weishampel et al., *The Dinosauria*, pp. 32–44; Padian, K., 1997, "Physiology," in Curry and Padian, *Encyclopedia of Dinosaurs*, pp. 552–59.

131 ". . . skeleton of *Barosaurus* is shown rearing up . . ." Norell et al., *Discovering Dinosaurs*.

132 "And there is evidence, some scientists now argue, that certain dinosaurs were endotherms." This is still a controversial issue. A primary advocate is Bakker, *The Dinosaur Heresies*. Arguments pro and con appear in Thomas, R.D.K., and Olson, E. C., 1980, *A Cold Look at Hot-Blooded Dinosaurs*, Westview Press, Boulder, CO. A more cautious review considers the evidence ambiguous: Padian, K., 1997, "Physiology."

132 "Another clue comes from the examination of the fine structure . . ." Ibid. and de Ricqlès, A. J., 1974, "Evolution of endothermy: histological evidence," *Evolutionary*

Theory 1:51–80; Reid, R.E.H., 1997, "Histology of bones and teeth," in Curry and Padian, *Encyclopedia of Dinosaurs*, pp. 329–39.

132　"... the match... is not perfect." Reed, "Histology of bones and teeth."

132　"Living birds, undisputed endotherms, belong to a branch of dinosaurs..." Norell, M. A., et al., 1995, "A nesting dinosaur," *Nature* 378:774–76; Novacek, *Dinosaurs of the Flaming Cliffs*; Dingus and Rowe, *The Mistaken Extinction*; Padian, K., and Chiappe, L. M., 1997, "Bird origins," in Curry and Padian, *Encyclopedia of Dinosaurs*, pp. 71–79.

132　"... Liaoning dinosaurs..." Zhou, Z., Barrett, P. M., and Hilton, J., 2003, "An exceptionally well preserved Lower Cretaceous ecosystem," *Nature* 421:807–14; Chang, M.-M, et al. (eds.), 2004, *The Jehol Biota: The Emergence of Feathered Dinosaurs, Beaked Birds and Flowering Plants*, Shanghai Scientific and Technical Publishers, Shanghai; Norell, M., 2005, *Unearthing the Dragon: The Great Feathered Dinosaur Discovery*, Pi Press, New York.

132–33　"... the feathers in these animals are not flight feathers..." Norell, *Unearthing the Dragon*.

133　"... fossils of theropod dinosaurs with feathers..." Ibid.

133　"... duck-billed hadrosaurs as well as sauropod embryos, had a pebbly skin..." Forster, C. A., 1997, "Hadrosauridae," in Curry and Padian, *Encyclopedia of Dinosaurs*, pp. 293–300; Chiappe, L. M., et al., 1998, "Sauropod dinosaur embryos from the Late Cretaceous of Patagonia," *Nature* 396:258–61.

133　"Some researchers have returned directly to observations of the bones..." Stevens, K. A., and Parrish, M. J., 1999, "Neck posture and feeding habits of two Jurassic sauropod dinosaurs," *Science* 284:798–800.

135　"... a near-vertical head position for *Brachiosaurus*... would be physiologically impossible..." Seymour, R. S., and Lillywhite, H. B., 2000, "Hearts, neck posture and metabolic intensity of sauropod dinosaurs," *Proceedings of the Royal Society of London* 267:1883–87.

136　"... examples of evidence of what dinosaurs ate." Taggart, R. E., and Cross, A. T., 1997, "The relationship between land plant diversity and productivity and patterns of dinosaur herbivory," *Dinofest, International Proceedings* 1:403–16.

136　"... remains of small mammals in their macerated stomach contents." Zhou et al., "An exceptionally well preserved Lower Cretaceous ecosystem."

136　"... the alleged cannibalistic behavior... refuted." Nesbitt, S. J., et al., 2006, "Prey choice and cannibalistic behavior in the theropod *Coelophysis*," *Biology Letters*, First Cite Early Online Publishing, doi:10.1098/rsbl.2006.0524.

136　"... *Edmontosaurus*, has a mash of fruit seeds..." Krausel, R., 1922, "Die Nahrung von Trachedon," *Paleontologische Zeitschrift* 4:80.

137　"... these dinosaurs probably grazed on softer plants, such as ferns." Krassilov, V. A., 1981, "Changes of Mesozoic vegetation and the extinction of dinosaurs," *Palaeogeography, Palaeoclimatology, and Palaeoecology* 34:207–24; Taggart and Cross, "The relationship between land plant diversity."

137　"... ferns can grow and regenerate..." Taggart and Cross, "The relationship between land plant diversity."

137　"... the impact of feeding by large mammals, such as elephants..." Sukumar, *The Living Elephants*; Owen-Smith, *Megaherbivores*.

138　"Large herbivores... are bulk feeders..." Ibid. and Taggart and Cross, "The relationship between land plant diversity."

138 "... sauropod ... daily food intake requirements ..." Coe, M. J., et al., 1987, "Dinosaurs and land plants," in Friis, E. M., Chaloner, W. G., and Crane, P. R. (eds.), 1987, *The Origins of Angiosperms and Their Biological Consequences*, Cambridge University Press, Cambridge, UK, pp. 225–58.

138 "... an herbivore biomass of 93,000 kilograms per hectare ..." Ibid.

138–39 "... these giants relied on ... ferns ..." Taggart and Cross, "The relationship between land plant diversity."

139 "... the tundra ... sustains huge herds ..." Ibid.

139 "... trackways of the gargantuan sauropods ..." Lockley, M., 1995, "Track records," *Natural History* 104:46–51.

139 "... very like migrating herds of large mammals today." Ibid.

9: A FLOWER IN THE FOREST

140 "... a quiet revolution ..." Beck, C. B. (ed.), 1976, *Origin and Early Evolution of Angiosperms*, Columbia University Press, New York; Crane, P. R., 1985, "Phylogenetic analysis of seed plants and the origin of angiosperms," *Annals of the Missouri Botanic Garden* 72:716–93; Dilcher, D. L., 1986, "Origin of Flowering Plants," *1987 McGraw-Hill Yearbook of Science and Technology*, McGraw-Hill Publishers, New York, pp. 339–43; Doyle, J. A., and Donoghue, M. J., 1987, "The origin of angiosperms: a cladistic approach," in Friis, E. M., Chaloner, W. G., and Crane, P. R. (eds.), *The Origin of Angiosperms and Their Biological Consequences*, Cambridge University Press, Cambridge, UK, pp. 17–49; Friis, E. M., Chaloner, W. G., and Crane, P. R., 1987, "Introduction to angiosperms," ibid., pp. 1–15; Crane, P. R., Friis, E. M., and Pederson, K. R., 1995, "The origin and early diversification of angiosperms," *Nature* 374:27–33; Krassilov, V. A., 1997, *Angiosperm Origins: Morphological and Ecological Aspects*, Pensoft, Sofia, Bulgaria; Crane, P. R., Herendeen, P., and Friis, E. M., 2004, "Fossils and plant phylogeny," *American Journal of Botany* 91:1683–99. Tudge, C. *The Variety of Life*; Soltis, P. S., and Soltis D. E., 2004, "The origin and diversification of angiosperms," *American Journal of Botany* 91:1614–26. Ingrouille and Eddie, *Plants: Diversity and Evolution*.

140 "... small, unfertilized eggs were prepared for the arrival of pollen ..." Tudge, *The Variety of Life*.

140 "Pollen and perhaps some sweet nectar ..." Ibid.

141 "... a tiny tube that penetrated the protective crypt ... of the flower ..." Ibid.

141 "Tens of millions of years passed before this elaborate system took hold ..." Crane et al., "Fossils and plant phylogeny."

141 "Some studies ... suggest that flowers first appeared in moist, dimly lit forests ..." Field, T. S., Arens, N. C., and Dawson, T. E., 2003, "The ancestral ecology of angiosperms: emerging perspectives from extant basal lineages," *International Journal of Plant Sciences* 164:5129–42; Field, T. S., Arens, N. C., Doyle, J. A., Dawson, T. E., and Donoghue, M. J., 2004, "Dark and disturbed: a new image of early angiosperm ecology," *Paleobiology* 30:82–107.

141 "... pollen grains from Cretaceous rocks about 130 million years old ..." Friis et al., *The Origins of Angiosperms and Their Biological Consequences*.

141 "Angiosperms ... were the last major group of plants to appear on Earth." Ibid.

141 "Early angiosperm pollen grains ..." Hughes, N. F., 1994, *The Enigma of Angiosperm Origins*, Cambridge University Press, Cambridge, UK.

141 ". . . no . . , angiosperm pollen or plants prior to the Cretaceous." Friis, E. M., Pederson, K. R., and Crane, P. R., 2001, "Origin and radiation of angiosperms," in Briggs and Crowther, *Palaeobiology II*, pp. 97–102.

141 ". . . *Archaefructus* . . ." Sun, G., et al., 1998, "In search of the first flower: a Jurassic angiosperm, *Archaefructus*, from northeast China," *Science* 282:1692–95; Sun, G., et al., 2002, "Archaefructaceae, a new basal angiosperm family," *Science* 296:899–904.

141–42 ". . . is now believed to be much younger . . ." Friis et al., "Origin and radiation of angiosperms."

142 ". . . fossils tell us only the minimum date of origin for a given group." Novacek and Wheeler, *Extinction and Phylogeny*.

142 "The oldest . . . angiosperm fossils . . . are . . . from the western Portuguese Basin." Friis et al., "Origin and radiation of angiosperms."

142 ". . . between 90 and 100 million years ago, we have evidence of marked increase . . ." Ibid. and Friis et al., *The Origins of Angiosperms and their Biological Consequences*.

142 "Early Cretaceous flowers lacked the wondrous complexity . . ." Ibid.

142 ". . . the basic components of the flower . . ." Friis et al., "Introduction to angiosperms"; Tudge, *The Variety of Life*; Ingrouille and Eddie, *Plants*.

143 ". . . 'the abominable mystery' . . ." Darwin, C. R., 1879, letter to J. D. Hooker, July 22, 1879, in Darwin, F., and Seward A. C. (eds.), 1903, *More Letters of Charles Darwin: A Record of His Work in a Series of Hitherto Unpublished Papers*, vol. 2, John Murray, London, pp. 20–21; Crepet, W. L., 1998, "The abominable mystery," *Science* 282:1653–54.

143 ". . . most mysterious is the derivation of the angiosperm carpel . . ." The numerous hypotheses largely concern the tubular development of the leaf-bearing structures that become encapsulated; see comments in Friis et al., "Origin and radiation of angiosperms," Soltis, P. S., et al., 2004, "The diversification of flowering plants," in Cracraft, J., and Donoghue, M. J. (eds.), *Assembling the Tree of Life*, Oxford University Press, Oxford, UK, pp. 154–67.

143 "Two competing theories . . ." Meyen, S. V., 1986, "Hypothesis of the origin of angiosperms from Bennettitales by gamoheterotophy: transition of characters from one sex to another," *Zhurnal Obshchei Biologii* 47:291–309; Friis et al., "Origin and radiation of angiosperms"; Pryer, K. M., Schneider, H., and Magallón, S., 2004, "The radiation of vascular plants," in Cracraft and Donoghue, *Assembling the Tree of Life*.

143 "Many bennettitalean species . . ." Tudge, *The Variety of Life*.

144 ". . . bennettitaleans . . . certainly at least as old as the Late Triassic." Ibid.

144 "Gnetophytes . . ." Ibid.

145 ". . . double fertilization." Friis et al., "Introduction to angiosperms"; Soltis et al., "The diversification of flowering plants."

146 "Angiosperms and gnetophytes, and perhaps bennettitaleans, share other features . . ." Doyle, J. A., and Donoghue, M. J., 1986, "Seed plant phylogeny and the origin of angiosperms: an experimental cladistic approach," *Botanical Review* 52:331–429.

146 ". . . two major events are then implied . . ." Friis et al., "Origin and radiation of angiosperms"; Soltis et al., "The diversification of flowering plants"; Pryer et al., "The radiation of vascular plants."

147 ". . . phylogeny usefully points out . . ." Novacek and Wheeler, *Extinction and Phylogeny*.

147 ". . . the anthophyte hypothesis . . . has come under fire." For various analyses, reviews,

and opinions, see Donoghue, M. J., and Doyle, J. A., 2000, "Seed plant phylogeny: Demise of the anthophyte hypothesis?," *Current Biology* 10:R106–109; Frohlich., M. W., and Parker, D. S., 2000, "The mostly male theory of flower evolutionary origins: from genes to fossils," *Systematic Biology* 25:155–70; Pryer et al., "The radiation of vascular plants"; Soltis et al., "The diversification of flowering plants."

147 "... plants ... have been studied for their DNA sequences ..." Goremykin, V., et al., 1996, "Noncoding sequences from the slowly evolving chloroplast inverted repeat in addition to rbcL data do not support gnetalean affinities of angiosperms," *Molecular Biology and Evolution* 13:383–96; Soltis, D. E., Soltis, P. S., and Zanis, M. J., 2002, "Phylogeny of seed plants based on evidence from eight genes," *American Journal of Botany* 89:1670–81.

147 "Most of these results are starkly different from those suggested by morphology ..." Goremykin et al., "Noncoding sequences"; but some support morphology—for example, Rydin, C., Källersjö, M., Friis, E. M., 2002, "Seed plant relationships and the systematic position of Gnetales based on nuclear and chloroplast DNA: conflicting data, rooting problems, and the monophyly of conifers," *International Journal of Plant Science* 163:197–214.

147 "... many of the original arguments based on morphology have been either rejected or modified)." For example, see Loconte, H., and Stevenson, D. W., 1990, "Cladistics of the Spermatophyta," *Brittonia* 42:197–211.

147 "A study combining both morphology and molecular data ..." De la Torre, J.E.B., et al., 2006, "Estimating plant phylogeny: lessons from partitioning," *BMC Evolutionary Biology* 6:48.

147 "... oldest remains of known gymnosperms ..." DiMichele, "Carboniferous coal-swamp forests."

147–148 "Most botanists and paleobotanists have not been quick to embrace the gene-based story ..." Soltis et al., "The diversification of flowering plants."

148 "... gene studies have ... clarified relationships in groups ..." Pace, N. R., 2004, "The early branches of the tree of life," in Cracraft and Donoghue, *Assembling the Tree of Life*, pp. 76–85.

148 "... gene-based phylogenies ... are in step with the solid results ..." Soltis et al., "The diversification of flowering plants."

148 "The angiosperms' several innovations ..." Ibid. and Friis et al., "Introduction to angiosperms"; Tudge, *The Variety of Life*. For current and frequently updated phylogenies and classifications, see Soltis, P., Soltis, D., and Edwards, C., 2004, Tree of Life Web Project: Angiosperms, www.tolweb.org/angiosperms.

149 "Other features ... are ... well expressed in angiosperms." Tudge, *The Variety of Life*.

149 "Angiosperms have the most diverse of habits ..." Ibid.

149 "... sex is a very big deal in evolution." Stebbins, G. L., 1974, *Flowering Plants: Evolution Above the Species Level*, Belknap Press of Harvard University, Cambridge, MA; Williams, G. C., 1975, *Sex and Evolution*, Princeton University Press, Princeton, NJ; Smith, J. M., 1978, *The Evolution of Sex*, Cambridge University Press, Cambridge, UK; Ridley, M., 1995, *The Red Queen: Sex and the Evolution of Human Nature*, Penguin, New York; Colegrave, N., 2002, "Sex releases the speed limit on evolution," *Nature* 420:664–66.

149 "... reproduction through sex is the norm in life." Williams, *Sex and Evolution*.

150 "Hybrids ..." Stebbins, G. L., 1958, "On the hybrid origin of angiosperms," *Evolution*

12:267–70; Arnold, M. L., 1997, *Natural Hybridization and Evolution*, Oxford University Press, Oxford, UK.

150 "Darwin . . . took the principle one step further." Darwin, C. R., 1861, *On the Origin of Species*, 3rd ed., John Murray, London.

150 "As Darwin himself remarked . . ." Ibid, pp. 345–46.

151 ". . . internal fertilization . . ." King, T. J., Bliss, M., Roberts, V., 1987, *Biology: A Functional Approach*, Nelson Thornes, Cheltenham, UK; Tarin, J. J., and Cano, A., (eds.), 2000, *Fertilization in Protozoa and Metazoan Animals: Cellular and Molecular Aspects*, Springer, Berlin; Barnes, R.S.K., et al., 2000, *The Invertebrates: A Synthesis*, 3rd ed., Blackwell Publishing, New York; Hyde, K. M., 2004, *Zoology: An Inside Look at Animals*, Kendall Hunt, Dubuque, IA.

151 ". . . praying mantids . . ." Prete, F. R., Wells, H., Wells, P. H., and Hurd, L. E. (eds.), 1999, *The Praying Mantids*, Johns Hopkins University Press, Baltimore; Hurd, L. E., 2003, "Mantodea (praying mantids)," in Resh, V. H., and Cardé R. T. (eds.), 2003, *Encyclopedia of Insects*, Academic Press, San Diego, pp. 675–77.

152 ". . . sex in many species, including insects, can be ceremonial and prolonged . . ." Alcock, J., 2005, *Animal Behavior: An Evolutionary Approach*, 8th ed., Sinauer Associates, Sunderland, MA.

152 ". . . the lock and key fit of their male and female genitalia." Tuxen, S. L., 1970, *Taxonomist's Glossary of Genitalia in Insects*, 2nd ed., Munksgaard, Copenhagen; Hendrick, D. H., and Gordh, G., 2003, "Anatomy: head, thorax, abdomen, and genitalia," in Resh and Cardé, *Encyclopedia of Insects*, pp. 12–26.

152 ". . . by skewing influence . . . in the gene pool of the population." Alcock, *Animal Behavior*.

152 "Mammals are famous for the frequent instances of polygamy in species . . ." Vaughan, T. A., Ryan, J. M., and Capzaplewski, N. J., 2000, *Mammalogy*, 4th ed., Saunders College Publishing, Philadelphia.

152 "The northern elephant seal, *Mirounga angustirostris* . . ." Le Boeuf, B. J., 1974, "Male-male competition and reproductive success in elephant seals," *American Zoologist* 14:163–76.

152–53 "polygamy . . . in mammals . . . Monogamy . . . Polyandry . . ." Vaughan, Ryan, and Copzaplewski, *Mammalogy*.

153 "That big hyena I encountered was probably a female . . ." Estes, R.D.E., 1991, *The Behavior Guide to African Mammals*, University of California Press, Los Angeles.

153 ". . . females actually select for sperm . . ." Birkhead, T. R., 1998, "Sperm competition in birds," *Reviews of Reproduction* 3:123–29; Birkhead, T. R., 2000, *Promiscuity: An Evolutionary History of Sperm Competition*, Harvard University Press, Cambridge, MA.

153 ". . . animal pollination of flowering plants!" Proctor, M., Yeo, P., and Lack, A., 1996, *The Natural History of Pollination*, Timber Press, Portland, OR.

154 "Although a mere thirty species of mammals facilitate plant pollination . . ." Ibid.

10: THE GARDEN OF DELIGHTS

155 "It will be a long, arduous day for a male *Campsoscolia ciliata*." Proctor, M., Yeo, P., and Lack, A., 1996, *The Natural History of Pollination*, Timber Press, Portland, OR.; Pouyanne, A., 1917, "La fécondation des *Ophrys* par les insects," *Bulletin de la Société de l'Afrique du Nord* 8:6–7.

156 "... more than 40 percent of *Ophrys speculum* ... is pollinated ... in this manner." Ibid. and Correvon, H., and Pouyanne, A., 1923, "Nouvelles observations sur le mimétisme et la fécondation chez les *Ophrys speculum* et *lutea*," *Journal de la Société Nationale d'Horticulture de France* 1923:372–77; Kullenberg, B., 1961, "Studies in *Ophrys* pollination," *Zoologiska Bidrag Uppsala* 34:1–340.

156 "... exquisite timing in the life cycles ..." Proctor et al., *The Natural History of Pollination*, p. 206.

157 "The misnamed fly orchid, *Ophrys insectivora* ..." Kullenberg, "Studies in *Ophrys* pollination."

157 "... *Orchis mascula*, studied intensively by Darwin ..." Darwin, C. R., 1862, *On the Various Contrivances by Which British and Foreign Orchids Are Fertilised by Insects*, 2nd ed., John Murray, London.

157 The orchids ... twenty thousand strong ... epitomize extreme ... approaches to attracting insect pollinators." Arditti, J., 1992, *Fundamentals of Orchid Biology*, John Wiley and Sons, New York.

157 "Other flower groups offer strategies for attracting ... pollinators," Proctor et al., *The Natural History of Pollination*.

158 "... false promise of a brooding site ..." Knoll, F., 1956, *Die Biologie der Blüte*, Springer, Berlin; Nilsson, L. A., 1979, "Anthecological studies on the lady's slipper, *Cypripedium calceolus* (Orchidaceae)," *Botaniska Notiser* 132:329–47; Proctor et al., *The Natural History of Pollination*, pp. 188, 299.

158 "Some actually provide brooding sites." Anderson, A. B., and Overal, W. L., 1988, "Pollination ecology of the forest-dominant palm (*Orbignya phalerata* Mart.)," *Biotropica* 20:192–205.

158 "... brood-site mutualism between the edible fig ... and the fig wasp ..." Grandi, G., 1961, "The hymenopterous insects of the superfamily Chalcidoidea developing within the receptacles of figs. Their life-history, symbioses and morphological adaptations," *Bollettino dell'Istituto di Entomologia della Università degli Studi di Bologna* 26:I–XIII.

159 "... other plants have established complex mutualistic relationships ..." Powell, J. A., 1992, "Interrelationships of yuccas and yucca moths," *Trends in Ecology and Evolution* 7:10–15.

160 "... pollination in the carnivorous plants ..." Darwin, C. R., 1875, *Insectivorous Plants*, John Murray, London; Slack, A., 1988, *Carnivorous Plants*, Alphabooks, London; Zamora, R., Gomez, J. M., and Hodar, J. A., 1988, "Fitness responses of a carnivorous plant in contrasting ecological scenarios," *Ecology* 79:1630–44.

160 "The Venus flytrap ..." Hodick, D., and Sievers, A., 1988, "On the mechanism of closure of Venus flytrap (*Dionaea muscipula* Ellis)," *Planta* 179:32–42.

160 "... carnivorous plants ... apprehend and eat in different ways." Ibid.

160 "... How do the carnivorous plants manage to avoid killing and consuming the insect visitors they require for pollination?" Slack, A., 1988, and Brewer, J. S., 2002, "Why don't carnivorous pitcher plants compete with non-carnivorous plants for nutrients?" *Ecology* 84:451–62.

160–61 "... certain groups of flowers ... suited ... particular groups of insects." Proctor et al., *The Natural History of Pollination*, pp. 178–84.

161 "Similar strategies ... occur in distantly related flower groups." Ibid., pp. 145–86.

161 "... many species are less fastidious ..." Ibid., pp. 173.

162 "More than three hundred insect species visit carrot flowers ..." Hawthorn, L. R.,

Bohart, G. E., and Toole, E. H., 1956, "Carrot seed yield and germination as affected by different levels of insect pollination," *Proceedings of the American Society for Horticultural Science* 67:384–89.

162 "Composites thrive and dominate in many habitats." Burtt, B. L., 1961, "The Compositae and the study of functional evolution," *Transactions of the Botanical Society of Edinburgh* 39:216–32; Raven, P. H., Evert, R. F., Eichhorn, S.E., 2004, *Biology of Plants*, 7th ed., W. H. Freeman, San Francisco.

162 "... orchids ... often hang out ... in marginal or restricted habitats ..." Arditti, J., 1992, *Fundamentals of Orchid Biology*, John Wiley and Sons, New York.

162 "There are twenty-five thousand species of Asteracea ..." Proctor et al., *The Natural History of Pollination*, p. 174; Judd, W. S., et al., 1999, *Plant Systematics: A Phylogenetic Approach*, Sinauer Associates, Sunderland, MA.

162 "... insects ... have evolved in ways that have enhanced their role as pollinators ..." Frankie, G. W., and Thorp, R. W., 2003, "Pollination and pollinators," in Resh and Cardé, *Encyclopedia of Insects*, pp. 919–28.

163 "... intense pollinator behavior ... is the mark of only certain insect groups ..." Ibid. and Proctor et al., *The Natural History of Pollination*, p. 51.

163 "... insect pollinators ... share basic tools ..." Proctor et al., *The Natural History of Pollination*.

163 "... Thysanoptera ..." Kirk, W.D.J., 1984, "Pollen feeding in thrips (Insecta: Thysanoptera)," *Journal of Zoology, London* 204:107–15.

164 "... the beetles, represent a giant leap forward in refinement." White, R. E., 1983, *Beetles*, Houghton Mifflin, New York; Lippok, B., et al., 2000, "Pollination by flies, bees, and beetles of *Nuphar ozarkana* and *N. advena* (Nymphaeaceae), *American Journal of Botany* 87:898–902.

164 "They are in the main 'lazy visitors' to flowers ..." Kugler, 1984, "Die Bestäubung von Bläten durch den Schmalkäfer *Oedemera* (Coleoptera)," *Berichte der deutschen botansichen Gesellschaft* 97:383–390; Proctor et al., *The Natural History of Pollination*, p. 56.

164 "... mess and soil pollinators ..." Frankie and Thorp, "Pollination and pollinators," in Resh and Cardé, *Encyclopedia of Insects*, p. 920.

164 "... the Diptera, ... show an even greater elaboration of adaptations for pollination ..." Ibid., p. 921.

164 "... extreme development of its proboscis ..." Hammond, A., 1874, "The mouth of the crane fly," *Science Gossip* 1874:155–60; Gouin, F., 1949, "Recherches sur la morphologie de l'appareil buccal des Diptères," *Mémoires du Museum National d'Histoire Naturelle, Paris* N. S. 28:167–269.

165 "... Lepidoptera ... rival and even surpass the flies." Scoble, M. J., 1995, *The Lepidoptera: Form, Function and Diversity*, Oxford University Press, Oxford, UK; Kristensen, N. P. (ed.), 1999. *Lepidoptera, Moths and Butterflies*, vol. 1, *Evolution, Systematics, and Biogeography. Handbuch der Zoologie. Eine Naturgeschichte der Stämme des Tierreiches Band / Arthropoda: Insecta Teilband* 35: Walter de Gruyter, Berlin.

165 "... the meganosed fly ..." Sessions, L., and Johnson, S., 2005, "The flower and the fly," *Natural History*, March 2005, 58–63.

165 "The proboscis in the lepidopteran ..." Hepburn, H. R., 1971, "Proboscis extension and recoil in Lepidoptera," *Journal of Insect Physiology* 17:637–56.

166 "This coevolutionary one-upmanship has been carried to absurd extremes." Proctor et al., *The Natural History of Pollination*, p. 81, Table 4.1.

166–67 "Darwin observed . . ." Ibid., p. 382.

167 "Many of the flowers that attract lepidopterans . . ." Ibid., p. 85, Table 4.3; p. 182.

167 ". . . the most specialized and spectacular pollinators of all, the hymenopterans." Goulet, H., and Huber, J. T., 1993, "Hymenoptera of the world: an identification guide to families," *Agriculture Canada Research Branch Monograph*, no. 1894E; Grimaldi, D., and Engel, M. S., 2005, *Evolution of the Insects*; Tree of Life Project, *Hymenoptera, Wasps, Ants, Bees, and Sawflies,* http://tolweb.org/Hymenoptera.

167–68 ". . . Apocrita . . . Aculeata . . ." Goulet and Huber, "Hymenoptera of the world."

168 ". . . *Campsoscolia*, famously duped into pseudosex . . ." Pouyanne, "La fécondation des *Ophrys*"; Correvon, and Pouyanne, "Nouvelles observations sur le mimétisme et la fécondation"; Kullenberg, "Studies in *Ophrys* pollination."

168 "Vespids are common pollinators of many orchids . . ." Proctor et al., *The Natural History of Pollination*, p. 106, Table 5.3.

168 ". . . flowers highly adapted to pollination by ants . . ." Ibid., pp. 107–108.

168 ". . . many plants have developed defenses against voracious ant feeding . . ." Feinsinger, P., and Swarm, L. A., 1978, "How common are ant-repellent nectars?" *Biotropica* 10:238–39.

168 "Bees have an enormous, unchallenged impact as plant pollinators . . ." Michener, C. D., 2000, *Bees of the World*, Johns Hopkins University Press, Baltimore; Frankie and Thorp, "Pollination and pollinators."

168–69 ". . . one or more cultivars of 66 percent of the world's fifteen hundred crop species . . ." McGregor, S. E., 1976, *Insect Pollination of Cultivated Crop Plants*, U.S. Department of Agriculture—Agricultural Research Service, Washington, D.C.; Kremen, C., Williams, N. M., and Thorp, R. W., 2002, "Crop pollination from native bees at risk from agricultural intensification," *Proceedings of the National Academy of Sciences* 99:16812–16.

169 ". . . an estimated value of five to fourteen billion dollars per year in the United States alone." Southwick, E. E., and Southwick, L., Jr. , 1992, "Estimating the economic value of honey bees (Hymenoptera: Apidae) as agricultural pollinators in the United States," *Economic Entomology* 85:621–33; Morse, R. A., and Calderone, N. W., 2000, "Beeculture," http://bee.airoot.com.

169 "The decline . . . of beekeeping in the United States . . ." Buchanna, S., and Nabhan, G. P., 1996, *The Forgotten Pollinators*, Island Press, Covelo, CA; Allen-Wardell, G., et al., 1998, "The potential consequences of pollinator declines in the conservation of biodiversity and the stability of food crop yields," *Conservation Biology* 12:8–17; Kremen et al., "Crop pollination from native bees"; Frankie and Thorp, "Pollination and pollinators."

169 "The special status of bees . . . relates to a peculiar feature of their life cycle." Proctor et al., *The Natural History of Pollination*, pp. 108–109.

169 "Some high-tuned equipment . . ." Ibid., pp 109–33.

170 ". . . buzz pollination . . ." Ibid., p. 125; Buchmann, S. L., 1983, "Buzz pollination in angiosperms," in Jones, C. E., and Little, R. J. (eds.), *Handbook of Experimental Pollination Biology*, Van Nostrand Reinhold, New York, pp. 73–113.

170 ". . . amazing powers of communication . . ." Proctor et al., *The Natural History of Pollination*, pp. 133–42; Frisch, K. von, 1993, *The Dance Language and Orientation of Bees*, Harvard University Press, Cambridge, MA.

171 ". . . the last major group of insect pollinators to appear in the fossil record." Grimaldi and Engel, *The Evolution of Insects*, p. 431, Figure 11.33.

11: TOWARD A NEW ECOSYSTEM

172 "A clay pit in Sayreville, New Jersey . . ." This urban locality has received little in the way of scientific coverage. The geologic unit at the site, the Raritan Formation, is described in Kümmel, H. B., and Volney, J., 1940, *The Geology of New Jersey*, Bulletin 14, Geology Series, State of New Jersey Publication, pp. 1–203. See also Gallagher, W. B., 1997, *When Dinosaurs Roamed New Jersey*, Rutgers University Press, Piscataway, N.J. A good website relating the history and plunder of the locality is Cornet, B., 2003, "Upper Cretaceous Facies, Fossil Plants, Amber, Insects and Dinosaur Bones, Sayreville, New Jersey," www.sunstar-solutions.com/sunstar/Sayreville/Kfacies.htm. Web page created November 29, 2003; updated February 13, 2007.

172 "Ancient insects were engulfed in the resin of ancient conifers." Grimaldi, D., Shedrinsky, A., and Wampler, T., 2000, "A remarkable deposit of fossiliferous amber from the Upper Cretaceous (Turonian) of New Jersey," in Grimaldi, D. (ed.), *Studies on Fossils in Amber, with Particular Reference to the Cretaceous of New Jersey*, Backhuys, Leiden, Netherlands, pp. 1–76.

173 ". . . the earliest-known fossil bee, *Cretotrigona*." Engel, M. S., 2000, "A new interpretation of the oldest fossil bee (Hymenoptera, Apidae)," *American Museum Novitates*, no. 3296:1–11.

173 "There is a certain bitter irony to the urban sprawl engulfing Sayreville . . ." Cornet, "Upper Cretaceous Facies."

174 ". . . history behind the rise of flowers and of bees . . ." Friis et al., 2001, "Origin and radiation of angiosperms"; Grimaldi and Engel, *The Evolution of Insects*.

174 "These resemble the living genus *Hedyosmum* . . ." Friis et al., "Origin."

174 ". . . monocots . . . dicots . . ." A lucid explanation is in Tudge, *The Variety of Life*.

174 ". . . suggested by John Ray in 1703 . . ." Ray, J., 1686–1704, *Historia plantarum*.

174 "But recent studies by Peter Crane and others . . ." Crane, P., Friis, E. M., and Pederson, K. R., 1995, "The origin and early diversification of angiosperms"; Soltis, P. S., and Soltis, D. E., 2004, "The origin and diversification of angiosperms"; Soltis et al., 2004, "The diversification of flowering plants."

175 ". . . evolution of specializations that encourage the pickup and dispersal of pollen." Ibid. and Crepet, W. L., 2001, "Plant-animal interactions: insect pollination," in Briggs and Crowther, *Palaeobiology II*, pp. 426–29.

175 "which came first . . . wind-pollinated or animal-pollinated flowers." Tudge, *The Variety of Life*, pp. 578–79.

175 "the fossil record itself is good enough to track the . . . relationship between pollinator and pollen provider." Crepet, "Plant-animal interactions"; Grimaldi and Engel, 2005, *Evolution of the Insects*.

176 ". . . what would attract the insect to the ovule?" Crepet, "Plant-animal interactions."

176 "Beetles appeared and diversified early in the fossil record . . ." Farrell, B. D., 1998, "'Inordinate fondness' explained: why there are so many beetles," *Science* 281:555–59.

176 "This asynchrony in timing has stimulated one of the most . . . heated debates . . ." Ibid.

176 "The latter hypothesis, firmly preferred by some . . ." Labandeira, C. C., and Sepkoski, J. J., Jr., 1993, "Insect diversity in the fossil record," *Science* 261:310–14; Labandeira, C. C., 1998, "How old is [*sic*] the flower and the fly?" *Science* 280:57–58.

176 "Others disagree." Crepet, W. L., 1996, "Timing in the evolution of derived floral char-

acters: Upper Cretaceous (Turonian) taxa with tricolpate and tricolpate derived pollen," *Review of Palaeobotany and Palynology* 90:339–59. Grimaldi, D., 1999, "The co-radiations of pollinating insects and angiosperms in the Cretaceous," *Annals of the Missouri Botanical Garden* 86:373–406.

176 ". . . current evidence seems to favor the intimate evolutionary syncopation . . ." Grimaldi, "The co-radiations of" and Crepet, "Animal-plant interactions"; Grimaldi and Engel, *Evolution of the Insects*.

177 ". . . the advantages of insect . . . pollination are clear." Tudge, *The Variety of Life*, p. 579.

178 "Some environments favor wind pollination . . ." Proctor et al., *The Natural History of Pollination*, p. 265.

178 ". . . years of plant and insect evolution involve refinements . . ." Ibid., pp. 365–83.

12: DINOSAUR CAMELOT

180 ". . . American Museum's paleontological expeditions to the Gobi Desert of Mongolia . . ." For background and early expeditions, see Novacek, *Dinosaurs of the Flaming Cliffs*; for later expeditions, see Novacek, *Time Traveler*.

181 "A team led by the colorful, gun-toting Roy Chapman Andrews . . ." Andrews, R. C., 1932, *The New Conquest of Central Asia*, Vol. 1. *Natural History of Central Asia*, The American Museum of Natural History, New York.

181 ". . . Ukhaa Tolgod . . ." Dashzeveg, D., et al., 1995, "Extraordinary preservation in a new vertebrate assemblage from the Late Cretaceous of Mongolia," *Nature* 374:446–49.

182 ". . . adult oviraptorids perfectly preserved, still sitting on their nests . . ." Norell, M. A., et al., 1995, "A nesting dinosaur," *Nature* 378:774–76.

182–83 "The bones at this rich locality are not preserved along with plants." Novacek, *Dinosaurs*.

183 "These places represent oases in a burning desert." Ibid.

183 ". . . our geologists concluded . . ." Loope, D. B., et al., 1998, "Life and death in a Cretaceous dune field," *Geology* 26:27–30; Dingus, L., and Loope, D., 2000, "Death in the dunes," *Natural History*, vol. 109, No. 6, pp. 50–55.

184 ". . . another Cretaceous locality of much recent fame and generosity." Ji, Q., et al., 1998, "Two feathered theropods from the Upper Jurassic/Lower Cretaceous strata of northeastern China," *Nature* 393:753–61; Ji, Q., et al., 2001, "The distribution of integumentary structures in a feathered dinosaur," *Nature* 410:1084–88; Zhou, Z., Barret, P. M., and Hilton, J., 2003, "An exceptionally well preserved Lower Cretaceous ecosystem," *Nature* 421:807–14; Chang, M.-M., et al. (eds.), 2004, *The Jehol Biota: The Emergence of Feathered Dinosaurs, Beaked Birds and Flowering Plants*, Shaghai Scientific and Technical Publishing, Shanghai; Norell, M., 2005, *Unearthing the Dragon: The Great Feathered Dinosaur Discovery*, Pi Press, New York.

184 "The Jehol . . . about 128 million to about 110 million years in age." Zhou, Barrett, and Hilton, "An exceptionally well preserved Lower Cretaceous ecosystem."

184 "What Jehol brings . . . is . . . a whole ecosystem." Ibid.

185 ". . . fossils of theropod dinosaurs . . . have attracted the most attention." Ibid.

187 "Jehol is also enriched with . . . fully fledged . . . birds." Ibid.

187 "The most eye-popping of all fossils . . . one where the hunted became the hunter." Hu, Y., Meng, J., Wang, Y., Li, C., 2005, "Large Mesozoic mammals fed on young dinosaurs," *Nature* 433:149–52.

188 ". . . a largish mammal . . . with a flat, beaverlike tail . . ." Ji, Q., et al., 2006, "Swimming mammaliaform from the Middle Jurassic and ecomorphological diversification of early mammals," *Science* 311:1123–27.

188 ". . . a fossil mammal with impressions of a winglike flap of skin . . ." Meng, J., et al., 2006, "A Mesozoic gliding mammal from northeastern China," *Nature* 444:889–93.

188 "What natural forces produced such amazing fossils?" Zhou et al., "An exceptionally well preserved Lower Cretaceous ecosystem."

188 "Collecting the Jehol fossils . . ." Norell, *Unearthing the Dragon.*

189 ". . . the world of Cretaceous rocks still has much to offer." For an overview, see Novacek, *Dinosaurs of the Flaming Cliffs.*

189 ". . . the clay pits of New Jersey . . ." Cornet, "Upper Cretaceous Facies."

189 "Time takes its toll . . ." Novacek, *Dinosaurs.*

190 "The more remote the geologic time, the more elusive . . ." Ibid.

190 ". . . decline of the massive, long-necked sauropods after the Jurassic." Barrett, P. M., and Willis, K. J., 2001, "Did dinosaurs invent flowers? Dinosaur-angiosperm coevolution revisited," *Biological Reviews* 76:411–47.

190 ". . . the changeover from a conifer-dominated plant community . . ." Friis et al., *The Origins of Angiosperms and Their Biological Consequences.*

190 ". . . in the late 1970s, prominently Robert Bakker, proposed . . ." Bakker, R. T., 1978, "Dinosaur feeding behaviour and the origin of flowering plants," *Nature* 274:661–63.

190 "The logic of their argument is as follows." Ibid.

191 "In support of this scenario, David Weishampel and David Norman proposed . . ." Weishampel, D. B., and Norman, D. B., 1989, "Vertebrate herbivory in the Mesozoic: Jaws, plants, and evolutionary metrics," *Geological Society of America, Special Paper* 238:87–100.

191 ". . . Scott Wing and Bruce Tiffney calculated . . ." Wing, S. L., and Tiffney, B. H., 1987, "The reciprocal interaction of angiosperm evolution and tetrapod herbivory," *Reviews in Palaeobotany and Palynology* 50:179–210.

191 ". . . large herbivores tend to be generalists . . ." Taggart and Cross, "The relationship between land plant diversity."

191 ". . . new studies of sauropod neck architecture . . ." Stevens and Parrish, "Neck posture and feeding habits of two Jurassic sauropod dinosaurs."

191 ". . . the very trees that Bakker argued were safe . . ." Bakker, "Dinosaur feeding."

192 ". . . the change from sauropod- to ornithischian-dominated communities was not the same in all places." Barrett and Willis, "Did dinosaurs invent flowers?"

192 ". . . dinosaur distribution might actually provide a welcome test . . ." Wing, S. L., and Sues, H. D., 1992, "Mesozoic and Early Cenozoic terrestrial ecosystems," in Behrensmeyer, A. K., et al. (eds.), *Terrestrial Ecosystems Through Time,* University of Chicago Press, Chicago, pp. 327–416.

192 ". . . dinosaurs must have helped to disperse seeds of cycads and other plants . . ." Tiffney, 1997, "Plants and Dinosaurs," in Currie and Padian, *Encyclopedia of Dinosaurs,* pp. 557–59.

192 "But the theory . . . is irreparably damaged." Barrett and Willis, "Did dinosaurs invent flowers?"

192 "... marked change in atmospheric carbon dioxide over time." Berner and Kothavala, "Geocarb III: a revised model of atmospheric CO_2 over Phanerozoic time"; Scotese, www.scotese.com/climate.htm.

193 "... CO_2 infusion would increase warmer ... land areas." Otto-Bliesner, B. L., and Upchurch, G. R., Jr., 1997, "Vegetation-induced warming of high-latitude regions during the Late Cretaceous period," *Nature* 385:804–807.

193 "CO_2 has been cited as enhancing water and nutrient transport ..." Woodward, F. I., 1992, "A review of the effects of climate on vegetation: ranges, competition, and composition," in Peters and Lovejoy, *Global Warming and Biological Diversity*, pp. 105–23.

193 "... early studies described a warm 'greenhouse climate' ..." Barron, E. J., 1983, "A warm equable Cretaceous: the nature of the problem," *Earth Sciences Review* 19:305–38.

193 "... but recent work has demonstrated just the opposite." Kuypers, M.M.M., Pancost, R. D., and Damsté, J.S.S., 1999, "A large and abrupt fall in atmospheric CO_2 concentration during Cretaceous times," *Nature* 399:342–45.

193 "... the rise of flowering plants and specialized pollinating insects ... had a lot to do with each other." Grimaldi, D., 1999, "The co-radiations of pollinating insects and angiosperms in the Cretaceous"; Crepet, "Animal-plant interactions"; Grimaldi and Engel, *Evolution of the Insects*.

193–94 "The Late Cretaceous saw an all-time high in dinosaur diversity." See Fig. 2 in Barrett and Willis, "Did dinosaurs invent flowers?"

13: A PUZZLING CATASTROPHE

197 "The town of Gubbio ..." The official url is www.comune.gubbio.pg.it/.

197 "... another sacred tablet, this one dating back 65 million years." Alvarez, W., 1997, *T. Rex and the Crater of Doom*, Princeton University Press, Princeton, NJ (softcover 1998, Vintage Books, New York).

198 "The beds near the road are tilted at about forty-five degrees ..." Ibid.

198 "... the hard tests of foraminiferans ..." Ibid.

198 "The Scaglia Rossa was not laid down in a shallow ocean bottom offshore ..." Ibid. and Luterbacher, H. P., and Premoli Silva, I., 1962, "Note préliminaire sur une revision du profil de Gubbio, Italia," *Rivista Italiana di Paleontologia e Stratigrafia* 68:253–88.

199 "The K/T boundary ... has been dated ... at about 65 million years before the present." Pillmore, C. L., and Miggins, D. P., 2001, "A new 40Ar/39Ar Age determination on the K/T boundary interval: possible constraints on the timing of the K/T Event and sedimentation rates of the K/T sequence," International Conference on Catastrophic Events and Mass Extinctions: Impacts and Beyond, July 9–12, 2000, Vienna, Austria, abstract no. 3154.

199 "Paleontologists have recognized for many decades 65 million years before the present as the most infamous ... mass extinction event ..." Alvarez, L.W., 1983, "Experimental evidence that an asteroid impact led to the extinction of many species 65 million years ago," *Proceedings of the National Academy of Sciences* 80:627–42; Florentin, J. M., Maurrasse, R., Sen, G., 1991, "Impacts, tsunamis, and the Haitian Cretaceous-Tertiary boundary layer," *Science* 252:1690–93; Glen, W., 1990, "What killed the dinosaurs?" *American Scientist* 78:354–70; Hunter, J., 1994, "Lack of a high body count at the K-T boundary," *Journal of Paleontology* 68:1158; Sheehan, P. M., et al., 1991,

"Sudden extinction of the dinosaurs: Late Cretaceous, upper Great Plains, U.S.A.," *Science* 254:835–39; Stanley, S. M., 1987, *Extinctions*, Scientific American Books, New York; Williams, M. E., 1994, "Catastrophic versus noncatastrophic extinction of the dinosaurs: testing, falsifiability, and the burden of proof," *Journal of Paleontology* 68:183–90; Archibald, J. D., 1996, *Dinosaur Extinction at the End of an Era: What the Fossils Say*, Columbia University Press, New York; Novacek, *Dinosaurs of the Flaming Cliffs*; Dingus and Rowe, *The Mistaken Extinction*.

200 "This is a detective story, as engagingly told in Walter Alvarez's book . . ." Alvarez, *T. Rex*.

200 ". . . Earth's magnetic field had flip-flopped . . ." Cox, *Plate Tectonics and Geomagnetic Reversals*.

201 ". . . found themselves in geologic nirvana." Alvarez, *T. Rex*.

201 "The next phase of the investigation took a most surprising turn." Ibid.

201 ". . . Asaro shared his surprising and troubling results . . ." Ibid. and Asaro, F., 1987, "The Cretaceous-Tertiary iridium anomaly and the asteroid impact theory," in Twore, P. (ed.), *Discovering Alvarez—Selected Works of Luis Alvarez with Commentary by his Students and Colleagues*, University of Chicago Press, Chicago, pp. 240–42.

202 ". . . a wild but brilliant explanation." Alvarez, *T. Rex*.

202 "Was the spike of iridium . . . widespread . . . ?" Ibid.

202 "A number of influential paleontologists then (and now) . . . argued that . . . extraterrestrial impact was irrelevant." The arguments, primarily from a skeptical attitude, are reviewed by Archibald, *Dinosaur Extinction at the End of an Era*.

203 ". . . most of them were too small . . ." Alvarez, *T. Rex*.

203 ". . . peculiar composition called spherules . . ." Ibid.

203 ". . . the spherule sample was a weird mixture of chemicals . . ." Ibid.

203 ". . . shocked quartz, had for some time been associated with impact craters." Ibid.

204 ". . . several geologists claimed that the shocked quartz was a sign of . . . volcanic eruptions . . ." Carter, N. L., Officer, C. B., and Drake, C. L., 1990, "Dynamic deformation of quartz and feldspar: clues to the causes of some natural crises," *Tectonophysics* 171:373–91.

204 ". . . Alan Hildebrand . . . proposed that a strange-looking Cretaceous deposit . . . pointed to the location of the crater." The series of investigations summarized here and described in further detail by Alvarez, *T. Rex*, resulted in the landmark paper by Hildebrand, A. R., et al., 1991, "Chicxulub Crater: a possible Cretaceous/Tertiary boundary impact crater on the Yucatán Peninsula, Mexico," *Geology* 19:867–71.

204 ". . . in 1991, a paper entitled . . ." Ibid.

204 ". . . the crater . . . had actually been discovered ten years before . . ." Ibid.

205 "The Chicxulub depression . . ." Ibid.

206 ". . . age estimates . . . correlate to a date around 64.98 +/- 0.05 million years." Swisher, C. C., III, et al., 1992, "Coeval 40Ar/39Ar Ages of 65.0 million years ago from Chicxulub crater melt rock and Cretaceous-Tertiary boundary tektites," *Science* 257:954–58.

206 "What of the colliding object, the thing itself?" Kring, D. A., 1995, "The dimensions of the Chicxulub impact crater and impact melt sheet," *Journal of Geophysical Research* 100:16,979–86.

206 ". . . scientists have spent a good deal of time trying to reconstruct the devastation." A paper that accounts for many earlier reconstructions and offers its own is Robertson, D. S., et al., 2004, "Survival in the first hours of the Cenozoic," *Geological Society of America Bulletin* 116:760–68.

206 "The ejecta sprayed out . . ." Ibid.

206 ". . . temperatures in the upper atmosphere might have hovered around 700°C (1,300°F) for several hours." Melosh, H. J., 1990, "Reentry of fast ejecta: the global effects of large impacts [abstract]." *Transactions of the American Geophysical Union* 71:1429.

206 "Organisms . . . would have absorbed this thermal radiation from the entire visible sky." Ibid. and Robertson et al., "Survival in the first hours of the Cenozoic."

206 "The impact released other destructive forces too." Robertson et al., "Survival in the first hours of the Cenozoic."

207 "A recent study . . . shows a . . . high surge of CO_2 concentration . . ." Beerling, D. J., et al., 2002, "An atmospheric pCO_2 reconstruction across the Cretaceous-Tertiary boundary from leaf megafossils," *Proceedings of the National Academy of Sciences* 99:7836–40.

207 "Other nasty effects . . ." Robertson et al., "Survival in the first hours of the Cenozoic."

207 ". . . claim . . . that the extinction of . . . dinosaurs was already marked for some millions of years before this event . . ." Sloan, R. E., et al., 1986, "Gradual dinosaur extinction and simultaneous ungulate radiation in the Hell Creek formation," *Science* 232:629–33.

207 "Some, prominently Gerta Keller . . . argue that the Chicxulub impact occurred three hundred thousand years before the K/T extinction event . . ." Keller, G., et al., 2004, "Chicxulub impact predates the K-T boundary mass extinction," *Proceedings of the National Academy of Sciences* 101:3753–58; Keller, G., et al., 2004, "More evidence that the Chicxulub impact predates the K/T mass extinction," *Meteoritics and Planetary Science* 39:1127–44.

207 ". . . she and others argue that the huge amounts of volcanic activity . . ." Keller et al., "More evidence that the Chicxulub impact predates the K/T mass extinction," and Keller, G., 2003, "Biotic effects of impacts and volcanism," *Earth and Planetary Science Letters* 215:249–64.

208 "Smit and others have a decent alternative explanation . . ." Smit, J., van der Gaast, S., and Lustenhouwer, W., 2004, "Is the transition impact to post-impact rock complete? Some remarks based on XRF scanning, electron microprobe, and thin section analyses of the Yaxcopoil-1 core in the Chicxulub crater," *Meteoritics and Planetary Science* 39:1113–26.

208 "Moreover, the impact date is 65 million plus or minus about 50,000 years . . ." Swisher et al., "Coeval 40Ar/39Ar Ages of 65.0 million years."

208 ". . . scientists still argue that climate change . . . may have promoted extinction . . ." Sloan et al., "Gradual dinosaur extinction." For a pluralistic view on the effects of climate change, volcanism, and extraterrestrial impact, see Archibald, *Dinosaur Extinction at the End of an Era*; Archibald, J. D., 1997, "I. Extinction, Cretaceous," in Currie and Padian, *Encyclopedia of Dinosaurs*, pp. 221–30.

209 ". . . that most marine organisms did not suffer significant extinction . . ." Kiessling, W., and Claeys, P., 2002, "A geographic database approach to the KTB," in Buffetaut, E., and Koeberl, C. (eds), *Geological and Biological Effects of Impact Events*, Springer, Berlin and Heidelberg, pp. 83–140.

209 ". . . declines in cycads and ferns during the Late Cretaceous were accompanied by a presumably related increase in angiosperm species." Wing and Sues, "Mesozoic and Early Cenozoic terrestrial ecosystems."

209 "Conifers did not show any significant decrease . . ." Ibid.

209 ". . . major decline in angiosperm . . . diversity in the high latitudes . . ." Spicer, R. A., and Parrish, J. T., 1986, "Paleobotanical evidence for cool north polar climates in Middle Cretaceous (Albian-Cenomanian) time," *Geology* 14:703–706.

209–10 ". . . angiosperm diversity in these floras was persistently lower . . ." Ibid.

210 ". . . no falloff in floral diversity in Antarctica . . ." Askin, R. A., and Jacobson, S. R., 1996, "Palynological change over the Cretaceous-Tertiary boundary on Seymour Island, Antarctica: environmental and depositional factors," in MacLeod, N., and Keller, G. (eds.), *Cretaceous-Tertiary Mass Extinctions: Biotic and Environmental Changes*, W. W. Norton, New York.

210 ". . . for insects, the trend through the Cretaceous . . ." Grimaldi and Engel, *Evolution of the Insects*.

210 ". . . no significant decline in diversity for the last phase of the Cretaceous leading up to the extinction event, or the data are insufficient to muster a conclusion." Sheehan et al., "Sudden extinction of the dinosaurs"; Archibald, *Dinosaur Extinction at the End of an Era*; Archibald, "I. Extinction, Cretaceous."

210 ". . . the only localities with enriched terrestrial vertebrate faunas that straddle the K/T boundary are in western North America . . ." Archibald, *Dinosaur Extinction at the End of an Era*.

210 ". . . we find no significant falloff in diversity . . ." Ibid.

210 ". . . no statistical difference in diversity and abundance of dinosaurs . . ." Ibid.

211 ". . . the 70 percent extinction factor . . . doesn't typify *all* groups." Kiessling and Claeys, "A geographic database approach to the KTB."

211 ". . . ammonites, were roundly whacked . . ." Ibid.

211 "Species losses were nearly this devastating . . . for . . . plankton . . ." Ibid.

211 ". . . some groups, such as gastropods . . . lost less than 60 percent of their species." Ibid.

211 ". . . more than 22,000 plant specimens representing 353 species . . ." Wilf, P., and Johnson, K. R., 2004, "Land plant extinction at the end of the Cretaceous: a quantitative analysis of the North Dakota megafloral record," *Paleobiology* 30:347–68.

212 ". . . study of pollen, or palynology, has its own rules . . ." Ibid.

212 "Studies of fossil pollen . . . show less devastating losses . . ." Nichols, D. J., and Johnson, K. R., 2002, "Palynology and microstratigraphy of Cretaceous-Tertiary boundary sections in southwestern North Dakota," in Hartman, J. H., Johnson, K. R., and Nichols, D. J. (eds.), *The Hell Creek Formation and the Cretaceous-Tertiary Boundary in the Northern Great Plains—An Integrated Continental Record at the End of the Cretaceous*, Geological Society of America Special Paper 361, pp. 95–143.

212 "High concentrations of ferns, or fern spikes, are known from . . . North America. But this pattern does not hold up everywhere." Askin, R. A., and Jacobson, S. R., 1996, "Palynological change over the Cretaceous-Tertiary boundary on Seymour Island, Antarctica: environmental and depositional factors," in MacLeod, N., and Keller, G. (eds.), *Cretaceous-Tertiary Mass Extinctions: Biotic and Environmental Changes*, W. W. Norton, New York; Flannery, T., 2001, "North American devastation or global cataclysm?" *Science* 294:1668–69.

212 ". . . fern spikes . . ." Vajda, V., Raine, J. I., and Hollis, C. J., 2001, "Indication of global deforestation at the Cretaceous-Tertiary boundary by New Zealand fern spike," *Science* 294:1700–2.

212 "It suggested . . . mass extinction may have not been so marked outside North America." Askin and Jacobson, "Palynological change."

213 ". . . a post-K/T fern spike . . ." Vajda et al., "Indication of global deforestation."

213 "But an ingenious series of studies of the damage done to fossil leaves by insect feeding . . ." Wilf, P., Labandeira, C. C., Johnson, K. R., and Ellis, B., 2006, "Decoupled plant and insect diversity after the end-Cretaceous extinction," *Science* 313:1112–15.

213 ". . . relied all too heavily on the limited geographic coverage of western North America . . ." Archibald, *Dinosaur Extinction*.

213 ". . . where the samples reveal a striking selectivity in extinction." Ibid.

213 ". . . three other groups . . . suffered more than 75 percent of species extinction." Ibid.

214 ". . . 50 to 100 percent of species of certain groups survived into the earliest Paleocene." Ibid.

214 "Why this extreme bias in victimization?" Novacek, M. J., 1999, "100 million years of land vertebrate evolution: the Cretaceous–Early Tertiary transition," *Annals of the Missouri Botanical Gardens* 86:230–58.

214 "A more involved explanation offered by D. S. Robinson and several coauthors . . ." Robinson et al., "Survival in the first hours of the Cenozoic."

215 ". . . the post-Cretaceous world had its own spectacular events . . ." Novacek, "100 million years of land vertebrate evolution."

215 ". . . a certain bipedal relative of the greater apes . . ." Stringer and Andrews, *The Complete World of Human Evolution*.

215 ". . . ice age extinctions . . ." MacPhee, *Extinctions in Near Time*.

14: THE ERA AFTER

217 ". . . chaparral." See www.blueplanetbiomes.org/chaparral.htm; www.nceas.ucsb.edu/nceas-web/kids/biomes/chaparral.htm.

217 "Many plants . . . require intense heat to germinate." Hanes, T. L., 1971, "Succession after fire in the chaparral of southern California," *Ecological Monographs* 41:27–52.

218 "I was astonished to see the rebound of the habitat . . ." Ibid.

218 "After the K/T apocalypse not all organisms were so resilient as chaparral . . ." Novacek, "100 million years of land vertebrate evolution."

218 "Planktonic foraminifera . . . so devastated that only a single species . . . is found in . . . the Paleocene above the K/T boundary." Olsson, R. K., et al., 1992, "Wall texture classification of planktonic foraminifera genera in the lower Danian," *Journal of Foraminiferal Research* 22:195–213; Koutsoukos, E.A.M., 1996, "Phenotypic experiments into new pelagic niches in Early Danian planktonic foraminifera: aftermath of the K/T boundary event," in Hart, M. B. (ed.), *Biotic Recovery from Mass Extinction Events*, Geological Society, Special Publication 102, London, pp. 319–35.

218 "Remarkably, this and one other species gave rise to all of the hundreds of younger . . . species." Koutsoukos, "Phenotypic experiments."

218 "A similar pattern is recorded for benthic foraminifera . . ." Speijer, R. P., and Van der Zwaan, G. J., 1996, "Extinction and survivorship of southern Tethyan benthic foraminifera across the Cretaceous/Palaeogene boundary," in Hart, *Biotic Recovery from Mass Extinction Events*, pp. 343–71.

218 "The first few meters above the K/T boundary preserve only a few species of crinoids . . ." Håkansson, E., and Thomsen, E., 1999, "Benthic extinction and recovery patterns at the K/T boundary in shallow water carbonates, Denmark," *Palaeogeography, Palaeoclimatology, Palaeoecology* 154:67–85.

218 ". . . more diverse fossils . . ." Ibid.

218 "A similar pattern . . . in the Gulf coastal plain . . ." Hansen, T. A., 1988, "Early Tertiary radiation of marine molluscs and the long-term effects of the Cretaceous-Tertiary extinction," *Paleobiology* 14:37–51.

219 ". . . the famous fern spikes . . ." Vajda et al., "Indication of global deforestation."

219 "It took conifers and angiosperms longer . . ." Tschudy, R. H., and Tschudy, B. D., 1986, "Extinction and survival of plant life following the Cretaceous Tertiary boundary event, western interior, North-America," *Geology* 14:667–70; Arens, N. C., and Jahren, A. H., 2000, "Carbon isotope excursion in atmospheric CO_2 at the Cretaceous-Tertiary boundary: evidence from terrestrial sediments," *PALAIOS* 15:314–22.

219 "So did insects . . . and . . . birds." Wing and Sues, "Mesozoic and Early Cenozoic terrestrial ecosystems"; Novacek, "100 million years"; Wilf et al., "Decoupled plant and insect diversity after the end-Cretaceous extinction."

219 ". . . it takes at least 10 million years for the recovery . . ." Kirchner and Weil, "Delayed biological recovery."

219 "Others . . . question this . . ." For data and references cited therein, see Erwin, "Lessons from the past."

219 ". . . productivity . . . recovered within a few hundred thousand years . . ." Ibid.

219 ". . . recovery of the terrestrial carbon cycle . . . occurred within 130,000 years." Arens and Jahren, "Carbon isotope excursion."

220 "Analyses of mammalian diversity . . ." Alroy, J., 1999, "The fossil record of North American mammals: evidence for a Paleocene evolutionary radiation," *Systematic Biology* 48:107–18.

220 "(These figures are now outdated . . .)" Novacek, *Time Traveler*; Chang et al., *The Jehol Biota*.

220 "Body size in North American mammals sharply increased . . ." Alroy, "The fossil record of North American mammals."

220 "It was not until . . . 20 million years after the K/T event . . . that truly large mammals evolved." Wing and Sues, "Mesozoic and Early Cenozoic terrestrial ecosystems"; Novacek, "100 million years."

220 ". . . the oddity of this 30-million-year lag in trophic opportunism." Wing and Sues, "Mesozoic and Early Cenozoic terrestrial ecosystems."

220 ". . . a time when biological diversity is at a high point." Wilson, *The Diversity of Life*.

220 "A computer simulation analysis showed . . ." Nee, S., and May, R. M., 1997, "Extinction and the loss of evolutionary history," *Science* 278:692–94.

221 ". . . the Cenozoic Era . . ." (ICS), 2005, International Stratigraphic Chart.

221 ". . . about 210 million years in age." For an overview of mammalian history, see Novacek, *Time Traveler*.

221 ". . . tiny one-celled algae called diatoms." Round, F. E., and Crawford, R. M., 1990, *The Diatoms. Biology and Morphology of the Genera*, Cambridge University Press, Cambridge, UK.

222 ". . . diatom expert Edward Theriot . . ." Theriot, E., and Stoermer, E. F., 1984, "Morphological and ecological evidence for two varieties of the diatom *Stephanodiscus niagarae*," *Eighth Diatom Symposium*, Richard, M. (ed.), Koeltz, Scientific Books, Koenigstein, Taurus, Germany; pp. 385–94.

222 ". . . at about 50 million years something important happened to them." Falkowski, P. G., et al., 2004, "The evolution of modern eukaryotic phytoplankton," *Science* 305:354–60.

222 "The proliferation of oxygen-producing diatoms . . ." Ibid.

222 ". . . oxygen, just like CO_2 content, has gone up and down over time . . ." Berner and Canfield, "A new model for atmospheric oxygen"; Berner, "The rise of plants and their effect on weathering and atmospheric CO_2"; Berner and Kothavala, "Geocarb III"; Scotese, Paleomap Project; Huey and Ward, "Hypoxia."

222 ". . . a paper that ties this Eocene oxygen surge to a major change in the ecosystem . . ." Falkowski, P. G., et al., 2005, "The rise of oxygen over the past 205 million years and the evolution of large placental mammals," Science 309:2202–4.

224 ". . . the very slow increase in body mass in mammals through time." Alroy, "The fossil record of North American mammals."

224 ". . . another upward surge in body mass." Ibid.

224 ". . . when many of the modern placental mammals . . . first appear in the fossil record." Novacek, "100 million years."

225 "A . . . dip . . . about 3 million years ago." Alroy, "The fossil record."

225 ". . . extinction events between fifty thousand and ten thousand years ago . . ." Ibid.

225 ". . . Paleocene-Eocene Thermal Maximum . . ." Wing, S. L., et al., 2005, "Transient floral change and rapid global warming at the Paleocene-Eocene Boundary," Science 310:993–96; Novacek, "100 million years"; Janis, C. M., 1993,"Tertiary mammal evolution in the context of changing climates, vegetation, and tectonic events," Annual Reviews of Ecology and Systematics 24:467–500.

225 "During the PETM, warm tropical to subtropical conditions extended nearly to . . . Arctic regions like the Ellesmere Islands . . ." Hickey, L. J., et al., 1983, "Arctic terrestrial biota: paleomagnetic evidence of age disparity with mid northern latitudes during the Late Cretaceous and Early Tertiary," Science 221:1153–56; Novacek, M. J., et al., 1991, "Wasatchian (early Eocene) mammals and other vertebrates from Baja California, Mexico, the Lomas Las Tetas de Cabra Fauna," Bulletin of the American Museum of Natural History 208:1–88.

225 ". . . cooling was accompanied by increased aridity." Novacek, "100 million years"; Janis, C. M., 1993,"Tertiary mammal evolution in the context of changing climates, vegetation, and tectonic events," Annual Reviews of Ecology and Systematics 24:467–500.

226 ". . . prompt some paleontologists to claim that the inauguration of widespread grasslands marks the origin of the true modern ecosystem." Collinson, M. E., 2001, "Rise of modern land plants and vegetation," in Briggs and Crowther, Palaeobiology II, pp. 112–15.

226 ". . . their history goes back to those 90-million-year old . . . clay pits . . ." Cornet, "Upper Cretaceous Facies."

226 "Grasses are . . . well suited to the reliably energetic winds of the open plains . . ." Tudge, The Variety of Life.

226–27 "The Early Cenozoic was a time . . . for . . . mammals that spent their lives in or above trees." Wing and Sues,"Mesozoic and Early Cenozoic terrestrial ecosystems"; Novacek, "100 million years."

227 "Today . . . bats are the main reason why the tropics have a much greater diversity . . ." Wilson and Reeder, Mammal Species of the World: A Taxonomic and Geographic Reference.

227 "Bats . . . can be traced back to the Early Eocene." Novacek, M. J., 1985, "Evidence for echolocation in the oldest known bats," Nature 315:140–41.

227 "And what an opportunistic group bats have turned out to be!" Fascinating aspects of bat behavior and ecology are treated in Hill, J. E., and Smith, J. D., 1984, Bats: A Nat-

ural History, University of Texas Press, Austin; Nowak, R. M., 1994, *Walker's Bats of the World*, Johns Hopkins University Press, Baltimore and London.

228 ". . . other arboreal mammals." Walker, E. P., and Paradiso, J. L., 1975, *Mammals of the World*, Johns Hopkins University Press, Baltimore.

228 ". . . a mammal group of special interest to us." Ibid.

229 ". . . as Queen Victoria noted haughtily, 'so disagreeably human.'" www.amnh.org /exhibitions/darwin/.

229 ". . . primates evolved in trees for most of their history." Among the many volumes on primate history and evolution, see Ciochon, R. L., and Fleagle, J. R. (eds.), 1987, *Primate Evolution and Human Origins*, Aldine de Gruyter, New York.

230 ". . . *Australopithecus afarensis* . . . was probably adept at climbing trees." Stringer and Andrews, *The Complete World of Human Evolution*.

230–31 "Early humans were, pound for pound, much weaker . . ." Ibid.

231 ". . . they were easy prey for . . . the leopard and the hyena." Hart, D., and Sussman, R. W., 2005, *Man the Hunted: Primates, Predators, and Human Evolution*, Westview Press, Boulder, CO.

15: WHO THEY WERE

235 "Olduvai Gorge . . ." The history of this site is enriched with personal accounts and biographical works. Among them are Leakey, L.S.B., 1974, *By the Evidence: Memoirs, 1932–1951*, Harcourt, Brace, Jovanovich, New York; Leakey, M. D., 1985, *Disclosing the Past*, Doubleday, New York; Morell, V., 1996, *Ancestral Passions: The Leakey Family and the Quest for Humankind's Beginnings*, Touchstone Books, New York.

235 ". . . Richard Fortey might call an oxymoronic sacred place for evolution . . ." Fortey, *Earth: An Intimate History*.

235 ". . . dogged paleontological labor . . ." Morell, *Ancestral Passions*.

236 ". . . what do we mean when we say *human* evolution?" Diamond, J. M., 1992, *The Third Chimpanzee: The Evolution and Future of the Human Animal*; Tattersall, I., 2002, *The Monkey in the Mirror*, Harcourt, New York; Stringer and Andrews, *The Complete World*.

236 ". . . 1.3 percent of the 3 billion DNA nucleotides in the genomes of chimps and humans differ." Chimpanzee Sequencing and Analysis Consortium, "Initial sequence of the chimpanzee genome and comparison with the human genome."

236 ". . . scientists' modern classifications of humans and their nearest primate relatives." Stringer and Andrews, *The Complete World*.

236 ". . . skeletal features that show our antecedents . . . as having been bipedal . . ." Ibid. and Tattersall, *The Fossil Trail*.

236 "However . . . chimpanzees . . . use grass stems and other objects as tools . . ." Stringer and Andrews, *The Complete World*.

236 "Another oft-cited feature is the development of language." Johanson, D. C., 1996, *From Lucy to Language*, Simon & Schuster, New York.

237 "One aspect of the brain . . . shows a clear trend in human evolution . . ." Holloway, R. L., 1974, "The casts of fossil hominid brains," *Scientific American*, July 1974, pp. 106–15.

237 ". . . the relative brain size was hardly larger than that of present-day apes." Stringer and Andrews, *The Complete World*.

237 "... complex language and intricate social organization are at least indirectly indicated by this increased brain capacity." Ibid.

237 "Instead, human history was ... a branching bush comprising numerous species that appeared and eventually died out ..." Ibid. and Tattersall, *The Fossil Trail.*

237 "Olduvai Gorge is important because in it are preserved some key branches ..." Tattersall, I., et al., 1999, *Encyclopedia of Human Evolution and Prehistory,* 2nd ed., Garland Publishing, New York.

237–38 "... the Flaming Cliffs of Mongolia's Gobi Desert." Andrews, *New Conquest.*

238 "... Olduvai Gorge proved stubbornly resistant to disclosure of its primary treasure ..." Leakey, L.S.B., *By the Evidence: Memoirs, 1932–1951;* Morell, *Ancestral Passions.*

238 "But it was not until 1959 ... that Mary Leakey found ..." Leakey, M. D., *Disclosing the Past.*

238 "This inaugural skull of what is now called *Paranthropus boisei* was later joined by a variety of fossils ..." Leakey, L.S.B, 1959, "A new fossil skull from Olduvai," *Nature* 184:491–93; Leakey, L.S.B., 1960, "Recent discoveries at Olduvai Gorge," *Nature* 188:1050–52.

238 "... *Homo habilis* ..." Stringer, C., 1992, "*Homo habilis* closely examined," *Current Anthropology* 33:338–40; Tattersall, I., 1992, "The many faces of *Homo habilis,*" *Evolutionary Anthropology* 1:33–37.

239 "Olduvai is not the only crucible for early human evolution." For a concise, updated survey of the major sites and fossil discoveries of early humans in Africa, see Stringer and Andrews, *The Complete World.*

239 "... *Sahelanthropus tchadensis* ..." Brunet, M., et al., 2002, "A new hominid from the upper Miocene of Chad, central Africa," *Nature* 418:145–51.

239 "... *Orrorin tugenensis* ..." Senut, B., et al., 2001, "First hominid from the Miocene (Lukeino Formation, Kenya)," *Comptes rendus de l'Académie de Sciences* 332:137–44.

239 "... *Ardipithecus kadabba* ..." White, T. D., Suwa, G., and Asfaw, B., 1994, "*Australopithecus ramidus,* a new species of early hominid from Aramis, Ethiopia," *Nature* 371:306–12.

239 "... including Lucy, of *Australopithecus afarensis.*" Johanson, D. C., and Edey, M. A., 1981, *Lucy: The Beginnings of Humankind,* Simon & Schuster, New York; see also Tattersall, *The Fossil Trail;* Stringer and Andrews, *The Complete World.*

239 "... 3.5-million-year-old footprints in the ash beds of Laetoli ..." Hay, R. L., and Leakey, M. D., 1982, "Fossil footprints of Laetoli," *Scientific American,* February 1982, pp. 50–57.

239 "Localities farther south in Africa have been particularly critical ..." Stringer and Andrews, *The Complete World.*

239 "... Darwin's intuition that Africa was the homeland of humans ..." Darwin, C. R., 1871, *The Descent of Man, and Selection in Relation to Sex,* John Murray, London.

239 "... *Australopithecus africanus* ..." Dart, R. A., 1925, "*Australopithecus africanus:* the man-ape of South Africa," *Nature* 115:195–99.

239 "... famous false fossils as the Piltdown Man from England." Spencer, F., 1990, *Piltdown: A Scientific Forgery,* Oxford University Press, London.

240–41 "The skeleton, known as Little Foot ..." Clarke, R. J., and Tobias, P. V., 1995, "Sterkfontein member 2 foot bones of the oldest South African hominid," *Science* 269:521–24.

241 "... *Paranthropus robustus* ..." Broom, R., 1938, "The Pleistocene anthropoid apes of South Africa," *Nature* 142:377–79.

241 "... *Homo erectus.*" Leakey, L.S.B., 1966, "*Homo habilis, Homo erectus* and the australopithecines," *Nature* 209:1279–81; Leakey, R. E., and Walker, A. C., 1985, "*Homo erectus* unearthed," *National Geographic* 168 (November): 624–29; de Chardin, T. P., 1930, "*Sinanthropus pekinensis*: an important discovery in human palaeontology," *Revue des questions scientifiques*, July 20.

241 "... the striking early diversity and experimentation in our ancestry ..." A more current count is summarized in Stringer and Andrews, *The Complete World*.

241 "... first 6 million years of that history is locked in the rocks of Africa." Ibid.

241 "... Dmanisi site in the Caucasus region ..." Gabunia, L., et al., 2000, "Earliest Pleistocene hominid cranial remains from Dmanisi, Republic of Georgia: taxonomy, geological setting, and age," *Science* 288:1019–25; Vekua, A., et al., 2002, "New Skull of Early *Homo* from Dmanisi, Georgia," *Science* 297:85–89.

241 "New research on human fossils from sub-Saharan Africa ..." Grine, F. E., et al., 2006, "Late Pleistocene skull from Hofmeyr, South Africa, and modern human origins," *Science* 315:226–29.

241 "Even most studies of genes ... fit well with ... African origins." Stringer, C. B., and Andrews, P., 1988, "Genetic and fossil evidence for the origin of modern humans," *Science* 239:1263–68; Vigilant, L., et al., 1991, "African populations and the evolution of human mitochondrial DNA," *Science* 253:1503–7; Bowcock, A. M., et al., 2001, "High resolution of human evolutionary trees with polymorphic microsatellites," *Nature* 368:455–57; Ke, Y., et al., 2001, "African origin of modern humans in East Asia: a tale of 12,000 Y chromosomes," *Science* 292:1151–53.

241 "... humans lived in a terrifying world." Hart and Sussman, *Man the Hunted*; Stringer and Andrews, *The Complete World*.

242 "The dispersal of humans from Africa ... was followed by a number of other milestones ..." Tattersall, *The Fossil Trail*; Stringer and Andrews, *The Complete World*.

243 "... *Homo floresiensis* ..." Brown, P., et al., 2004, "A new small-bodied hominin from the Late Pleistocene of Flores, Indonesia," *Nature* 431:1055–61. Morwood, M. J., et al., 2004, "Archaeology and age of a new hominin from Flores in eastern Indonesia," *Nature* 431:1087–91; Martin, R. D., et al., 2006, "Comment on 'The Brain of LB1, *Homo floresiensis.*'" *Science* 312:999.

243 "... a complex and uncertain pattern." Tattersall, *The Fossil Trail*.

243 "... (*Homo neanderthalensis*) ..." Trinkaus, E., and Shipman, P., 1992, *The Neandertals: Changing the Image of Mankind*, Knopf, New York; Tattersall, I., 1995, *The Last Neanderthal: The Rise, Success and Mysterious Extinction of Our Closest Human Relatives*, Macmillan, New York.

243 "... artistic Cro-Magnon *Homo sapiens*." Leakey, R. E., 1994, *The Origin of Humankind*, Basic Books, New York.

243 "... a genetic bottleneck ..." Harpending, H. C., et al., 1998, "Genetic traces of ancient demography," *Proceedings of the National Academy of Sciences* 95:1961–67; Cann, R. L., 2001, "Genetic clues to dispersal in human populations: retracing the past from the present," *Science* 291:1742–48.

244–45 "... at least four humanlike species were living in Africa ..." Stringer and Andrews, *The Complete World*.

245 "Today we are alone ..." Ibid.

245 "Nonetheless, we see in *Homo sapiens* a refinement in the tools . . ." Leakey, R. E., *The Origin of Humankind.*

245 ". . . at least thirteen thousand years ago, when humans likely reached the Americas . . ." Alroy, J., 1999, "Putting North America's end-Pleistocene megafaunal extinction in context."

245 ". . . new species arise most often from isolated populations . . ." Darwin, *Origin;* Mayr, *Systematics and the Origin of Species.*

16: THE EXTERMINATORS

246 "Australia was at one time connected with other huge landmasses . . ." Scotese, 2002, Paleomap Project.

246 "More than 80 percent of its native flowering plants and inshore temperate zone fish are endemic." For general information, see Australian Government Department of Environment and Heritage, National Biodiversity Hotspots, www.deh.gov.au/biodiversity /hotspots/national.html.

246 ". . . fifty thousand years ago . . . Australia was a land of giants." Murray, P. F., 1991, "The Pleistocene megafauna of Australia," in Rich, P. V., et al., *Vertebrate Palaeontology of Australasia,* Pioneer Design Studio, Lilydale, Australia, pp. 1071–164.

247 "It is today devoid of the large browsers and carnivores . . ." Strahan, R. (ed.), 1995, *The Mammals of Australia,* Reed Books, Chatswood, Australia.

247 ". . . a synchrony between the appearance of humans in Australia and the demise of some of its charismatic denizens." Much of this evidence is reviewed in Flannery, T. F., and Roberts, R. G., 1999, "Late Quaternary extinctions of Australasia," in MacPhee, *Extinctions in Near Time,* pp. 239–55. Also see Flannery, T. F., 1999, "Debating extinction," *Science* 283:182–83; Field, J., and Dodson, J., 1999, "Late Pleistocene megafauna and archaeology from Cuddie Springs, South-eastern Australia," *Proceedings of the Prehistoric Society* 65:275–301.

247 ". . . first arrival of humans in Australia is between fifty and sixty thousand years." Flannery and Roberts, "Late Quaternary extinctions of Australasia." Also see Roberts et al., 2001, "New ages for the last Australian megafauna: continent-wide extinction at 46,000 years," *Science* 292:1888–92.

247 ". . . locality called Cuddie Springs . . ." Flannery and Roberts, "Late Quaternary extinctions of Australasia."

247 "Other places in the world reveal . . ." For summaries of events, places, and casualties, see Martin, P., and Steadman, D. W., 1999, "Prehistoric extinctions on islands and continents," in MacPhee, *Extinctions in Near Time,* pp. 17–55.

247 "In North America . . . the earliest-known human sites—about 13,400 years before present . . ." Fiedel, S. J., 1999, "Older than we thought: implications of corrected dates for paleoindians," *American Antiquity* 64:95–115.

248 ". . . a detailed book, *The Hippos* . . ." Eltringham, S. K., 1999, *The Hippos: Natural History and Conservation,* Princeton University Press, Princeton, NJ.

248 "The Pleistocene epoch . . . was besieged by intervals of enormous glacial expansion from the northern ice cap . . ." Imbrie, J., and Imbrie, K. P., 1986, *Ice Ages: Solving the Mystery,* Harvard University Press, Cambridge, MA; Lourens, L., et al., 2004, "The

Neogene Period," in Gradstein, F., Ogg, J., and Smith, A. G. (eds.), *A Geologic Time Scale 2004*, Cambridge University Press, Cambridge, UK.

249 "Climate instability was . . . the standard explanation." This theory is still upheld by many, including Graham, R. W., 1976, "Late Wisconsin mammalian faunas and environmental gradients of the eastern United States," *Paleobiology* 2:343–50; Graham, R. W., 1986, "Response of mammalian communities to environmental changes during the late Quaternary," in Diamond. J. D., and Case, T. J. (eds.), *Community Ecology*, Harper & Row, New York, pp. 300–13.

249 ". . . Paul Martin first presented in the 1960s a comprehensive theory showing that intense hunting by humans . . ." Martin, P. S., 1967, "Pleistocene overkill," in Martin, P. S., and Wright, H. E. (eds.), *Pleistocene Extinctions: The Search for a Cause*, Yale University Press, New Haven, CT, pp. 75–120; Martin, P. S., 1984, "Prehistoric overkill: the global model," in Martin, P. S. and Klein, R. G. (eds.), *Quaternary Extinctions: A Prehistoric Revolution*, University of Arizona Press, Tucson, pp. 354–403.

249 "Proponents of Martin's overkill theory, like John Alroy, have first focused on problems in the traditional climate, or ecological, model." The evidence and arguments for and against the overkill model are effectively treated in a detailed analysis by Alroy, "Putting North America's end-Pleistocene megafaunal extinction in context: large scale analyses of spatial patterns, extinction rates, and size distributions"; Alroy favors the overkill model. For a more pluralistic evocation of both overkill and climate, see Barnosky, A. D. et al., 2004, "Assessing the causes of Late Pleistocene extinctions on the continents," *Science* 306:70–74. Another paper that emphasizes a more complex interplay and a more extended period of extinction is Stuart, A. J., et al., 2004, "Pleistocene to Holocene extinction dynamics in giant deer and woolly mammoth," *Nature* 431:684–89.

249 "Extinction of large North American mammals occurred within 1,000 or 2,000 years of the arrival of humans . . ." Alroy, "Megafaunal extinction in context."

249 ". . . Clovis in New Mexico . . ." Dixon, E. J., 1999, *Bones Boats and Bison: The Early Archeology of Western North America*, University of New Mexico Press, Albuquerque.

249 "Masses of bones next to stone weapons . . ." Haynes, G., and Eiselt, B. S., 1999, "The power of Pleistocene hunter-gatherers: forward and backward searching for evidence about mammoth extinction," in MacPhee, *Extinctions in Near Time*, pp. 71–93.

249 ". . . that were either plausible prey species . . ." Alroy, "Megafaunal extinction in context."

250 "Finally, the extinctions of large vertebrates in South America . . . and New Zealand occurred at different times . . . but always when human hunters first appeared." Martin and Steadman, "Prehistoric extinctions on islands."

250 ". . . Africa and southern Asia were largely spared the exterminations . . ." Ibid.

250 "Darwin himself pondered this question as he stepped off the *Beagle* . . ." Darwin, C. R. 1845, *Journal of Researches into the Natural History and Geology of the Countries Visited During the Voyage of H.M.S. Beagle Round the World, Under the Command of Capt. Fitz Roy, R.N.*, 2nd ed., John Murray, London.

250 "There is a reasonable explanation . . ." Martin, "Prehistoric overkill: the global model."

250 ". . . paleontological evidence shows no further major extinctions of large mammals in North America in the last few thousand years . . ." This point is also addressed in Alroy, "Megafaunal extinction in context."

251   "How could a small population of early humans . . . devastate so many large animals so quickly?" Addressed ibid.

251   "One might readily envision such a possibility in smaller areas, such as islands . . ." Steadman, "Prehistoric extinctions of Pacific island birds"; Steadman, *Extinction and Biogeography of Tropical Pacific Birds*.

251   ". . . the spread of diseases from humans . . . was the primary reason for the destruction." MacPhee, R.D.E., and Marx, P. A., 1997, "The 40,000-year plague: humans, hyperdisease, and first-contact extinctions," in Goodman, S. M., and Patterson, B. D. (eds.), *Natural Change and Human Impact in Madagascar*, Smithsonian Institution Press, Washington, D. C., pp. 169–217.

251   ". . . armed hunters could . . . do significant damage." Alroy, J., 2001, "Multispecies overkill simulation of the end-Pleistocene megafaunal mass extinction," *Science* 292:1893–96.

251   ". . . claims for human sites in the Americas . . . as much as 50,000 years . . . are rejected by most experts." Stein, R., 2004, "Team says humans lived in North America earlier," *Washington Post*, November 18, 2004, www.washingtonpost.com/wp-dyn/articles/A58632-2004Nov17.html.

251   "Assuming a minimal colonization of North America, involving 100 humans and a modest population growth rate of 1.7 percent per year . . ." Alroy, "A multispecies overkill simulation."

251–52   ". . . would mean that there was a considerable overlap . . . between humans and the prey . . ." Ibid.

252   ". . . a million people by 13,250 years ago . . ." Ibid.

252   ". . . index for hunting ability . . ." Ibid.

252   "Circumstantial evidence suggests that the 'good hunter' assumption . . ." Haynes, G., and Eiselt, B. S., 1999, "The power of Pleistocene hunter-gatherers."

252   ". . . piles of mammoth skulls and bones at Mezhirich in Ukraine . . ." Pidoplichko, I. H., 1998, *Upper Palaeolithic Dwellings of Mammoth Bones in the Ukraine: Kiev-Kirillovskii, Gontsy, Dobranichevka, Mezin and Mezhirich*, J. and E. Hedges, Oxford, UK.

252   ". . . intensive fishing in coastal waters of northeastern North America extends back into aboriginal times." Jackson, J.B.C., et al., 2001, "Historical overfishing and the recent collapse of coastal ecosystems," *Science* 293:629–37.

253   ". . . early humans were capable within a millennium of eliminating most of the large mammals . . ." See also Alroy, "Multispecies overkill simulation."

253   ". . . these developments arose at least eleven thousand years ago in southwestern Asia and within about five thousand years thereafter in eight other regions . . ." Diamond, J., 1999, *Guns, Germs, and Steel*.

17: THE CULTIVATORS

254   ". . . 'nasty, brutish, and short' . . ." Hobbes, T., 1651, *Leviathan, or the Matter, Forme, and Power of a Commonwealth, Ecclesiasticall and Civill*.

254   ". . . a life portrayed by some prominent archeologists, such as Robert Braidwood . . ." Braidwood, R., 1964, *Prehistoric Men*, Chicago Natural History Museum, Chicago; also quoted in Winterhalder, B., 1993, "Work, resources, and population in foraging societies," *Man* 28:321–40.

254 "Yet this portrayal of bare hand-to-mouth existence is . . . grossly overstated." Jared Diamond deals with this and many other issues covered in this chapter in his well-known book *Guns, Germs, and Steel*. See especially his chapter 6: "To Farm or Not to Farm."

255 "Hunter-gatherer societies . . . of the Amazon . . . have an extraordinary knowledge of . . . thousands of species of plants and many animals . . ." Ibid.

255 "Early hunter-gatherers were taller, better nourished, and healthier . . ." Ibid.

255 "The best translation for the Wappo word *nappa* is 'plenty.'" Keller, T., 1999, *The French Laundry Cookbook*, Artisan/Workman Publishing, New York.

256 ". . . agricultural practice and its products . . ." As named in Diamond, *Guns, Germs, and Steel*, and many other works. Also see Rindos, D., 1984, *The Origins of Agriculture: An Evolutionary Perspective*, Academic Press, New York.

256 ". . . seed crops . . . root crops . . ." Diamond, *Guns, Germs, and Steel*.

257 "The origins of animal domestication . . ." Ibid. and Mason, I. L., 1984, *Evolution of Domesticated Animals*, Longman, London.

257 ". . . an impressive fourteen species of big herbivorous domesticated animals . . ." See Table 9.1 in Diamond, *Guns, Germs, and Steel*.

257 ". . . we must distinguish domesticated animals from merely tamed ones." Ibid.

257 "Attempts to domesticate very temperamental zebras are dangerous and the African buffalo . . . nearly suicidal." Ibid. Also see Nowak, *Walker's Mammals of the World*, vols. 1 and 2.

258 ". . . Jared Diamond and others regard as *the* singular event in postagricultural human history . . ." Diamond, *Guns, Germs, and Steel*.

258 "The scholarly literature offers many proposals . . ." Saur, C., 1952, *Agricultural Origins and Dispersals*, American Geographical Society, New York; Reed, C. (ed.), 1977, *Origins of Agriculture*, Mountan, The Hague; Rindos, *Origins of Agriculture*; MacNeish, R., 1992, *The Origins of Agriculture and the Settled Life*, University of Oklahoma Press, Norman.

258 ". . . climate change is deemed a critical element . . ." Richerson, P. J., Boyd, R., and Bettinger, R. L., 2001, "Was agriculture impossible during the Pleistocene but mandatory during the Holocene? A climate change hypothesis," *American Antiquity* 66:387–411.

259 ". . . population pressures compelled foragers to make the switch . . . [to agriculture]." Influential exponents of this theory are Boserup, E., 1965, *The Conditions of Agricultural Growth: The Economics of Agrarian Change Under Population Pressure*, Aldine, Chicago; Cohen, M. N., 1977, *The Food Crisis in Prehistory: Overpopulation and the Origins of Agriculture*, Yale University Press, New Haven, CT; Glassow, M. A., 1978, "The concept of carrying capacity in the study of culture process," in Schiffer, M. B. (ed.), *Advances in Archaeological Method and Theory*, vol. 1, pp. 31–48, Academic Press, New York.

259 ". . . there may have been as many as a million people on the continent . . ." Alroy, "Multispecies overkill simulation."

259 ". . . Native American populations at the time of Columbus were . . . twenty million.)" Diamond, *Guns*.

260 "Consider the tendency of humans to aggregate in communities." Perhaps the most influential exponent is Saur, *Agricultural Origins and Dispersals*.

260 "As the burgeoning local populations became sedentary, they developed new . . . farming methods . . ." Ibid.

260 "And there is a self-reinforcing, mutualistic relationship between cultivator and culti-

vated." The so-called evolutionary theory for the origins of agriculture is prominently the work of David Rindos; see Rindos, D., 1980, "Symbiosis, instability, and the origins and spread of agriculture: a new model," *Current Anthropology* 21:751–72; Rindos, *Origins of Agriculture*; Rindos, D., 1989, "Darwinism and its role in the explanation of domestication," in Harris, D. R., and Hillman, G. C. (eds.), *Foraging and Farming*, Unwin Hyman, London, pp. 27–41.

260 ". . . humans themselves have been powerful selective agents." Rindos, *Origins of Agriculture*.

260 "How did the system work?" Ibid.

261 "The selection of plants for the quality of their products . . ." Diamond, *Guns, Germs, and Steel*.

261 ". . . almonds, are actually poisonous to humans . . ." Ibid.

261 "Last to come were . . . fruit trees that required grafting . . ." Ibid.

261 ". . . pollination of plants through the activity of insects . . . was not understood until millennia after humans first domesticated plants and animals for their own benefit." An excellent history of this delayed discovery is given in chapter 1 of Proctor et al., *The Natural History of Pollination*.

262 ". . . Darwin published his important observations of flower pollination . . ." Darwin, C. R., *On the Various Contrivances by Which British and Foreign Orchids Are Fertilised by Insects*.

263 "Today beekeeping is a practice . . . for the managed pollination of . . . important crops . . ." See McGregor, *Insect Pollination of Cultivated Crop Plants*.

263 "Despite the apparent inevitability of agriculture . . . some cultures and regions developed faster and more elaborately than others." This is the primary focus of Diamond's *Guns, Germs, and Steel*.

263 "Why this checkered geographic pattern of haves and have-nots?" Ibid.

263 "Another geographic factor may have been critical." Ibid.

265 "Mongolia . . . has only 2.5 million people but more than 33.6 million cattle, goats, sheep, horses, and camels." University of Pennsylvania Museum of Archaeology and Anthropology website, Modern Mongolia: Reclaiming Genghis Khan, www.museum .upenn.edu/Mongolia/section3.shtml.

18: LAND RUSH

267 "About a hundred thousand years ago, the Laurentide ice sheet began to form." A good website on this last major glaciation phase is by the National Park Service, Wisconsin's Glacial Legacy, www.nps.gov/archive/iatr/expanded/history.htm.

268 ". . . modern fertilizer, the new wonder substance for crops . . ." Stoltzenberg, *Fritz Haber*.

269 "Croplands and pastures now occupy 40 percent of Earth's land surface." Foley, J. A., et al., 2005, "Global consequences of land use," *Science* 309:570–74.

269 ". . . 350 million years ago, dense fern, cycad, and gymnosperm forests blanketed the megacontinent of Pangea . . ." DiMichele, "Carboniferous coal swamp forests."

269 "Likewise in the Early Eocene . . . angiosperm-dominated forests extended from the equator to the poles." Wing, "Transient floral change and rapid global warming at the Paleocene-Eocene Boundary,"

269 ". . . 11,000 years ago, forests occupied nearly 70 percent of the land's surface." My-

gatt, E., 2006, "World's forests continue to shrink," Earth Policy Institute, www.earth-policy.org/Indicators/Forest/index.htm; FAO, *Global Forest Resources Assessment 2005*, Rome, 2006, www.fao.org/forestry/site/32038/en.

269 "But in subsequent centuries energetic cultivation thinned out those forests." Hughes, J. D., 1975, *Ecology in Ancient Civilizations*, University of New Mexico Press, Albuquerque. An excellent website with thorough documentation is Kovarik, W., *Environmental History Timeline*, www.radford.edu/~wkovarik/envhist/timeline.text.html. An excellent historical review of trends in land use is McNeill, J. R., and Winiwarter, V., 2004, "Breaking the sod: humankind, history, and soil," *Science* 304:1627–29.

269 "But these historical losses pale in comparison with those caused by the spread of agriculture . . . during the past two hundred years." Mygatt, "World's forests continue to shrink."

270 "The pace of destruction for tropical rainforests . . . is worse." Ibid.

270 ". . . world grain harvests, for example, now weigh in at more than two billion tons . . ." For the statistics on agricultural production given in this paragraph, see Foley et al., "Global consequences of land use."

270 "Our effort to maximize yield from the land has been in lockstep with the rise of civilizations." McNeill and Winiwarter, "Breaking the sod."

270 ". . . about 845 million people, 14 percent of the global human population, are . . . malnourished." Sanchez, P. A., and Swaminathan, M. S., 2005, "Cutting world hunger in half," *Science* 307:357–59.

271 "The long-ago development of great river settlements . . . made for . . . an increased rate of soil erosion." McNeill and Winiwarter, "Breaking the sod."

271 ". . . Loess Plateau . . ." Ibid.

271 "But then this changed radically between the sixteenth and the nineteenth centuries." Ibid.

272 ". . . accompanied the rise in the use of artificial fertilizers in the 1950s." Smil, V., 2001, *Enriching the Earth*, MIT Press, Cambridge, MA.

272 ". . . dust bowls of the Great Plains . . ." Egan, T., 2006, *The Worst Hard Time: The Untold Story of Those Who Survived the Great American Dust Bowl*, Houghton Mifflin, New York.

272 ". . . stimulated improvements in North America and Europe, such as the U.S. Department of Agriculture's Soil Conservation Service . . ." McNeill and Winiwarter, "Breaking the sod."

272 ". . . 17 percent was degraded by erosion." This and related statistics are given in Foley et al., "Global consequences of land use." Also see Kaiser, J., 2004, "Wounding Earth's fragile skin," *Science* 304:1616–18.

272 ". . . 11 billion tons of human-induced erosion yearly . . . at least eleven billion U.S. dollars annually in lost crop production." Ibid.

272 "Dire predictions . . . of Malthusian proportions have not been borne out." Kaiser, "Wounding Earth's fragile skin."

272 "In Haiti only 3 percent of the nation's once lush tropical forests remain . . ." Ibid.

273 "Demand for food and other resources is very unevenly distributed on a global scale." Sanchez and Swaminathan, "Cutting world hunger in half."

274 "The water table in the Gobi and all Central Asia is sinking." Brown, "Water deficits growing in many countries—water shortages may cause food shortages." Great Lakes Directory, www.greatlakesdirectory.org/zarticles/080902_ water_shortages.htm.

274 "Global statistics illustrate why such a decline in freshwater is not just a local phenom-enon." This is the subject of a comprehensive volume by Pearce, F., 2006, *When the Rivers Run Dry Water: The Defining Crisis of the Twenty-first Century*, Beacon Press, Boston.

274 "Mexico City . . . has sunk thirty feet over the last hundred years." Dillon, S., 1998, "Journal: Mexico City sinking into a depleted aquifer." *The New York Times*, January 29, 1998, www.greatdreams.com/cities.htm.

274 ". . . the Aral Sea . . ." Bissell, T., 2000, "Eternal winter: lessons of the Aral Sea disaster," *Harper's Magazine*, April 2002, pp. 41–56; Ferguson, R., 2003, *The Devil and the Disappearing Sea*, Raincoast Books, Vancouver, BC; Bendhun, F., and Renard, P., 2004, "Indirect estimation of groundwater inflows into the Aral sea via a coupled water and salt mass balance model," *Journal of Marine Systems* 47:35–50.

274 ". . . Europe's cities are using groundwater at unsustainable rates. More than thirteen African countries suffer from water . . . scarcity . . ." Global Water Outlook to 2025: Averting an Impending Crisis, www.ifpri.org/media/water_countries.htm.

274 "North America's largest aquifer . . . is being depleted . . ." Northern Plains Ground-water Conservation District (survey), www.npwd.org/Ogallala.htm.

274 "Worldwide, agriculture accounts for 70 percent of the total consumption of freshwa-ter, while industry comes in at about 22 percent, and domestic use at only 8 percent." UNESCO World Water Assessment Programme, 2003, The United Nations World Water Development Report: Water for People, Water for Life (Executive Summary), www.unesco.org/water/wwap/wwdr1/ex_summary/index.shtml.

274 "Freshwater is of course only a small fraction of the water on the planet." Ibid.

274 "Most of the available freshwater belongs to a fortunate few large countries . . ." Ibid.

274 "Water-deprived countries include rich ones . . . and poor ones . . ." Ibid.

275 ". . . global consumption of water increased sixfold . . ." Ibid.

275 ". . . bottled water . . . currently accounting for sales of $4 billion a year." National Resources Defense Council (NRDC), *Bottled Water: Pure Drink or Pure Hype?*, www.nrdc.org/water/drinking/bw/exesum.asp.

275 "Stark differences in consumption exist even among well-endowed countries." Interna-tional Food Policy Research Institute, Global Water Outlook to 2025: Averting an Im-pending Crisis, www.ifpri.org/media/water_countries.htm.

275 "Freshwater ecosystems . . . are highly sensitive to environmental disruption." Dud-geon, D., et al., 2005, "Freshwater biodiversity: importance, threats, status and conser-vation challenges," *Biological Reviews* 81:2, 163–82.

275 ". . . tropical rainforests are being rapidly destroyed . . ." Mygatt, "World's forests con-tinue to shrink."

275 "The simple rate of destruction of these habitats is what gives our current estimates of extinction their heft." Pimm et al., "The future of biodiversity."

276 "It is fair to recognize mitigating trends." Foley et al., "Global consequences of land use."

276 "Here we confront the worldwide problem concerning those most proficient of pollina-tors the bees." Buchmann and Nabhan, *The Forgotten Pollinators*; Allen-Wardell et al., "The potential consequences of pollinator declines in the conservation of biodiversity and the stability of food crop yields"; Kremen et al. "Crop pollination from native bees at risk from agricultural intensification"; Frankie and Thorp, "Pollination and pollinators."

276–77 ". . . pollinate one or more of the cultivars of 66 percent of the world's fifteen hun-dred crop species . . ." Southwick and Southwick, 1992, "Estimating the economic value of honey bees (Hymenoptera: Apidae) as agricultural pollinators in the United States"; Morse and Calderone, *Beeculture*, http://bee.airoot.com/content/Pollination Reprint07.pdf.

277 ". . . honeybee *Apis mellifera*, has an estimated annual value of five to fourteen billion . . ." Ibid.

277 "In May 2007 came news . . ." Stokstad, E., 2007, "The case of the empty hives," *Science* 316:970–72.

277 ". . . a ready insurance . . . comes from our surprisingly strong dependence on wild species of bees for agriculture." Kremen et al., "Crop pollination."

277 "A 2002 study . . ." Ibid.

277 ". . . native bees could provide . . . pollination service . . ." Ibid.

277 "Unfortunately, numerous areas . . . show a decline in native bee populations because of degraded habitats . . ." Ibid.

278 ". . . Ricketts and several coauthors showed that coffee farms . . . benefited substantially from wild pollinators . . ." Ricketts, T. H., et al., 2004, "Economic value of tropical forest to coffee production," *Proceedings of the National Academy of Sciences* 101:12579–82.

19: DARK FORCES

279 "This began nearly fifty thousand years ago with the human invasion of Australia . . ." MacPhee, *Extinctions in Near Time*.

280 "Many large marine creatures . . . had been nearly exterminated by the early 1800s." Jackson et al., "Historical overfishing."

280 ". . . resurgence of the elephant is not uniform across Africa." 2006 IUCN Red List.

280 "The horn of the rhinoceros has long been an item treasured . . ." Ellis, R., 2005, *Tiger Bone and Rhino Horn: The Destruction of Wildlife for Traditional Chinese Medicine*, Shearwater Books/Island Press, Washington, D.C.

280 ". . . Yemen was importing about 6,000 pounds of black rhino horns a year." Ibid.

281 "The demand . . . reduced the African black rhino populations by more than 90 percent . . ." Milliken, T., Nowell, K., and Thomsen, J. B., 1993, *The Decline of the Black Rhino in Zimbabwe*, TRAFFIC International.

281 ". . . in Ghana . . . [r]egional bushmeat trade . . . deviated significantly . . . depending on the availability of fish." Brashares, J. S., et al., 2004, "Bushmeat hunting, wildlife declines, and fish supply in West Africa," *Science* 306:1180–83.

281 "Thus declining fish stocks . . . could have a ripple effect on land . . ." Ibid.

282 "Foreign enterprise has artificially increased the profitability of fishing . . ." Ibid.

282 "Fish is the primary food source for as much as one-fifth of the world's population." FAO, 2000, *The State of World Fisheries and Aquaculture 2000*, FAO, Rome, Italy.

283 ". . . overfishing has been far more devastating to marine ecosystems than either habitat destruction or pollution." Jackson et al., "Historical overfishing."

283 "Surveys of world fisheries . . . show a steady decline, followed by a precipitous fall." Pauly et al., "Fishing down marine food webs"; Pauly et al., "The future for fisheries."

283 ". . . underestimates of decline . . ." Ibid.

283 ". . . haddock has slowly climbed back . . ." Safina, C., Andrew, A., Rosenberg, A. A., Myers, R. A., Quinn, T. J., II, and Collie, J. S., 2005, "U.S. ocean fish recovery: staying the course," *Science* 309:707–708.

283 ". . . Sustainable Fisheries Act of 1996 . . ." Ibid.

284 "The once-populous Atlantic cod continues to decline . . ." Ibid.

284 ". . . there is much pressure to emaciate the Sustainable Fisheries Act." Ibid.

284 "In December 2006 the Senate adopted a measure that reauthorized the Magnuson-Stevens Act . . ." Stokstad, E., 2006, "Ocean policy: fisheries bill gives bigger role to science—but no money," *Science* 314:1857.

284 ". . . Japan continues to be unmatched in its drain upon the global marine ecosystem . . ." Data in FAO, 2001, *The State of World Fisheries and Aquaculture 2001*, FAO, Rome, Italy.

284 ". . . average annual per capita consumption of fish and shellfish in Japan was an impressive 145.7 pounds . . ." Ibid.

285 ". . . more than 70 percent of the world's fish species are either fully exploited or depleted." Ibid.

286 "The zebra mussel . . ." A thorough profile of this invasive species that includes images, history, maps, and other resources is the U.S. Department of Agriculture (USDA) National Invasive Species Information Center website, www.invasivespeciesinfo.gov/aquatics/zebramussel.shtml. Also see Nalepa, T. F., and Scholesser, D. W. (eds.), 1993, *Zebra Mussels: Biology, Impacts, and Control*, Boca Raton Lewis Publishers, Boca Raton, FL.

287–88 ". . . a grand isolated bestiary on South America . . . was wiped out by invading species from the north." See Simpson, G. G., 1980, *Splendid Isolation: The Curious History of South American Mammals*, Yale University Press, New Haven, CT. See also an updated general overview in Novacek, *Time Traveler*.

289 "The presence of the dog, probably the first domesticated animal, doubtlessly thinned small wild prey . . ." Diamond, *Guns, Germs and Steel*.

289 "Diseases that spread from these domesticates . . . promoted the extinction of nearly two thousand native Pacific island birds . . ." Steadman, *Extinction and Biogeography of Tropical Pacific Birds*.

289 "A list of the hundred worst invasive species . . ." Lowe, S., et al., 2000, *100 of the World's Worst Invasive Alien Species: A Selection from the Global Invasive Species Database*, ISSG.

289 ". . . virtually all invasive species share these qualities: . . ." Some of the examples here are in ibid.

291 "Classically, biologists have ascribed to these invaders some quality of robustness . . ." This is how Simpson explained the devastation of the South American mammal fauna with the invasion of the supposedly more robust mammals from the north. Simpson, *Splendid Isolation*.

291 "Some ecologists have mustered evidence that . . . may not explain fully the intricacies of invasion . . ." Callaway, R. M., and Aschehoug, E. T., 2000, "Invasive plants versus their new and old neighbors: a mechanism for exotic invasion," *Science* 290:521–23.

291–92 ". . . in its Eurasian habitat *Centaurea* . . . did not harm the other plants." Ibid.

292 "In North America . . . chemicals released by *Centaurea* detrimentally affected some of the resident grasses." Ibid.

292 "Around Chesapeake Bay are huge expanses of feathery reeds known to botanists as *Phragmites australis* . . ." Marris, E., 2005, "Shoot to kill," *Nature* 438:272–73.

292 ". . . strike teams have been on a *Phragmites* search and destroy mission." Ibid.

292 "Some ecologists . . . question the . . . obliteration program . . ." Ibid.

293 ". . . United States spends more than a billion dollars spraying *Phragmites* . . ." Ibid.

293 "... invasion is often a two-way street." Examples here are from Normile, D., 2004, "Expanding trade with China creates ecological backlash," *Science* 306:968–69.

293 "... drain on the U.S. economy of about $137 billion a year." Ibid.

293 "A leaf beetle ... effective in attacking those destructive salt cedars ..." Ibid.

293–94 "In the United States the one-billion-dollar program in cargo and baggage inspections ..." Ibid.

20: THE WASTE OF A WORLD

295 "California ... still has one of its worst air pollution problems." American Lung Association, *State of the Air: 2006*. For the full report see the association website http://lung action.org/reports/sota06_full.html.

295 "Air pollution is also the scourge of many smaller cities." Ibid.

295 "... pollution has been a human contribution to the environment for millennia ..." Kovarik, W., *Environmental History Timeline*, www.radford.edu/~wkovarik/envhist /timeline.text.html; Hughes, J. D, 1975, *Ecology in Ancient Civilizations*, University of New Mexico Press, Albuquerque; Nriagu, J, 1983, *Lead and Lead Poisoning in Antiquity*, Wiley Interscience, New York.

296 "Romans called their polluted air *gravioris caeli* (heavy heaven) or *infamis aer* (infamous air) ..." Kovarik, *Environmental History Timeline*.

296 "... they set new standards for public health." Ibid.

296 "During medieval times the effects of pollution in Europe, mostly smoke from burning wood and wastes dumped into water ..." Kovarik, *Environmental History Timeline*.

296 "... England banned coal burning in smoky, sooty London." Ibid. and Brimblecombe, P., 1988, *The Big Smoke*, Routledge, London; Fitter, R.S.R., 1946, *London's Natural History*, Collins, London.

297 "A study ... of copper impurities preserved in Greenland ice layers ..." Hong, S., Candelone, J.-P., Patterson, C. C., and Boutron, C. F., 1996, "History of ancient copper smelting recorded in Roman and medieval times recorded in Greenland ice," *Science* 272:246–49.

297 "Similar patterns are seen for profiles of lead (Pb) concentrations as recorded in Arctic ice layers." Nriagu, J. O., 1996, "History of global metal pollution," *Science* 272:223.

297 "... clean air acts in the United States and elsewhere were inspired by a series of atmospheric disasters." Markham, A., 1994, *A Brief History of Pollution*, St. Martin's, New York; Kovarik, *Environmental History Timeline*; Snyder, L. P., 1994, "The death-dealing smog over Donora, Pennsylvania: industrial air pollution, public health policy and the politics of expertise, 1948–1949," *Environmental History Review*, Spring 1994, 117–38; Halliday, S., 2001, *The Great Stink of London: Sir Joseph Bazelgette and the Cleansing of the Victorian Metropolis*, Phoenix Mill, Sutton, UK.

298 "Modern pollution disasters come in all forms ..." For listing of recent environmental accidents, see Kovarik, *Environmental History Timeline*.

298 "... Baiyangdian Lake ..." AFP, April 29, 2006, "Pollution slowly choking north China's largest lake to death," www.todayonline.com/articles/115456.asp.

298 "Lack of fish is not the only blow to the quality of life ..." Ibid.

298–99 "'When we were kids, we used to drink the water ...'" Ibid.

299 "China now releases forty to sixty billion tons of wastewater . . ." Lim, L., March 23, 2005, "China warns of water pollution," BBC News, http://news.bbc.co.uk/1/hi/world /asia-pacific/4374383.stm.

299 ". . . the U.S. tanker *Exxon Valdez* hit an undersea reef . . ." Alaska Oil Spill Commission, 1990, *Spill: The Wreck of the Exxon Valdez, Final Report*, state of Alaska, Juneau.

299 ". . . the resultant kill-off . . ." Piper, E., 1993, *The Exxon Valdez Oil Spill: Final Report, State of Alaska Response*, Alaska Dept. of Environmental Conservation, Anchorage.

299 "There have been worse oil spills . . ." A description of the major oil spill accidents of the years between 1967 and 2004 can be found at The Mariner Group, Oil Spill History, www.marinergroup.com/oil-spill-history.htm.

300 ". . . hotspots for ground pollution." LaGrega, M. D., 2000, *Hazardous Waste Management*, McGraw-Hill, New York; Trevors, J. T. (ed.), *Water, Air, and Soil Pollution: An International Journal of Environmental Pollution*. Springer Verlag, Berlin; Stegmann, R., 2001, *Treatment of Contaminated Soil: Fundamentals, Analysis, Applications*, Springer Verlag, Berlin.

300 "Substances that contaminate the substrate . . ." Trevors, *Water, Air, and Soil Pollution*; Carson, *Silent Spring*; American Geological Institute (AGI), Groundwater and Soil Contamination Database, www.agiweb.org/georef/onlinedb/gscweb.html.

300 "The health hazards of dioxins and other contaminants have been well documented." U.S. Environmental Protection Agency (EPA), Hazardous Wastes, www.epa.gov /epaoswer/osw/hazwaste.htm#hazwaste.

300 ". . . (although recent reports call for a better scientific assessment . . ." National Research Council of the National Academies, 2006, *Health Risks from Dioxin and Related Compounds: Evaluation of the EPA Reassessment*, National Academies Press, Washington, D.C. www.nap.edu/catalog/11688.html.

300 ". . . recent DNA studies have shown much lower levels of microbial diversity . . ." Gand, J., et al., 2005, "Computational improvements reveal great bacterial diversity and high metal toxicity in soil," *Science* 309:1387–90.

300 ". . . EPA's National Priority List as Superfund sites . . ." U.S. Environmental Protection Agency (EPA), Superfund: Cleaning Up America's Hazardous Waste Sites, www.epa.gov/superfund/.

300 "More than sixty-five million Americans . . . live within four miles of a Superfund site." Ibid.

300 ". . . women living within a quarter mile of a Superfund site were at significantly high risk . . ." California Birth Defects Monitoring Program, 1999, Birth Defects and Hazardous Waste Sites, www.cbdmp.org/ef_waste.htm.

300 "Then, in 1995, the Superfund tax was suspended . . ." A criticism of this suspension is Mess, D., 2003, *Renewal of Superfund's Polluter Pays Tax: The Need to Lighten the Economic and Health Burdens of Toxic Waste Sites on the Shoulders of Individual Taxpayers*, John Glenn Institute for Public Policy and Public Service, Ohio State University Press, Columbus, http://glenninstitute.osu.edu/washington/DavidMess.htm.

301 "REACH is more ambitious than any regulatory program . . ." European Chemicals Bureau (ECB), REACH (registration, evaluation, and authorization of Chemicals), http://ecb.jrc.it/reach/.

301 "The Union Carbide pesticide plant . . ." Kurzman, D., 1987, *A Killing Wind: Inside Union Carbide and the Bhopal Catastrophe*, McGraw-Hill, New York.

301 "The killing fields remain . . ." Stringer, R., et al., 2002, "Chemical Stockpiles at the Union Carbide India Limited in Bhopal: An Investigation," Greenpeace Research Laboratories, www.greenpeace.org/raw/content/international/press/reports/chemical -stockpiles-at-union-c.pdf

301 ". . . surveys locate five of the world's most polluted places . . ." Blacksmith Institute, *World's Worst Polluted Places*, www.blacksmithinstitute.org/ten.php.

302 "Japan has the biggest landfill complex of all . . ." MacIntyre, D., and Tashiro, H., 2000, "Japan's dirty secret: as deadly toxins poison the environment, the government is doing its best to avoid the issue," *Time Asia*, May 29, 2000, vol. 156, no. 21.

302 ". . . surrounding the city of Kathmandu, Nepal, are several stockpiles of . . . pesticides . . ." Neupane, F. P. (ed.), 2003, *Integrated Pest Management in Nepal: Proceedings of a National Seminar, Kathmandu, Nepal 25–26 September 2002*, Himalayan Resources Institute, Kathmandu.

302 "These problems are common in Africa and South America too, of course." Blacksmith Institute, *World's Worst Polluted Places*.

302 ". . . developing countries . . . have been compelled to use . . . discarded pesticides . . ." Environment News Service, "Pesticides sent as aid to Nepal now toxic waste," http://ens-newswire.com/ens/oct2001/2001-10-18-03.asp; FAO, 2002, *The Legacy of Obsolete Pesticide Stocks in Africa*, www.mindfully.org/Pesticide/2002/African-Obsolete -Pesticides FAO2002.htm.

302–303 ". . . air pollution was their country's biggest problem . . ." "The most polluted city in the world: sixteen of the 20 most polluted cities in the world are in China," *Epoch Times*, June 10, 2006, www.theepochtimes.com/news/6-6-10/42510.html.

303 "Beijing may now have the worst smog of any capital in the world . . ." Watts, J., 2005, "Satellite data reveals Beijing as air pollution capital of world," *Guardian*, October 31, 2005, www.guardian.co.uk/china/story /0,7369,1605146,00.html.

303 ". . . Mexico City was no longer the world's smoggiest . . ." For a table of the world's worst cities for air pollution, see American Association for the Advancement of Science (AAAS) Atlas of Population and Environment, *Air Pollution*, http://atlas.aaas.org/index .php?part=2&sec=atmos&sub=air.

303 "Air pollution is composed of a complex, variant mixture of gases . . ." There are many general texts on atmospheric emissions, including notably Stern, A. C., 1984, *Fundamentals of Air Pollution*, 2nd ed., Academic Press, Orlando; Lutgens, F. K., and Tarbuck, E. J., 1995, *The Atmosphere*, 6th ed., Prentice Hall, Upper Saddle River, NJ; Wayne, R. P., 2000, *Chemistry of Atmospheres*, Oxford University Press, Oxford, UK.

304 "In the early 1980s satellite images showed emissions . . ." Akimoto, H., 2003, "Global air quality and pollution," *Science* 302:1716–19.

304 "Pollutants in the atmosphere don't last forever." Ibid.

304 ". . . ozone . . . ozone 'hole' . . . CFCs . . ." Cagin, S., and Dray, P., 1993, *Between Earth and Sky: How CFCs Changed Our World and Endangered the Ozone Layer*, Pantheon, New York; Weatherhead, E. C., and Andersen, S. B., 2006, "The search for signs of recovery of the ozone layer," *Nature* 441:39–45.

304 ". . . the Montreal Protocol . . ." The Montreal Protocol on Substances that Deplete the Ozone Layer as Adjusted and/or Amended in London 1990[,] Copenhagen 1992[,] Vienna 1995[,] Montreal 1997[,] Beijing 1999[,] UNEP, http://hq.unep.org /ozone/Montreal-Protocol/Montreal-Protocol2000. shtml.

305 "The atmospheric lifetime of tropospheric ozone is about one or two weeks . . ." Akimoto, "Global air quality."

305 "Another pollutant, carbon monoxide (CO) . . . also 'lives' long enough in the atmosphere . . ." Ibid.

305 "Other long-distance travelers, the nitrogen oxides . . ." Ibid.

305 "The modern controls of such emissions in the United States . . . Europe . . . Asia . . ." Ibid.

306 ". . . the Southern Hemisphere will contribute more and more to the decline of global air quality." Ibid.

306 ". . . toxic release inventories . . ." EPA, Toxic Release Inventories (TRI) www.epa.gov/tri/.

306 ". . . some of the most polluted areas in the United States are not near major cities . . ." Scorecard, *The Pollution Information Site*, www.scorecard.org/index.tcl.

306 "The most populous region high in the rankings of chemical pollution is Salt Lake City . . ." Ibid.

306 ". . . one major mining polluter in North America is named U.S. Ecology Idaho Inc." Ibid.

306 ". . . a revised rule that puts limitations on the amount of information . . . in its TRI reports." EPA, *Toxic Release Inventories (TRI), Burden Reduction Proposed Rule (Phase II)*, www.epa.gov/tri/tridata/modrule/phase2/index.htm.

307 ". . . pollutants can also profoundly transform the *global* environment." Kiehl, J. T., and Trenberth, K. E., 1997, "Earth's annual global mean energy budget," *Bulletin of the American Meteorological Society* 78:197–208; Houghton et al. (eds.), *Climate Change 2001: The Scientific Basis Contribution of Working Group I to the Third Assessment Report of the Intergovernmental Panel on Climate Change (IPCC)*, Cambridge University Press, Cambridge, UK. IPCC Working Group II contribution to Fourth Assessment Report. Climate Change 2007: Impacts, Adaptation, and Vulnerability, www.ipee.ch.

307 "We have not yet sorted out all the reasons why CO_2 concentrations spiked at certain times in the past . . ." Berner and Kothavala, "Geocarb III."

307 ". . . we know that present levels in the atmosphere are unusually high, and . . . we know why." Houghton et al., (eds.), *Climate Change 2001*. IPCC Working Group II, Climate Change 2007.

21: HEAT WAVE

308 ". . . Antarctica . . ." An excellent resource for the physical environment and biota of the part of the continent we visited is Moss, S., 1988, *Natural History of the Antarctic Peninsula*, Columbia University Press, New York.

308 ". . . the great explorer Ernest Shackleton . . ." The story, brilliantly retold, is in Alexander, C., 1998, *The Endurance: Shackleton's Legendary Antarctic Expedition*, Knopf, New York.

309 "When . . . Shackleton launched his audacious expedition . . ." Ibid.

309 "Little pink flowers (*Colobanthus quietensis*) . . ." Moss, *Natural History of the Antarctic Peninsula*.

310 ". . . CFC emissions . . ." Cagin and Dray, *Between Earth and Sky*.

310 ". . . a series of articles . . . glacier movement in Greenland and Antarctica." Joughin, I., 2006, "Climate change: Greenland rumbles louder as glaciers accelerate," *Science* 311:1719–20; Bindschadler, R., 2006, "Climate change: hitting the ice sheets where it

hurts," *Science* 311:1720–21; Overpeck, J. T., Otto-Bliesner, B. L., Miller, G. H., Muhs, D. R., Alley, R. B., and Kiehl, J. T., 2006, "Paleoclimatic evidence for future ice-sheet instability and rapid sea-level rise," *Science* 311:1747–50; Velicogna, I., and Wahr, J., 2006, "Measurements of time-variable gravity show mass loss in Antarctica," *Science* 311:1754–56.

310 "Glaciers draining the Greenland ice sheet are speeding up alarmingly . . ." Ibid.

310–11 ". . . loss of hundreds of cubic kilometers of ice annually in West Antarctic . . ." Ibid.

311 ". . . an article . . . from the British Antarctic Survey . . ." Turner, J., Lachlan-Cope, T. A., Colwell, S., Marshall, G. J., and Connolley, W. M., 2006, "Significant warming of the Antarctic winter troposphere," *Science* 311:1914–17.

312 "The average increase in temperature per decade . . ." Ibid.

312 ". . . since the 1860s shows an increase of between 0.4° and 0.8°C . . ." Hulme, M., 2005, "Recent climate trends," in Lovejoy and Hannah, *Climate Change and Biodiversity*, pp. 31–40.

312 ". . . increase in global mean surface temperatures by 2030 to 2050 somewhere between 1.5° and 4.5°C . . ." Houghton et al., *Climate Change 2001*; Karl, T. R., and Trenbeth, K. E., 2005, "What is climate change?" in Lovejoy and Hannah, *Climate Change and Biodiversity*, pp. 15–28.

312 ". . . different models vary considerably in projections for 2080 to 2100." Houghton et al., *Climate Change 2001*; Hannah, L., Lovejoy, T. E., and Schneider, S. H., 2005, "Biodiversity and climate change in context," in Lovejoy and Hannah, *Climate Change and Biodiversity*, pp. 3–14.

313 "Global mean temperatures that wobbled by only about 5° to 6°C during the Pleistocene epoch . . . were enough to promote dramatic and sweeping shifts . . ." Huntley, B., 2005, "North temperate responses," in Lovejoy and Hannah, *Climate Change and Biodiversity*, pp. 109–41.

313 "Antarctica is clearly warming up, but they could not definitely say why." Turner et al., "Significant warming."

314 "Yet another article in *Science* was more assertive." Kerr, R. A., 2006, "Global change: no doubt about it, the world is warming," *Science* 312:825.

314 ". . . a report commissioned by the Bush administration . . ." www.climatescience.gov /Library/sap/sap1-1/finalreport/default.htm.

314 "In February 2007, the Intergovernmental Panel on Climate Change (IPCC) issued the conclusion . . ." IPCC Working Group II, Climate Change 2007.

314 "Even so, some scientiests protested . . ." Kerr, R. A., 2007, "Scientists tell policymakers we're all warming the world," *Science* 315:754–57.

314–15 "In 1992 the American Museum opened an exhibit on global warming . . . almost too early." A perception of the exhibit and the mood at the time are revealed in Liounis, A., 1992, "Global warming—mixed opinions about threat of global warming," *Omni*, May 1992.

315 "Efforts to answer these questions are pinned on several lines of evidence . . ." Hannah et al., "Biodiversity and climate change in context"; Overpeck, J., Cole, J., and Bartlein, P., 2005, "A 'paleoperspective' on climate variability and change," in Lovejoy and Hannah, *Climate Change and Biodiversity* pp. 91–108.

315 "We need to distinguish . . . climate change from climate variation." Karl and Trenbeth, "What is climate change?" in Lovejoy and Hannah, *Climate Change and Biodiversity*.

316 "The climate variation can naturally oscillate . . ." Ibid.

317 "We know of other climate variations on even longer timescales." Ibid.

317 ". . . one major trend might be for climate through time to become more variable . . ." Ibid. and Hulme, "Recent climate trends."

317 ". . . rewind time and then fast forward, but here do it in terms of changing climates." Many references, including those cited in earlier chapters, are relevant here. Among them are Gould, *The Book of Life*; Fortey, *Life: An Unauthorized Biography*; Berner and Canfield, "A new model for atmospheric oxygen over Phanerozoic time"; Scotese, Paleomap Project; Overpeck et al., "A 'paleoperspective' on climate variability and change."

320 ". . . we would expect to become icebound again in about ten thousand years." Huntley, "North temperate responses."

320–21 ". . . an increase in greenhouse gases . . ." Kiehl and Trenberth, "Earth's annual global mean energy budget"; Houghton et al., *Climate Change 2001*.

321 ". . . very warm periods in Earth's history are associated with higher concentrations of greenhouse gases." Berner and Kothavala, "Geocarb III"; Overpeck et al., "Paleoperspective."

321 ". . . 55 to 60 million years ago . . . CO_2 levels may have exceeded one thousand parts per million . . ." Kennedy, D., and Hanson, B., "Ice and history," *Science* 311:1673.

321 "The removal of CO_2 cooled things off considerably . . ." Ibid.

321 "In the current postglacial phase the CO_2 level stayed . . . at 290 ppmv . . ." Ibid.

321 "Even 130,000 years ago, when climates were warmer and sea levels higher . . ." Ibid.

322 ". . . double the preindustrial level of 290 ppmv, to nearly 600 ppmv by the middle of this century." *Climate Change 2001*.

322 "Then what next?" Ibid. and Karl and Trenberth, "What is climate change?"

322 "Only 5°C is what makes the difference between . . ." see Bush, M. B., and Hooghiemstra, H., 2005, "Tropical biotic responses to climate change," in Lovejoy and Hannah, *Climate Change and Biodiversity*, pp. 125–37.

322 ". . . dramatically shift many physical aspects of the planet's system." Karl and Trenberth, "What is climate change?" and Hulme, "Recent climate trends."

322 "Or could they?" Ibid.

323 "At higher latitudes the pattern is ambiguous." Ibid.

323 ". . . El Niño events have become more frequent and intense . . ." Ibid. and Glantz, M., 2000, *Currents of Change: Impacts of El Niño and La Niña on Society*, 2nd ed., Cambridge University Press, Cambridge, UK.

323 ". . . returning flow of subsurface waters in the northern loop between Greenland and Scotland . . ." Hanson, B., Turrell, W. R., and Osterhuis, S., 2001, "Decreasing overflow from the Nordic seas into the Atlantic Ocean through the Faroe Bank channel since 1950," *Nature* 411:927–30.

323–24 "It makes no sense to develop strategies for preserving biodiversity . . . on the assumption that the climate is stable." Lovejoy and Hannah, *Climate Change and Biodiversity*.

324 "The causes of other past extinction events are enigmatic, but climate change is invariably proposed as one cause . . ." Erwin, "Lessons from the past"; Archibald, *Dinosaur Extinction*; Briggs and Crowther, *Palaeobiology II*.

324 "In the Tertiary Period . . . the climates grew hotter, then cooler and then frigid . . ." See Overpeck et al., "A paleoperspective."

324–25 "Climates were also affected by major tectonic events . . ." Zachos, J., Pagani, M., Sloan, L., Thomas, E., and Billups, K., 2001, "Trends, rhythms, and aberrations in global climate 65 Ma to present," *Science* 292:686–93.

325 ". . . Panamanian bridge allowed reciprocal invasions . . ." Simpson, *Splendid Isolation*; Novacek, *Time Traveler*.

325 "In Australia changing climates in the Late Tertiary . . ." Markgraf, V., and McGlone, M., 2005, "Southern temperate ecosystem responses," in Lovejoy and Hannah, *Climate Change and Biodiversity*, pp. 142–56.

325 "A critical phase is . . . between 2.5 and 1.8 million years ago." Huntley, "North temperate responses."

325 "Organisms . . . had to cope with wildly shifting environments . . ." Ibid.

325 "Marine habitats were, if anything, even harder hit . . ." Ibid. and Jackson, J.B.C., 1995, "Constancy and change of life in the sea," in Lawton, J. H. , and May, R. M., *Extinction Rates*, Oxford University Press, Oxford, UK, pp. 45–54.

325 "The effect of this climate schizophrenia . . . is not clear . . ." Huntley, "North temperate responses."

326 "Extinction events connected with the ice ages were few." Ibid.

326 "At low latitudes and in the tropics the effects are subtler." Bush, M. B., and Hooghiemstra, H. , 2005, "Tropical biotic responses to climate change," in Lovejoy and Hannah, *Climate Change and Biodiversity*, pp. 125–37.

326 "The warming trend at . . . the beginning of the Holocene caused sea levels to rise 125 meters . . ." Ibid.

327 "One would expect that the Southern Hemisphere . . . would mimic the . . . Northern Hemisphere, but this is not the case." Markgraf and McGlone, "Southern temperate ecosystem responses."

327 ". . . species reacted to climate changes individualistically . . ." Hannah, Lovejoy, and Schneider, "Biodiversity and climate change in context"; Huntley, "North temperate responses."

328 ". . . phenology . . ." Jeffree, E. P., 1960, "Some long-term means from the phenological reports (1891–1948) of the Royal Meteorological Society," *Quarterly Journal of the Royal Meteorological Society* 86:95–103.

328 "In 1736, Robert Marsham . . . began making notes in his ledger . . ." Sparks, T. H., and Carey, P. D., 1995, "The responses of species to climate over two centuries: an analysis of the Marsham phenological record, 1736–1947," *Journal of Ecology* 83:321–29.

328 ". . . the first date of frog calling . . . was tightly correlated with temperature." Ibid.

328 ". . . more refined studies reaffirmed that frogs are cued . . . by seasonal shifts in temperature." These are cited in Root, T. L., and Hughs, L., 2005, "Present and future phenological changes in wild plants and animals," in Lovejoy and Hannah, *Climate Change and Biodiversity*, pp. 61–69.

328 "Two frog species . . . spawned two to three weeks earlier," Ibid. and Bebee, T.J.C., "Amphibian breeding and climate," *Nature* 374:219–20.

328 "The current warming trend influences myriad species." Root and Hughs, 2005, "Present and future phenological changes"; Thomas et al., 2004, "Extinction risk from climate change," *Nature* 427:145–48; Bradshaw, W. E., and Holzapfel, C. M., 2006, "Evolutionary response to rapid climate change," *Science* 312:1477–78.

328 ". . . life history data in 694 species recorded between 1951 and 2001." Root, T. L., Price, J. T., Hall, K. R., Schneider, S. H., Rosenzweig, C., and Pounds, J. L., 2003, "Fingerprints of global warming on wild animals and plants," *Nature* 421:37–42.

328 "The current warming trend has set species in motion." Parmesan, C., 2005, "Biotic response: range and abundance changes," in Lovejoy and Hannah, *Climate Change and Biodiversity*, pp. 41–55.

329 "Finally . . . malaria-carrying mosquitoes are migrating . . . to higher latitudes." Rogers, D. J., and Randolph, S. E., 2000, "The global spread of malaria in a future, warmer world," *Science* 289:1763–66.

329 "On the other hand, some species have drastically contracted their ranges . . ." Parmesan, "Biotic response."

329 "Precipitation changes have caused major shifts in the Sonoran desert flora . . . for the vegetation and the kinds of ants, reptiles, and rodents . . ." Ibid. and Turner, R. M., 1990, "Long-term vegetation change at a fully protected Sonaran desert site," *Ecology* 71:464–77; Brown, J. H., Valone, T. J., and Curtin, C. G., 1997, "Reorganization of an arid ecosystem in response to recent climate change," *Proceedings of the National Academy of Sciences* 94:9729–33.

329 ". . . these weak but persistent forces may ultimately eclipse the others . . ." Hewitt, G. M., and Nichols, R. A., 2005, "Genetic and evolutionary impacts of climate change," in Lovejoy and Hannah, *Climate Change and Biodiversity*, pp. 176–92.

330 "Rapid alteration of a species's range can distend its populations and transform the nature of its genetic enrichment." Ibid.

330 "But these colonizers may have a very low level of gene diversity . . ." Ibid. and Hewitt, G. M., 1993, "Postglacial distribution and species substructure: lessons from pollen, insects and hybrid zones," in Lees, D. R., and Edwards, D., (eds.), *Evolutionary Patterns and Processes*, Linnaean Society Symposium Series 14:97–123, Academic Press, London.

330 ". . . a population's genetic makeup and evolution . . ." Ibid. and Thomas, C., 2005, "Recent evolutionary effects of climate change," in Lovejoy and Hannah, *Climate Change and Biodiversity*, pp. 75–88.

331 "Some genetic studies suggest that climate change has easily outrun the rate at which a given population can adjust to it . . ." Etterson, J. R., and Shaw, R. G., 2001, "Constraint to adaptive evolution in response to global warming," *Science* 294:151–54.

331 ". . . others indicate that local gene pools are preserved through numerous past climate fluctuations." Hewitt, G. M., 1996, "Some genetic consequences of the ice ages and their role in divergence and speciation," *Biological Journal of the Linnaean Society* 58:247–76; Avise, J. C., Walker, D., and Johns, G. C., 1998, "Speciation durations and Pleistocene effects on vertebrate phylogeography," *Proceedings of the Royal Society of London: Biological Sciences*, 265:1707–12.

331 ". . . some species, especially in mountainous and tropical regions, are buffered by the amount of genetic variation . . ." Hewitt, "Some genetic consequences . . ."

331 "Although there is some debate over the matter . . ." See opposing views in Aronson, R. B., et al., 2003, "The causes of coral reef degradation," *Science* 302:1502; Pandolfi, J. M., et al., 2003, "Response," ibid.: 1502–3; Hughes, T. P., et al. , 2003, "Response," ibid.: 1503.

331 ". . . the drastic decline of coral reefs in many regions of the world resulted from a clear sequence of events." Jackson et al., 2001, "Historical overfishing"; Hoegh-Guldberg, O., 2005, "Climate change and marine ecosystems," in Lovejoy and Hannah, *Climate Change and Biodiversity*, pp. 256–73.

331 ". . . increases in . . . temperature . . . will result in environmental stresses not seen during the last five hundred thousand years of coral reef evolution." Hughes, T. P., et al.,

2003, "Climate change, human impacts, and the resilience of coral reefs," *Science* 301:929–33. See also Roy, K., and Pandolfi, J. M., 2005, "Responses of marine species and ecosytems to past climate change," in Lovejoy and Hannah, *Climate Change and Biodiversity*, pp. 160–75.

331 "... the sea's richest ecosystem ..." Knowlton, N., 2001, "The future of coral reefs," *Proceedings of the National Academy of Sciences*, 98:5419–25.

332 "More than a million named species ... may inhabit about four hundred thousand square kilometers of coral reef." Hoegh-Guldberg, "Climate change and marine ecosystems."

332 "Darwin ... noted that coral reefs ..." Darwin, C., 1842, *The Structure and Distribution of Coral Reefs*, Smith, Elder and Company, London.

332 "Tropical oceans are 0.5 to 1.0°C warmer than they were one hundred years ago ..." Hoegh-Guldberg, O., 1999, "Coral bleaching, climate change, and the future of the world's coral reefs," *Marine and Freshwater Research* 50:839–66.

332 "... coral bleaching." Ibid.

332 "... mortality rates ... of up to 90 percent ..." Hoegh-Guldberg, "Climate change and marine ecosystems."

332 "... 10 to 16 percent died off ..." Ibid.

332 "In the 1980s in lagoons off Belize ..." Ibid. and Aronson, R. B., et al., 2000, "Ecosystems: coral bleach out in Belize," *Nature* 405:36.

333 "The destruction of reefs worldwide can be plotted on a blacklist ..." Pandolfi et al. 2005, "Are U.S. coral reefs on the slippery slope to slime?" *Science* 307:1725–26.

333 "Another project, conducted by scientists at the American Museum ..." Mumby, P. J., et al., 2006, "Fishing, trophic cascades, and the process of grazing on coral reefs," *Science* 311:98–101.

333 "... massively affected ... from the increase in atmospheric carbon dioxide." Doney, S., 2006, "Ocean acidification," *Scientific American*, March 2006, pp. 58–65.

334 "... marine organisms depend on carbonate ions ..." Ibid.

334 "... hydrogen ions ... readily lower the pH of seawater ..." Ibid.

334 "... bad news for the organisms that need calcium carbonate ..." Ibid.

334 "... acidification of the seas could rampantly disrupt the marine ecosystem ..." Ibid.

334 "Organisms that live in ... freshwater are among the most endangered on Earth." Allen et al., "Climate change and freshwater ecosystems"; Dudgeon et al., "Freshwater biodiversity."

335 "... species respond independently ..." Huntley, "North temperate responses."

335 "... but that habitat loss greatly constrains migrations ..." Lovejoy, T., 2005, "Conservation with a changing climate," in Lovejoy and Hannah, *Climate Change and Biodiversity*, pp. 325–28.

335 "... global warming threatens ... species at high latitudes ..." Huntley, "Northern temperate responses."

335 "... species reactions to climate change are idiosyncratic." Hannah et al., "Biodiversity and climate change in context."

335 "Computer models ... take into account many factors ..." Midgley, G. F., and Miller, D., 2005, "Modeling species range shifts in two biodiversity hotspots," in Lovejoy and Hannah, *Climate Changes and Biodiversity*, pp. 229–31.

335–36 "... 25 percent of the *protea* flower species ... will be extinct ..." Ibid.

336 "Native Alaskans and Aleuts ... are worried about the rising sea level ..." Mac-

Cracken, M., 2004, "Arctic region noticeably changing," Climate.org: A Project of the Climate Institute, www.climate.org/topics/climate/arctic_change.shtml.

336 "Inhabitants of many small low-lying Pacific islands..." Marks, K., 2006, "Global warming threatens Pacific island states," *Independent*, October 27, 2006, www.countercurrents.org/cc-marks271006.htm.

336 "... Venice, New Orleans, and many other coastal communities are living on borrowed time ..." Pilkey, O. H., and Cooper, J.A.G., 2004, "Society and sea level rise," *Science* 303:1781–82.

336 "Venice is sinking ..." Bohannon, J., 2005, "A sinking city yields some secrets," *Science* 309:1878–1980.

336 "As a recent government report stated ..." Burkett, V. R., Zilkoski, D. B., and Hart, D. A., 2005, "Sea-level rise and subsidence: implications for flooding in New Orleans, Louisiana," U.S. Geological Survey (USGS), www.nwrc.usgs.gov/hurricane/Sea-Level-Rise.pdf.

336 "... Pleistocene climates did not cause widespread extinction ..." Huntley, "North temperate responses."

337 "Holocene warming did not raise the temperature range ..." Overpeck et al., "A 'paleoperspective.'"

337 "... restoration ... after a mass extinction ... takes from 3 to 15 million years ..." Kirchner and Weil, "Delayed biological recovery"; Erwin, "Lessons from the past."

337 "... Kyoto Protocol ..." http://unfccc.int/resource/docs/convkp/kpeng.html.

22: FUTURE WORLD

338 "... 300,000 people a year in China die prematurely ..." Economy, E., 2004, "Economic boom: environmental bust," Council of Foreign Relations, www.cfr.org/publication/7548/economic_boom_environmental_bust.html.

339 "... most polluted cities are in China.)" *Epoch Times*, June 10, 2006. www.theepochtimes.com/news/6-6-10/42510.html; The World Bank, 2007, *Cost of Pollution in China*, The State Environmental Protection Administration, Beijing, China; BBC News, July 3, 2007, "China 'buried smog death finding,'" www2.newsbbc.co.uk/2/hi/asia-pacific/6265098.stm.

340 "... Paul J. Crutzen ... has coined a new term, *Anthropocene* ..." Crutzen, P. J., 2002, "Geology of mankind," *Nature*, 415:23.

340 "... half will be extinct by the mid-twenty-first century." Wilson, *The Diversity of Life*; Pimm et al., "The future of biodiversity."

341 "... humans have been exterminating its components for forty thousand years." MacPhee, *Extinctions in Near Time*.

341 "In New Zealand ... flightless birds were reduced ..." Steadman, D. W., 1995, "Prehistoric extinctions of Pacific island birds: biodiversity meets zooarchaeology," *Science* 267:1123–31.

341 "'People won: native plants and animals lost.'" Steadman, D. W., 2002, "Paleontology: everything you want and moa," *Science* 298:2136–37.

341 "... extinction rate is soon to be as much as ten thousand times faster ..." Pimm et al., "The future of biodiversity."

341 "... mid-century loss of 30 to 50 percent ..." Ibid.

341 "... 90 percent loss at the end of the Permian ... but it approaches the Cretaceous ..." Erwin, "Lessons from the past."

341 "... the error range ... at 65 million years is ..." Swisher et al.,"Coeval 40Ar/39Ar Ages."

342 "'Death is one thing, an end to birth is something else.'" Soule, M. E., and Wilcox, B. A. (eds.), 1980, *Conservation Biology: An Evolutionary-Ecological Perspective*, Sinauer Associates, Sunderland, MA, pp. 150–71.

342 "... post-Cretaceous recovery of some species ... took hundreds of thousands of years ..." Erwin, "Lessons from the past."

342 "The endangerment ... of spider species in many areas ..." Platnick, N. I., 2000–2007, *The World Spider Catalog, Version 7.5*, American Museum of Natural History, http://research.amnh.org/entomology/spiders/catalog/BIB9.html

342–43 "An analysis of African ecosystems and human population distribution ..." Balmford, A., Moore, J. L., Brooks, T., Burgess, N., Hansen, L. A., Williams, P., and Rahbek, C., 2001, "Conservation conflicts across Africa," *Science* 291:2616–19.

343 "... bees ... pollinate a majority of the world's crop species ..." McGregor, *Insect Pollination*.

343 "But we can expect other biological effects ..." Myers, N., and Knoll, A. H., 2001, "The biotic crisis and the future of evolution," *Proceedings of the National Academy of Sciences* 98 (10): 5389–92.

343 "Species that reproduce explosively ... might be favored." Ibid.

344 "... evolutionary dead end of large vertebrates." Ibid.

344 "... novelties will invariably emerge ..." Ibid.

345 "... estimate of global human populations by mid-century is somewhat lower than originally predicted." Cohen, J. E., "Human population: the next half century."

345 "... a perfect ecological storm is brewing in West Africa ..." Brashares et al., "Bushmeat hunting, wildlife declines, and fish supply in West Africa."

345 "Many important areas rich in biodiversity lie on international borders ..." Balmford et al., "Conservation conflicts across Africa."

345 "... Holy Grail ..." www.intriguing.com/mp/_scripts/grail.asp.

346 "The IUCN Red List of 2006 documented a few reversals in disturbing trends ..." 2006 IUCN Red List, www.iucnredlist.org/.

347 "... dwindling populations of butterflies on the British Isles ..." Thomas et al., "Comparative losses of British butterflies."

347 "... and dragonflies in Sri Lanka ..." 2006 IUCN Red List.

347 "Creating corridors lacing through human-dominated regions ..." Western, "Human-modified ecosystems and future evolution."

348 "... land use that sustains ecosystems is possible ..." Foley, J. A., et al., 2005, "Global consequences of land use," *Science* 309:570–74.

348 "... a more global strategy of land use must distinguish among three options." Ibid.

349 "... extremely rapid climate changes ... preclude this kind of flexibility ..." Lovejoy and Hannah, *Climate Change and Biodiversity*.

349 "... countries in Asia ... are making huge contributions to air pollution ..." Akimoto, "Global air quality and pollution."

349 "... scientists ... have well stated the nature of the problem and the possible solutions." Lovejoy, T. E., and Hannah, L., 2005, "Global greenhouse gas levels and the future of biodiversity," in Lovejoy and Hannah, *Climate Change and Biodiversity*, pp. 387–95.

349 "... a higher concentration than at any time in the last 10 million years." Kennedy and Hanson, "Ice and history."

350 "... decrease the use of carbon fuels ... just to stay within a 'stabilized' range ..." Lovejoy and Hannah, "Global greenhouse gas levels."

350 "... three major strategies:" Ibid.

350 "... an efficiency savings of 50 percent ..." Ibid.

350 "... proposals to pump CO_2 into abandoned oil wells ..." Lovejoy and Hannah, "Global greenhouse."

350 "... gas released from Lake Nyos in Cameroon ..." Sano, Y., et al., 1987, "Helium isotope evidence for magmatic gases in Lake Nyos, Cameroon," *Geophysical Research Letters* 14:1039–41.

350 "... mineral carbonates could store fifty thousand gigatons of carbon ..." Lovejoy and Hannah, "Global greenhouse."

350 "... switch entirely to renewable energy sources ..." Ibid.

352 "... powerful nations refused to sign on to the Kyoto Protocol." Watson, R. T., "Emissions reductions and alternative futures."

352 "China, for its part ..." "China Climate Change," Pew Center on Global Climate Change, www.pewclimate.org/policy_center/international_policy/china.cfm.

352 "... enlisted David Legates ..." Kintisch, E., 2005, "Climate change: global warming skeptic argues U.S. position in suit," *Science* 308:482.

353 "Yet Bush was among those who rejected ..." Eilperin, J., "US trying to weaken G-8 climate change declaration," May 14, 2007, www.commondreams.org/archive/2007/05/14/1174/.

353 "... Stephen Hawking ... emphasized that the world as we know it is facing imminent catastrophe ..." Reuters UK, November 30, 2006, "Hawking says humans must colonise other planets," http://today.reuters.co.uk/news/.

354 "... the discoverer of the saola ..." Stone, R., 2006, "Wildlife conservation: the saola's last stand," *Science* 314:1380–83.

354–55 "... most optimistic counts hover around two hundred." Ibid.

355 "... Do Tuoc would prefer to fight for the saola ..." Ibid.

355 "... I have related success stories ..." 2006 IUCN Red List.

355 "... flew above North America in flocks of two billion strong ..." Schorger, A. W., *The Passenger Pigeon: Its Natural History and Extinction.*

355 "... were 'as beautiful as they were numerous.'" Cokinos, C., 2000, *Hope Is the Thing with Feathers*, Warner Books, New York.

355 "Henry David Thoreau described them ..." "Pigeons dart by on every side,—a dry slate color, like weather-stained wood (the weather-stained birds), fit color for this aerial traveller, a more subdued and earthly blue than the sky, as its field (or path) is between the sky and earth,—not black or brown, as is the earth, but a terrene or slaty blue, suggesting their aerial resorts and habits." *Thoreau's Journal*, September 15, 1859.

ACKNOWLEDGMENTS

I have described my motivations for writing this book in the prologue, but an attempt to account for 100 million years of ecosystem evolution and the current forces assaulting that ecosystem takes more than motivation—it takes the support, encouragement, and expert advice of loved ones, friends, and professional colleagues. For sharing their highly influential insights I thank Maureen O'Leary, Tom Lovejoy, Craig Morris, George Amato, Eleanor Sterling, John Flynn, Ian Tattersall, Rob DeSalle, Chris Raxworthy, Melanie Stiassney, Julie Novacek, and Travis Godsoe. I am so fortunate to work in a great scientific institution made all the greater because of the extraordinary leadership of President Ellen Futter, who has fostered an environment for both creativity and action. To her and to all of my dear friends on the President's Council I owe much. Among these, Gary Zarr, my fellow explorer of the Hoi An markets in chapter 1, was, as always, enthusiastic and encouraging. I am also deeply grateful to Chairman of the Board Lewis Bernard, and board chair emeriti Anne Eristoff and William Golden and many other members whose interests, devotion, and generosity have brought new and exciting opportunities to the institution.

I have benefited enormously from the expert readings of various chapters of the book by Ed Mathez, Eleanor Sterling, Peter Crane, David Grimaldi, Michael Engel, George Amato, Ian Tattersall, and Maureen O'Leary. They are not, of course, responsible for the unwanted shortcomings that may still persist. Superb original artwork was contributed by Mick Ellison, David Grimaldi, Doug Boyer, and Joyce A. Powzyk. Frank Ippolito provided his artistic expertise and a template for map illustrations while

Mick Ellison gave me useful tips for grappling with the complexities of Adobe Illustrator and Photoshop. Mary DeJong, Tom Baione, and the other Museum library staff were generous with their time in helping me locate and reproduce some important illustrations.

I am so honored to work with Elisabeth Sifton, my editor, who, through a combination of brilliance, fortitude, and patience, was once again indispensable for shaping the manuscript into something far more coherent and connective. I am as always indebted to my agent Al Zuckerman, whose superb judgment, persuasive powers, and faith in my effort have inspired me to try to write more and better. My mother, June Novacek, showed a strength and a continued passion for knowledge, even with the passing of my father, that once again demonstrated the healing powers of intellectual discovery even in times of great sorrow. Lastly, I must express my love and thanks to my wife, Maureen O'Leary, whose beauty, wit, and warmth has made life so constantly exciting to me.

INDEX

Page numbers in *italics* refer to illustrations.

ILLUSTRATION CREDITS

xvi Illustration based on charts in Christian, D., 2003, *Maps of Time: An Introduction to Big History*, University of California Press, Berkeley.

4 Illustration by Joyce A. Powzyk in Sterling, E. J., Hurley, M. M., and Le Duc Minh, 2006, *Vietnam: A Natural History*, Yale University Press, New Haven and London, American Museum of Natural History (AMNH). Reproduced with permission.

10 Illustration by Joyce A. Powzyk in Sterling et al., 2006, AMNH. Reproduced with permission.

19 Illustration from Wagler, J. G., 1830, *Natürliches System der Amphibian*, J. G. Cotta, Munich, courtesy AMNH.

23 Illustration by Violaine Martin from Gage, J. D., and Tyler, P. A., 1991, *Deep-Sea Biology, a Natural History of the Organisms at the Deep-Sea Floor*, Cambridge University Press, Cambridge. Reproduced with permission.

25 Illustration by the author based on data in Harrison, I., Laverty, M., and Sterling, E., 2004, "Species Diversity," *Connexions*, http://cnx.org/content/m12174/1.3/.

34 Illustration by David Grimaldi from Grimaldi, D., and Nguyen, T., 1999, "Monograph on the spittlebug flies, genus *Cladochaeta* (Diptera: Drosophilidae: Cladochaetini)," *Bulletin of the American Museum of Natural History* 241:1–326. Reproduced with permission.

36 Illustration by Grahame Chambers from Tudge, C., 2000, *The Variety of Life*, Oxford University Press, New York.

39 Image from Grimaldi, D., and Engel, M. S., 2005, *Evolution of Insects*, Cambridge University Press, Cambridge. Reproduced with permission.

42 Illustration by the author based on figures in Phillips, J., 1860, *Life on the Earth: Its Origin and Succession*, MacMillan, Cambridge and London (reprint edition, 1980, Arno Press, New York).

44 Illustration by Ed Heck, author's copyright.

49 Illustration by author after Ricketts, T. H., et al., 2005, "Pinpointing and preventing imminent extinctions," *Proceedings of the National Academy of Sciences U.S.A.* 102:18497–501.

50 Illustration modified from Plate 8 in Lovejoy, T. E., and Hannah, L. (eds.), 2005, *Climate Change and Biodiversity*, Yale University Press, New Haven. Reproduced with permission.

56 Illustration by the author based on Agricultural and Biological Engineering Department, University of Florida website; see Owens, J., "Biology for Engineers," www.agen.ufl.edu/~chyn/age2062/, and various sources.

62 Illustration by Chris Hunt from a photograph by P. B. Edwards from Edwards, P. B., and Aschenborn, H. H., 1988, "Male reproductive behavior of the African ball-rolling dung beetle, *Kheper nigroaeneus* (Coleoptera: Scarabaeidae)," *Coleopterists Bulletin* 42:17–27. Reproduced with permission.

63 Modified from illustration by Tom Prentis from Waterhouse, D. F., 1974, "The biological control of dung," *Scientific American*. Reproduced with permission.

66 Illustration by the author, after Cotgreave, P., and Forsyth, I., 2002, *Introductory Ecology*, Blackwell Publishing, Oxford.

70 Illustration by the author, after Cotgreave and Forsyth, ibid.

72 Illustration by the author after Christianson, G. E., 1999, *Greenhouse: The 200-Year Story of Global Warming*, Walker and Company, New York.

82 Durer's woodcut of the Indian rhinoceros from Gesner, C., 1551, *Historia animalium*, C. Froschauer, Zurich, courtesy AMNH special collections.

83 AMNH special collections. Reproduced with permission.

84 Haeckel, E., 1874, *Anthropogenie oder Entwicklungsgeschichte des Menschen. Gemeinverständliche wissenschaftliche Vorträge über die Grundzüge der menschlichen Keimes- und Stammes-Geschichte*, Wilhelm Engelmann, Leipzig.

85 Illustration by the author after various sources.

88 Modified from illustration by Ed Heck in Novacek, M. J., 1999, "100 million years of land vertebrate evolution: the Cretaceous–Early Tertiary transition," *Annals of the Missouri Botanical Gardens* 86:230–58. Author's copyright.

100, 101 Illustrations by Ed Heck. Author's copyright.

104 Illustration by the author after various sources, including Edwards, E., 1986, *Botanical Journal of the Linnaean Society*, 93:173–204 (*Aglaophyton*); and Edwards, E., 1980, *Review of Palaeobotany and Palynology* 29:177–88, Elsevier Science (*Rhynia*).

105 Illustration by D. Grimaldi from Grimaldi and Engel, 2005, *Evolution of the Insects*. Reproduced with permission.

109 Illustration modified from Ahlberg, E., and Clack, J. A., 2006, "A firm step from water to land," *Nature* 440:747–49. Reproduced with permission.

116 Illustration by author after Huey, R. B., and Ward, P. D., 2005, "Hypoxia, global warming, and terrestrial Late Permian extinctions," *Science* 308:398–401.

119, 128 Illustrations by D. Grimaldi from Grimaldi and Engel, 2005. Reproduced with permission.

134 Illustration (revised) from Stevens, K., and Parrish, M. J., 1999, "Neck posture and feeding habits of two Jurassic sauropod dinosaurs," *Science* 284:798–800. Reproduced with permission. Dinomorph™ images courtesy Kent Stevens, University of Oregon.

138 Modified from illustration in Lockley, M. G., 2001, "Trackways—dinosaur locomotion" in Briggs, D.E.G., and Crowther, P. R. (eds.), *Palaeobiology II*, Blackwell Publishing, Oxford, UK, pp. 408–412. Reproduced with permission.

142 Modified from illustration in Friis, E. M., Pedersen, K. R., and Crane, P. R., ibid., pp. 97–102. Reproduced with permission.

144, 145 Illustrations from Tudge, C., 2000, *The Variety of Life*, Oxford University Press, New York. Reproduced with permission.

148 Illustration by the author based on a figure in Pryer, K. M., Schneider, H., and Magallón, S., 2004, "The radiation of vascular plants," in Cracraft, J., and Donoghue, M. J. (eds.), *Assembling the Tree of Life*, Oxford University Press, New York.

151 AMNH special collections. Reproduced with permission.

155 Illustration from Proctor, M., Yeo, P., and Lack, A., 1996, *The Natural History of Pollination*, Timber Press, Portland.

159, 165 Illustrations by Patricia J. Wynne from Sessions, L. A., and Johnson, S. D., 2005, "The flower and the fly," *Natural History*, March 2005, pp. 58–63. Reproduced with permission.

166 Illustration from Maria Sibylla Merian, 1705, *Metamorphosis insectorum Surinamensium*, Amsterdam.

169 Illustration by Michael Rothman in Engel, M. S., 2000, "A new interpretation of the oldest fossil bee (*Hymenoptera: Apidae*)," *American Museum Novitates* 3296:1–11. Reproduced with permission.